Immunotoxicology of Environmental and Occupational Metals

Edited by

JUDITH T. ZELIKOFF

New York University School of Medicine

and

PETER T. THOMAS

Covance Laboratories Inc.

UK Taylor & Francis Ltd, 1 Gunpowder Square, London EC4A 3DE
USA Taylor & Francis Inc., 1900 Frost Road, Suite 101, Bristol, PA 19007

British Library Cataloguing in Publication Data
A catalogue record for this book is available from the British Library.
ISBN 0-7484-0390-6

Library of Congress Cataloging Publication Data are available

Cover design by

Typeset in Times 10/12pt by Best-set Typesetter Ltd, Hong Kong

Printed in Great Britain by T.J. International Ltd, Padstow, UK

Contents

Contents

Preface

Heavy metals are ubiquitous in the biosphere, where they occur as part of the natural background of chemicals to which human beings are constantly exposed. However, industrial uses of metals and other industrial and domestic processes have introduced substantial amounts of potentially toxic heavy metals into the atmosphere and into the aquatic and terrestrial environment.

The toxicological effects associated with high-dose exposure to many metals including arsenic, cadmium, and mercury are well known. However, adverse effects resulting from long-term subclinical levels of exposure, comparable with environmental conditions, are less clear. Within the last few decades, experimental data have shown that low-level exposure to certain metals induces subtle changes within a host, including altered immunological competence. Environmental stressors such as metals may act directly to kill the exposed organisms or indirectly to exacerbate disease states by lowering resistance and allowing the invasion of infectious pathogens or the progression of nascent tumors.

Although effects of metals are dependent upon such variables as host species and exposure parameters, including route, dose, and duration, the conclusion reached in most immunotoxicological studies is that heavy metals act to suppress immunocompetence. The most consistent finding in experimental and epidemiological studies evaluating the effects of metals on immune functions is a decreased host resistance to infectious agents. Immunotoxicity of metals may occur via direct effects upon a specific immune system component or, alternatively, via inhibition of immunoregulation, which can result in immunosuppression, hypersensitivity reactions, or autoimmune disorders. It has been postulated that metal toxicity may, at least in part, be due to autoimmunity, since an autoimmune disorder exists for all of the major target organs affected by heavy metals.

Preface

The purpose of this book is to provide information on the immunotoxicological effects of important environmentally and occupationally encountered metals, for investigators in the field and those interested in expanding their knowledge in this rapidly growing area. The chapters focus on the immunotoxicological effects from highly toxic, commonly encountered metals such as arsenic, beryllium, cadmium, chromium, lead, mercury, nickel, and vanadium, while others review the effects from the less well-studied metals indium and platinum. The final chapter reviews the immunotoxicological effects of the major essential metals: iron, zinc, and copper. In order to better understand the overall effects associated with each particular metal, individual chapters also review their history, use, occurrence, biology, and general toxicological properties. Lastly, a series of tables have been included to provide the reader with a comprehensive summary of the immunotoxicological effects associated with the most commonly encountered environmental/occupational metals. This book, as written, should serve as an important reference source for academics, policy makers, and regulators involved with matters related to the immunotoxicology of metals.

Judith T. Zelikoff
Peter T. Thomas

Acknowledgment

We would like to thank the individual authors for their time and effort on this project. We also gratefully acknowledge Dr Mitchell Cohen for his editorial comments and preparation of the immunotoxicology tables.

List of Contributors

LEIGH ANN BURNS
Dow Corning Corporation
Mail No. CO3101
2200 West Salzburg Road
Midland, MI 48686
USA

LEE S. NEWMAN
Division of Environmental and
    Occupational Health Sciences
National Jewish Medical and Re-
    search Center; and
Department of Medicine and
    Department of Preventative
    Medicine and Biometrics
University of Colorado School of
    Medicine
1400 Jackson Street
Denver, CO 80206
USA

LOREN D. KOLLER
College of Veterinary Medicine
Oregon State University
Corvallis, OR 97331-4801
USA

DARRYL P. ARFSTEN
Karch & Associates Inc.
1701 K Street N.W., Suite 1000
Washington, DC 20006
USA

LESA L. AYLWARD
Karch & Associates Inc.
1701 K Street N.W., Suite 1000
Washington, DC 20006
USA

NATHAN J. KARCH
Karch & Associates Inc.
1701 K Street N.W., Suite 1000
Washington, DC 20006
USA

MARK E. BLAZKA
Colgate E. Palmolive
909 River Road
Piscataway, NJ 08855-1343
USA

MICHAEL J. McCABE, J.R.
Institute of Chemical Toxicology
Wayne State University
2727 Second Avenue
Detroit, MI 48201-2654
USA

PIERLUIGI E. BIGAZZI
Department of Pathology
University of Connecticut Health
  Center
Farmington, CT 06030
USA

RALPH J. SMIALOWICZ
National Health and Environmental
  Effects Research Laboratory
US Environmental Protection
  Agency MD-92
Research Triangle Park, NC 27711
USA

KATHLEEN RODGERS
Livingston Research Center
University of Southern California
1321 North Mission Road
Los Angeles, CA 90033
USA

MITCHELL D. COHEN
Institute of Environmental Medicine
New York University Medical
  Center
Long Meadow Road
Tuxedo, NY 10987
USA

FELIX OMILA OMARA
TOXEN
University of Quebec at Montreal
Montreal, Quebec
Canada H3C 3P8

PAULINE BROUSSEAU
Concordia University
Montreal, Quebec
Canada H3G 1M8

BARRY RAYMOND BLAKLEY
Department of Veterinary
  Physiological Sciences
University of Saskatchewan
Saskatoon, Saskatchewan
Canada S7N 0W0

MICHEL FOURNIER
TOXEN
University of Quebec at Montreal
Montreal, Quebec
Canada H3C 3P8

1

Arsenic

LEIGH ANN BURNS

*Dow Corning Corporation, Mail No. CO3101, 2200 West Salzburg Road,
Midland, MI 48686, USA*

1.1 Introduction

This chapter will briefly review the history of arsenic as a medicinal and toxic
compound, its exposure, disposition, mechanism of action, toxicology, and
immunotoxicology in mammalian systems. It is, however, impossible to review
in depth all of these components within the confines of this chapter. This
review is intended to provide the reader enough information with which to
comprehend the more extensive review of the immunotoxicology of arsenic.
For more detailed information, the reader is referred to the following references (Squibb and Fowler, 1983; Dickerson, 1994; Burns *et al.*, 1995; Goyer,
1996) which were used to compile the material presented herein: *Occupational
Medicine 3ʳᵈ Edition, Casarett and Doull's Toxicology: The Basic Science of
Poisons 5ᵗʰ Edition*, and *Biological and Environmental Effects of Arsenic*. In
addition, the following may be used interchangeably throughout this chapter:
arsenite and inorganic trivalent arsenic (As^{3+}); arsenate and inorganic
pentavalent arsenic (As^{5+}).

1.2 History

Arsenic has held the attention of scientists and authors (e.g. *Arsenic and Old
Lace*) for hundreds of years due to its use as both a medicinal agent and a
poison. Thus, its history is rather complex. The use of arsenic compounds as
medicinal agents can be traced back to the time of Hippocrates (460–377 BC)
when realgar (AsS) was used to treat ulcers. In the Middle Ages arsenicals
possessed popularity as medicinal, suicidal, and homicidal agents. It was during this time that arsenous oxide (white arsenic) became recognized by many

as an effective poisoning agent. By the 1800s, arsenicals were widely used in prescribed tonics, with the therapeutic dosage for these tonics being determined by establishing a NOEL (no observable effect level) for toxic symptoms. It has been suggested by some scholars that instead of being deliberately poisoned by arsenic (as many historians have postulated), the French leader Napoleon was actually being *treated* with arsenical compounds.

Arsenic compounds, particularly the sulfides, have also been used commercially in the production of dyes and pigments. Realgar and orpiment possess orange-red and bright yellow pigments and were extensively employed in Chinese lacquers. In the textile industry, Paris green (cuproacetoarsonate) and Scheele's green (cupric arsenite) were used to dye fabrics. The dusts generated in cutting rooms from these agents represented potential arsenic intoxication hazards and in the late 1800s it was recommended that their use in this industry be eliminated. At approximately the same time that these dyes were being phased out of use, a description of skin cancer as a result of arsenic exposure made its way into the literature. Although the onset of symptoms of skin cancer was sometimes delayed (for reasons still not clear), this condition was attributed to the use of arsenic compounds in the treatment of several disease states.

Arsenicals were once widely used as chemotherapeutic agents until the development of more specific and less toxic agents with higher therapeutic indices. Around the turn of the century Erlich discovered the famous '606' (arsphenamine) and from that, literally thousands of arsenic derivatives were synthesized and evaluated for use. Although inorganic arsenic compounds were too toxic for human use, organic arsenicals possessed comparatively lower toxicities in mammalian systems and were relatively effective as chemotherapeutic agents in the treatment of syphilis, trypanosomiasis, amebic dysentery, trichomenal vaginitis, relapsing fever, and other diseases. Atoxyl was among the first arsenicals used successfully in treating trypanosomiasis and was later prescribed for syphilis. When the side effects of long-term atoxyl use became apparent, less toxic agents such as arsphenamine (salvarsan) and neoarsphenamine (neosalvarsan) began to be used. Other therapeutic arsenic-containing preparations include sodium cacodylate, Fowler's solution, Donovan's solution, and asiatic pills. These compounds were used to treat long-term dermatological conditions, pulmonary diseases, and leukemias. Although the use of arsenicals in medicine today has been essentially eliminated and replaced by less toxic chemotherapeutic agents and therapies, treatment of the advanced stages of trypanosomiasis with arsenic-containing compounds was an acceptable practice as recently as the 1980s, and some arsenicals may still be found occasionally as components of antiparasitic drugs.

Most recently, the use of arsenical compounds has been as toxic agents in pesticides, herbicides, fungicides, and in chemical warfare. During World War I, solid arsenical derivatives were combined and dispersed in smokes and arsine gases ($RAsCl_2$, R = methyl, ethyl, or phenyl). Most notably came the development of the arsenical gas, lewisite (2-chlorovinyl dichloroarsine).

Lewisite is a powerful vessicant which produces blistering lesions, local and respiratory irritation, and general systemic toxicity. Because of the insidious nature of Lewisite, the search for the mechanism of action and an antidote began almost immediately. This led to the development of BAL (British Anti-Lewisite; 2,3-dimercaptopropanol), a metal-binding agent effective for most arsenic poisonings. Today, arsenic is used in metal smelters, as a dessicant in the agricultural industry, as a silvicide in forestry, and in sheep dip. This represents the majority of the current uses of both inorganic and organic arsenicals and, thus, the primary means of occupational exposures.

Arsine (AsH_3) is a hydride of arsenic which is finding significant application in the semiconductor industry. Hybrid semiconductor materials consisting of silicon complexed with gallium arsenide (GaAs) or indium arsenide (InAs), or epitaxial gallium arsenide alone require reaction with AsH_3 gas. This is achieved when highly pure arsenic is reacted with pure hydrogen to produce the AsH_3 gas. This gas is then combined with gallium to produce GaAs, a material used in the electronics and telecommunications industries (sometimes complexed with silicon) as a more rapid conductor than the long-standing silicon chips. Arsine can also be used to 'dope' silicon and germanium devices, and is used with indium in infrared detectors and Hall effect applications (Dickerson, 1994).

1.3 Occurrence

Arsenic is a ubiquitous metal that occurs naturally in our environment. The concentration of arsenic in the soil is estimated to be less than 2 mg As/kg, although the value is dependent on the location and content of the soil/bedrock (NAS, 1977). Concentrations of arsenic in the drinking water also vary with location and bedrock content. Most sources of drinking water contain small quantities of arsenic (less than 0.01 mg As/L) (Squibb and Fowler, 1983). However, this concentration is dependent on the community in question as concentrations of 0.05 mg As/L have been reported in Nova Scotia, where arsenic content of the bedrock is high. Greater levels have been seen in communities around mineral springs such as those in California, the former Soviet Union, and New Zealand (0.4 to 1.3 mg As/L) (Schroeder and Balassa, 1966), with more extreme values observed in Japan (1.7 mg As/L) (Goyer, 1996) and Argentina (3.4 mg As/L) (Squibb and Fowler, 1983). Although the speciation of arsenic in the water source is largely unknown, one report indicated that in surveyed ground waters, 25 to 50 percent of the arsenic was in the trivalent form (Squibb and Fowler, 1983).

Arsenic is also present in many food items and an extensive list has been published (NAS, 1977). The content of arsenic in foodstuffs is generally below 0.25 mg As/kg (Jelinek and Comeliussen, 1977; Squibb and Fowler, 1983). This value is influenced by the natural content of arsenic in the soil, the species of plant and its uptake of arsenic, and by the quantity of arsenic-containing herbicides and insecticides that are used. Values for arsenic (primarily trivalent) of 0.5 mg As/L have been reported in wine and were attributable to the insecticides used to spray the grapes (Crecelius, 1977). However, by far the highest concentrations of arsenic in foodstuffs are found in consumable marine life and seaweed. Concentrations may vary from 10 mg As/kg in bony fishes to over 100 mg As/kg in crustaceans and other bottom-dwelling marine flora and fauna (Munro *et al.*, 1974; Jelinek and Comeliussen, 1977). The average daily intake of arsenic has been estimated to be 0.05 to 0.1 mg As per individual and is greatly influenced by the amount of seafood consumed in the diet.

Although the vast majority of arsenic exposure comes from dietary intake, exposure to arsenic may also occur in the occupational setting. Persons employed in copper, zinc, and lead smelters, in the manufacture and spraying of pesticides, and in the production of wood treated with chromated copper arsenate are at risk of occupational exposure to arsenic. The Occupational Safety and Health Administration (OSHA) has a recommended permissible exposure limit (PEL) for inorganic arsenic (except AsH_3) of 0.01 mg As/m^3 of air as a time-weighted average over an 8-h shift. In 1988, the National Institute for Occupational Safety and Health (NIOSH) issued a recommended exposure limit (REL) for arsenic and all its compounds of 2 μg As/m^3 as a ceiling in a 15-min air sampling. Finally, the threshold limit value (TLV) set by the American Conference of Government Industrial Hygienists for arsenic and its soluble compounds is 0.2 mg/m^3, and 0.05 ppm for AsH_3 gas (Dickerson, 1994).

Recently, individuals employed in the semiconductor industry, producing GaAs or InAs, have also been determined to be at risk. Although the manufacture of GaAs requires the use of AsH_3 gas, in the industrial setting the primary route of exposure to GaAs is inhalation of GaAs particles generated in the sawing and polishing of GaAs wafers. It has been estimated that between 50 and 60 percent of the GaAs crystals are converted into GaAs dust in this step of processing (Briggs and Owens, 1980). The toxicity of GaAs has only begun to be investigated in a limited number of laboratories in the past decade and thus the degree of risk that workers in these industries assume is not clear. In 1983, Willardson reported that manufacturing processes used 5 to 10 tons of arsenic in GaAs devices and that production was expected to increase three- to tenfold by 1990. A study by SRI International projected that in 1995 over 4 percent of total integrated circuit sales would be GaAs (Harrison, 1986). Presently, GaAs is regulated on the basis of inorganic arsenic toxicity data (Code of Federal Regulations, 1985) with an REL of 2 μg As/m^3 as a ceiling in a 15-min air sampling (NIOSH, 1982). Although it has been demonstrated recently that GaAs dissociates into its component metals, gallium and arsenic, following intratracheal instillation or oral exposure in rats (Webb *et al.*, 1984) and intratracheal exposure in mice (Burns *et al.*, 1991), the actual levels of arsenic exposure in GaAs-exposed workers remains unknown.

1.4   Essentiality and Metabolism

1.4.1   Toxicokinetics and metabolism of arsenic

Dermatological absorption of arsenic is possible, since systemic arsenic toxicity has been reported in workers exposed accidentally via the skin to arsenical compounds. Although this is relevant, occupational exposure to arsenic occurs primarily through the respiratory tract and usually enters the body in the form of arsenic trioxide. Because the particle size is often relatively large ($\geq 5\,\mu m$), these particles tend to settle in the upper airways. Smaller particles, capable of passing further into the alveolar space, represent a means of pulmonary absorption of arsenic. In the manufacture of GaAs, the primary route of exposure to GaAs is inhalation of particles generated in the sawing and polishing of GaAs wafers. Many of these particles are capable of penetrating deep into the lung (Briggs and Owens, 1980) and will be discussed in more detail later in this chapter.

Understanding absorption of arsenic from the lung is therefore important to the understanding of the disposition and toxicity of arsenic. Even so, most of the respired dose of arsenic (GaAs or others) consists of the larger particles, which are swept up the mucociliary escalator and subsequently swallowed, thus making gastrointestinal absorption critical in understanding the toxicokinetics of arsenic. The biological half-life of ingested inorganic arsenic is approximately 10 h and more than half (50–80 percent) of the arsenic that is excreted is eliminated in urine within days after exposure. Administration of radiolabeled arsenic (tri- or pentavalent forms, soluble or particulate) results in a very small amount of arsenic being eliminated in either urine or feces, suggesting arsenic is nearly completely absorbed from the digestive tract.

Following absorption, arsenic is rapidly cleared from the bloodstream and is distributed into a wide variety of tissues (Dickerson, 1994; Goyer, 1996). One exception is the rat, a species in which approximately 80 percent of the absorbed dose appears to concentrate in red blood cells (Vahter, 1981). In man, it does not cross the blood–brain barrier easily and can rarely be found in spinal fluid. Absorbed arsenic has a penchant for skin and other keratin-containing tissues such as nails and hair. In nails, arsenic produces the characteristic Mee's lines, transverse white lines across nails, which may appear up to 6 weeks after the onset of toxic symptoms. Arsenic deposition in hair has been used to reflect exposure history; however, it is difficult to distinguish between exposure via the diet (intrinsic exposure) and exposure from external sources. Elimination of arsenic can occur through the desquamation of skin and through the natural secretions of the sebaceous glands in the skin. It has also been detected in breast milk of humans (Dickerson, 1994). Placental transfer of arsenic has been demonstrated experimentally using high doses of sodium arsenate in hamsters. In the tissues of newborn babies and fetuses in Japan, it was shown that the total amount of arsenic increases throughout gestation, suggesting placental transfer (Dickerson, 1994).

Organic arsenicals, such as those found in seafood, are usually eliminated from the body unchanged. Trivalent arsenic may be converted into the pentavalent form, biotransformed, and then eliminated. In addition, some *in vivo* studies have suggested that pentavalent arsenic is converted into arsenic trioxide prior to detoxification and elimination, although this is still somewhat controversial. Within the digestive system, microorganisms may act to reduce and methylate arsenicals to monomethyl and dimethyl forms. This can be viewed as detoxification, since methylated forms of arsenic are less toxic. Once in the bloodstream, arsenic may pass through the liver and undergo similar detoxification. In cases of excessive external exposure, dimethyl arsenic acid is the primary biotransformation product which is formed and eliminated rapidly. This process, like many metabolic processes, can become a rate-limiting step in detoxification of inorganic arsenic. Interestingly, ingestion of arsenic-containing seafood does not result in enhanced elimination of inorganic, monomethyl, or dimethyl forms of arsenic. Instead, it appears to be transformed and eliminated as cacodylic acid (Dickerson, 1994; Goyer, 1996).

1.4.2 Mechanism of action

Much of the basis of our current understanding of the mechanism of arsenic toxicity comes from toxicity studies performed in the late 1800s, development of arsenical drugs, and the need to develop antidotes for arsenical warfare agents such as lewisite. What we have learned over these many years is that there is no single mechanism associated with arsenic toxicity (Squibb and Fowler, 1983; Dickerson, 1994; Goyer, 1996). Three associated mechanisms will be considered here: pentavalent arsenic toxicity, trivalent arsenic toxicity, and toxicity associated with AsH_3 gas exposure. Arsine, a potent hemolytic agent, has an affinity for glutathione (GSH), a peptide consisting of glycine, cysteine, and glutamic acid. Reduced glutathione is required to maintain the integrity of the red blood cell (RBC) erythrocyte-potassium pump. When the pump is disrupted, sodium ions may enter the RBC leading to an uptake of water and, ultimately, cellular lysis. The RBC 'ghosts' (associated with arsenic) can cause renal failure by blocking the renal tubules. Renal failure may also be enhanced by an AsH_3-induced decrease in oxygen uptake by the organ. This has been demonstrated in renal tissue slices *in vitro* (Dickerson, 1994).

Arsenic accumulates in mitochondria, where it has been linked to the uncoupling of oxidative phosphorylation resulting in a stimulation of mitochondrial ATPase activity. The arsenate ion (pentavalent arsenic) is both isosteric and isoelectric with phosphate, and numerous studies have been conducted which show that arsenate can substitute for phosphate in enzyme-catalyzed reactions, forming arsenic esters during the reactions. This mechanism of toxicity may be related to the fact that esters of arsenicals are inherently less stable than their phosphate counterparts, thus they may spontaneously decompose. 'Arsenolysis' refers to the spontaneous hydrolysis of arsenate esters, and occurs in biological systems. The result can be either a shut down of metabolic systems (e.g. succinate respiration) or an activation of enzyme systems (e.g. glyceraldehyde phosphate dehydrogenase). Additionally, arsenate may inhibit the energy-linked reduction of NAD^+ within cells (Squibb and Fowler, 1983).

Although the previously discussed ways in which arsenic may exert toxicity are significant, by far the most important and well-studied mechanism of arsenic toxicity involves the interaction of trivalent arsenic (arsenite) with sulfhydryl (-SH) groups associated with enzymes and other proteins. A comprehensive list of enzymes inhibited by arsenite has been published (Webb, 1966). Interestingly, and to continue to make the assessment of toxicity more complicated, this mechanism of action can be subdivided into the types of enzymes being inhibited. In some cases, the activities of the enzymes being inhibited can be restored by the addition of excess GSH, suggesting the involvement of a single thiol group in the inhibition of the enzyme (e.g. monoamine oxidase, transaminase). In other cases, such as lipoic acid dehydrogenase and pyruvate oxidase (which utilize lipoic acid as a coenzyme), restoration of activity cannot be achieved by the addition of excess GSH. Instead, dithiol compounds such as 2,3-dimercaptopropanol (BAL) are more effective. This suggests a more complex interaction of the arsenical with the enzyme, often resulting in the formation of a five- or six-membered ring-like structure. In the case of lipoic acid, the six-membered ring which is formed is particularly stable, thus contributing to the greater sensitivity of this enzyme system to arsenic exposure. Mono-substituted arsenicals react with two thiol groups while di-substituted arsenicals will react with a single thiol group.

In summary, there are several possible mechanisms of toxicity of arsenic. Arsenic may interact with critical enzyme thiol groups or persulfide residues and inhibit function (trivalent). It may act by the inhibition of substrate binding by competitive interactions with structurally similar organic arsenicals (pentavalent). It may interact with secondary molecules or enzyme substrates to inhibit metabolic reactions, and it may disrupt energy systems or ion balances within cells.

1.5 General Toxicology

As a single metallic entity the toxicity of arsenic is very difficult to characterize, primarily due to the fact that there are numerous organic and inorganic arsenical compounds and the chemistry of arsenic (as described previously) is relatively complex. It exists in both inorganic and organic forms, in both the trivalent and pentavalent species. Although the mechanisms of action of the various forms of arsenic are at least partially understood, characterizing their toxicity is complicated by metabolic interconversions to various chemical forms and oxidation states involved in the differential binding of these arsenical compounds in cells, and by the tissue distribution of arsenic in general. These studies have been reviewed extensively in the literature (Webb, 1966; Fowler *et al.*, 1979; Squibb and Fowler, 1983; Aposhian, 1989).

1.5.1 Acute effects

Symptoms of acute arsenic poisoning follow a pattern that is independent of the form of arsenic to which the individual was exposed and may appear minutes, hours, or even days after exposure (Dickerson, 1994). Vasodilation, hyperemia (particularly of the spleen), and plasma extrusion as a result of capillary damage result in edema, decreased blood pressure, and shock. There may also be centrally mediated paralysis of major muscle groups, including the respiratory muscles. Most often observed are signs of cardiac abnormalities and of acute gastrointestinal distress: stomach pain, forceful emesis, cramps, and gastrointestinal inflammation. Oliguria is characteristic of acute arsenic intoxication and the urine often contains protein, RBCs, and RBC ghosts. Neurological effects often appear several weeks following a significant exposure and may consist of reversible axonal degeneration. Reversible leukopenia and granulocytopenia have also been observed.

Arsine is a colorless gas with a slight garlic odor. It is a strong reducing agent and is chemically unstable in air (Klimecki and Carter, 1995). Arsine may be generated when nascent hydrogen contacts solutions or metals containing inorganic arsenic, or by reacting arsenic trioxide with reducing agents such as sodium borohydride. In humans, most workers exposed to AsH_3 gas are unaware of exposure since the gas is relatively non-irritating at low concentrations (Jenkins *et al.*, 1965). Inhalation of 250 ppm AsH_3 is fatal almost instantly; single exposures to concentrations as low as 10 ppm have been associated with symptoms in humans which may develop into serious illness within hours. Like acute arsenic exposure, acute AsH_3 toxicity is associated with headache, light headedness, vertigo, fever, tachycardia, abdominal pain and rigidity, and liver enlargement. Jaundice and a 'bronzing' of the skin color may develop. Other symptoms are a result of hemolysis, hypoxemia, and heme-pigment overload induced by AsH_3 exposure. However, by far the most significant target for AsH_3 toxicity is the erythrocyte (Jenkins *et al.*, 1965). Hemoglobin concentrations may drop below 10 µg/dl and may rise significantly in the plasma. RBC counts have been reported below 10^6/ml. Renal effects are likely the result of the significant amount of hemoglobin deposited in the nephrons and hematuria is a classic sign of acute AsH_3 poisoning.

1.5.2 Chronic effects

Symptoms of chronic arsenic exposure may develop over the course of a few weeks or months. These may include significant central nervous system effects such as neuritis, paresthesia (pins and needles), and muscle weakness.

Peripheral neuropathy is also possible, leading ultimately to the demyelination of nerves. Long-term chronic arsenic exposure may result in hepatic injury (jaundice and portal cirrhosis) and peripheral vascular disease. In addition, chronic arsenic exposure is reported to result in hair loss, as well as hyperpigmentation, hyperkeratinosis, and desquamation of the skin. Keratoses on the palms of the hands and soles of the feet have also been observed and are often related to the appearance of skin cancer. Immunotoxic effects (enhanced susceptibility to infection, immunosuppression, lymphopenia) have also been reported and will be discussed in the following sections.

There are several lines of evidence in the literature that indicate that GaAs represents a health hazard as a result of systemic arsenic exposure. Webb *et al.* (1984) reported partial dissolution of GaAs in buffers of various ionic strength; these investigations have been confirmed by other laboratories (Yamauchi *et al.*, 1986; Pierson *et al.*, 1989). Further investigations by Webb *et al.* (1986) confirmed GaAs dissolution 14 days following a single intratracheal administration of 100 mg GaAs/kg in rats. Gallium trioxide and arsenic trioxide were also evaluated and revealed that gallium trioxide produced little biological activity. Arsenic trioxide produced some of the same qualitative effects as those noted when GaAs was tested (increased dry lung weight, protein, DNA, and lipid content). In those same studies (Webb *et al.*, 1986), intratracheal administration of GaAs at 10, 30, and 100 mg GaAs/kg to rats resulted in detection of blood arsenic concentrations of 5.5, 14.3, and 53.6 ng As/ml, respectively, by flame atomic absorption spectrophotometry; oral administration of GaAs showed similar results. In those studies, no gallium was detected, in contrast to studies by Burns *et al.* (1991) who found gallium to accumulate over 14 days in the blood and spleen of mice exposed to intratracheal GaAs. It is noteworthy to remember that the RBCs of rats have a propensity for arsenic while those in mice apparently do not. These data suggest GaAs dissociates into gallium and arsenic species when administered orally or intratracheally. Qualitative and quantitative alterations in urinary porphyrins were reported as well (Webb *et al.*, 1984), and have also been noted in hamsters following subcutaneous InAs (100 mg InAs/kg) exposure (Conner *et al.*, 1995). Interestingly, the profile observed for InAs by Conner *et al.* (1995) was unique from that observed previously for GaAs. Similar alterations in urinary porphyrins were found when inorganic arsenic was administered to rats (Woods and Fowler, 1978).

Yamauchi *et al.* (1986) also showed that organic products of arsenic metabolism (dimethylarsenic acid and methylarsenic acid) were found in urine and other tissues of hamsters exposed either orally (10–1000 mg/kg) or subcutaneously (100 mg/kg) to GaAs or InAs (100 mg/kg) (Yamauchi *et al.*, 1992). Comparison of metabolism and excretion following intratracheal exposure to GaAs, arsenic (As^{+3}) oxide, and arsenic (As^{+5}) oxide at 5 mg/kg revealed that GaAs is metabolized in the hamster to the same compounds as arsenite and arsenate, showing a metabolic profile most similar to that observed for sodium arsenite (Rosner and Carter, 1987). Taken together, the toxicology data

9

collected on GaAs and InAs indicate the likelihood of systemic arsenic exposure due to the dissociation of these group III–V metal complexes *in vivo*. This conclusion is in agreement with other investigations evaluating the primary immunosuppressive component of GaAs (Burns *et al.*, 1991) and *in vivo* solubility (Webb *et al.*, 1984).

It is noteworthy to mention that although the evidence for systemic arsenic exposure following GaAs administration implicates arsenic as the primary immunosuppressive component of GaAs, a role for gallium in GaAs toxicity has been proposed by some investigators (Goering *et al.*, 1988). In this study, rats were exposed intratracheally to a single concentration of GaAs (50–200 mg GaAs/kg). Six days after exposure, serum δ-aminolevulinic acid dehydratase (ALAD) was dose-dependently decreased while urinary excretion of ALAD was enhanced. Using gallium nitrate, sodium arsenite ($NaAsO_2$), and sodium arsenate (Na_2HAsO_4), the authors concluded that the gallium component of GaAs was the primary inhibitor of ALAD following dissolution *in vivo*, and that competition for displacement of zinc from the active site of this enzyme may be involved in this inhibition. Since arsenic has a higher affinity for rat RBC membranes (Vahter, 1981), the lower sensitivity of ALAD to arsenite and arsenate may be a result of non-specific binding of the arsenic to the RBC membranes, thus reducing the pool of free arsenic available for enzyme inhibition. However, it is important to remember that compounds such as GaAs and InAs, which are intermetallic in nature, may possess multifaceted toxicities which are a combination of the toxic effects of each metal. The relative potency for each metal in producing a toxic effect may depend upon the system being evaluated and the relative affinity of each metal for enzymes or other proteins or target tissues.

1.5.3   Carcinogenicity

The evidence for occupational or accidental exposure to many metals, including arsenic, having a carcinogenic result is convincing and apparently dose-related (WHO, 1981; EPA, 1987; Snow, 1992). The carcinogenicity of arsenic is well-documented for the lung and skin, and this metal also appears to be involved in cancers of the liver and bladder (Squibb and Fowler, 1983; Hindmarsh and McCurdy, 1986; Wu *et al.*, 1987). The US Public Health Service has adopted a standard of 50 pg As/L in the drinking water. Based on epidemiological studies, there is a 1 in 5 million lifetime probability of developing cancer from a daily intake of approximately 30 ng.

It is interesting that studies designed to show the carcinogenicity of arsenic in animals have generally proved fruitless for reasons yet unknown, although differences in uptake and metabolism have been suggested (Snow, 1992). Others have proposed that arsenic may be a co-carcinogen (tumor promoter), potentiating the carcinogenic effect of proven carcinogens (Snow, 1992; Yamamoto *et al.*, 1995; Brown and Kitchin, 1996). The mechanism(s) by which arsenic is carcinogenic is/are not completely clear, but several possibilities exist. Trivalent arsenic can interfere with replication of deoxyribonucleic acid (DNA), with alkyl transferase and excision repair (Petres *et al.*, 1977; McCabe *et al.*, 1983; Squibb and Fowler, 1983; Aposhian, 1989), while pentavalent forms may substitute for phosphate and form unstable esters with DNA (Squibb and Fowler, 1983; Aposhian, 1989). Both of these have been implicated in chromosomal damage in human lymphocytes. Arsenic has also been implicated in alterations of gene expression such as induction and amplification (Squibb and Fowler, 1983). More importantly, exposure to arsenic may compromise immunological integrity and suppress natural immune responses. This issue is the primary focus of the remainder of this chapter.

1.6 Immunotoxicology

An understanding of the toxicity of occupational chemicals on the hematopoietic and immune systems is important due to the potential for daily exposure lasting up to 8 h. In humans occupational exposure to arsenic has been linked clinically with aplastic anemia, megaloblastic anemia, granulocytopenia, and generalized hemolytic reactions (Lisiewicz, 1993). The experimental literature contains many conflicting reports concerning the immunotoxic potential of arsenic. Several factors must be taken into account when evaluating these studies, including the various forms of arsenic, the variations in metabolism (which are not only related to the complexity of arsenic biotransformation, but may be related to the species being tested), the variety of species in which arsenic immunotoxicity has been studied, the variety of mechanisms of toxicity, different routes of exposure, and acute compared with chronic arsenic exposure regimes.

Specific immunotoxic consequences of metal exposure are well-documented in the literature and have been reviewed elsewhere (Burns *et al.*, 1995). It is evident that they may exert many immunomodulatory actions that are directly dependent on the exposure level. At high concentrations most metals exert immunosuppressive activity (Vos, 1977; Koller, 1980). However, at lower concentrations immunoenhancement may be seen. This is true for several metals including arsenic (Schrauzer and Ishmael, 1974; Kerkvliet *et al.*, 1980; McCabe *et al.*, 1983).

1.6.1 Inorganic arsenicals

Early investigations concerning the immunotoxic potential of arsenic primarily focused on arsenic-induced alterations in host resistance. In an elaborate series of studies involving various times of exposure of mice to sodium arsenite (NaAsO$_2$) in the drinking water (0.002 m) or subcutaneously (3.13–6.25 mg/kg), Gainer and Pry (1972) determined that exposed mice had a two- to ninefold increase in mortality rate to several viral pathogens including pseudorabies, encephalomyocarditis, and St Louis encephalitis viruses. In contrast, mortality to Western encephalitis virus was increased if mice received $NaAsO_2$ after intraperitoneal viral inoculation, but was decreased when arsenic exposure occurred prior to inoculation. Concomitant studies by Gainer (1972) showed that exposure to $NaAsO_2$, Na_2HAsO_4, roxarsone, and *p*-arsenilic acid inhibited the production and dose-dependently inhibited the activity of interferon (IFN) in mouse embryo cells. Additionally, the inhibition of activity was cell mediated and time dependent. These studies suggest that arsenic exposure increases susceptibility to viral disease by inhibiting IFN production and activity.

In contrast to these studies, other investigators (Schrauzer and Ishmael, 1974; Kerkvliet *et al.*, 1980) have reported different observations. C_3H/St mice exposed to $NaAsO_2$ in the drinking water (10 ppm for 15 months) demonstrated a 68 percent decrease in the development of spontaneous tumors (Schrauzer and Ishmael, 1974). Although there was decreased incidence, the tumors that did develop grew at a much faster rate, and mean survival time after the appearance of tumors was decreased 50 percent in arsenic-exposed mice when compared with controls. Similar results were noted for transplanted mammary tumors (injection into the gland). Kerkvliet *et al.* (1980) reported comparable observations in mice exposed to Na_2HAsO_4 in the drinking water (2.5 to 100 ppm for 10–12 weeks) and injected intramuscularly in the leg with Moloney sarcoma virus (MSV) sarcoma cells. In that study (Kerkvliet *et al.*, 1980), there was also a decreased incidence of tumors despite an increased mortality rate and enhanced growth rate of more mature tumors. No alterations in cell-mediated immunity could be demonstrated. These investigators proposed that arsenic-induced inhibition of viral transformation may account for the significant growth of late tumors since the sarcomas used by both groups have a viral etiology. This would not explain, however, the results observed by Gainer (1972).

Although Kerkvliet *et al.* (1980) could not demonstrate any changes in cell-mediated immunity following exposure to Na_2HAsO_4 in the drinking water (2.5 to 100 ppm for 10–12 weeks), other studies examining immunocompetence have shown that there are changes in immune function following exposure to arsenic compounds. In 1980, Blakely *et al.* (1980) exposed mice to $NaAsO_2$ in the drinking water for 3 weeks (0.5 to 10 ppm). *In vivo* evaluation of the primary (IgM) and secondary (IgG) immune responses showed significant immunosuppression (50 and 40 percent, respectively) at all concentrations; however, these results were not dose related.

Arsine gas is a potent hemolytic poison and a recognized industrial hazard that produces human toxicity at more than 30 ppm. Few studies have been conducted which address the toxicity associated with long-term exposure to AsH_3 gas. In general, rodents exhibit a peripheral erythrocyte regenerative response and extramedullary hematopoiesis has been reported (Klimecki and Carter, 1995). Interestingly, a single exposure of mice to 0.5 ppm AsH_3 (10 times the TLV) induced no changes in the hematopoietic system, while repeated exposures to 0.025 ppm produced profound hematotoxic effects (Blair *et al.*, 1990). Similar observations were made by other investigators. Immunotoxicology studies conducted in mice following inhalation exposure (0.5, 2.5, and 5 ppm AsH_3) of 6 h/day for 14 days demonstrated a dose-dependent increase in spleen weight (Hong *et al.*, 1989; Rosenthal *et al.*, 1989) and hemolysis in conjunction with fewer E-CFU (erythroid colony-forming units) in the bone marrow of mice repeatedly exposed to AsH_3 at 0.025 ppm, despite a significant peripheral RBC regenerative response (Hong *et al.*, 1989). Rosenthal *et al.* (1989) showed that, despite the observed splenomegaly, there was also significant depletion in splenic lymphocytes (83.4 percent in controls compared with 45.6 percent in animals dosed with 5 ppm AsH_3). T-lymphocytes appeared to be more sensitive than B-lymphocytes to the effects of AsH_3 gas. However, the *in vivo* T-lymphocyte-dependent antibody response was significantly enhanced by AsH_3 exposure. There were also dose-dependent decreases in natural killer (NK) cell and cytotoxic T-lymphocyte (T_C) functions but no remarkable differences in lymphoproliferative responses to phytohemagglutinin (PHA), concanavalin A (ConA), lipopolysaccharide (LPS), or in mixed lymphocyte reactions (MLR). Arsine gas exposure produced variable changes in host resistance including an increased susceptibility to *Listeria monocytogenes* and *Plasmodium yoelii*, but had no effect on resistance to influenza virus or B16F10 tumor cells.

As mentioned previously, inhalation of arsenicals represents a primary route of exposure to this metal. Although a significant amount may be transported up the mucociliary escalator and subsequently swallowed or be detoxified in the lung itself, the alveolar macrophage plays a critical role in recognition and elimination of foreign materials within the respiratory tract. Despite differences in proposed mechanisms of action between tri- and pentavalent arsenic, no distinction has been made between these chemical species in hazard identification following inhalation exposure. Lantz *et al.* (1994) therefore compared the *in vivo* (1 mg/kg, intratracheal, 24 h) and *in vitro* (0.1–300 µg/ml) toxicity of both oxidation states on rat alveolar macrophages. The results indicate that *in vivo* exposure to either form of arsenic decreased baseline and LPS-induced tumor necrosis factor-α (TNF-α) production and resulted in an inflammatory response in the lung (neutrophil influx). These findings are consistent with those of Aranyi *et al.* (1985) who showed that inhalation of arsenic trioxide increased susceptibility of mice to infection with *Streptococcus*. *In vitro* exposure of alveolar macrophages to trivalent arsenic results in suppression of phagocytic ability as well as a diminution of oxygen radical production (Fisher *et al.*, 1986; Labedzka *et al.*, 1989). Pentavalent arsenic has been shown to have no effect on oxygen radical production (Castranova *et al.*, 1984). Direct addition studies showed trivalent arsenic to be approximately 10-fold more potent than pentavalent arsenic in inhibiting oxygen radical production and LPS-induced TNF-α release (Lantz *et al.*, 1994).

Other studies examining *in vitro* exposure to arsenic have shown two dose-dependent effects on the immune system: immunoenhancement is generally seen at low concentrations and immunosuppression at higher concentrations. Sodium arsenite exposure augments the proliferative response of human and bovine peripheral blood lymphocytes to mitogens at arsenic concentrations of 1 μM, but suppresses proliferation at 4 μM (McCabe *et al.*, 1983). Exposure to Na_2HAsO_4 also showed similar results with enhancement at 5 μM (human) and 20 μM (bovine) and suppression at 40 μM (human) and 70 μM (bovine) Na_2HAsO_4 (McCabe *et al.*, 1983). Yoshida *et al.* (1986, 1987) have shown enhancement of the *in vitro*-generated antibody-forming cell (AFC) response to sheep red blood cells (SRBC) at low concentrations (50 ng/ml) and suppression at higher concentrations (100–500 ng/ml) of $NaAsO_2$. Also, addition of arsenic (50 ng As/ml) at the time of immunization or 96 h later enhanced the AFC response, while addition of arsenic 24, 48, or 72 h after immunization had no effect on antibody formation. These findings suggest arsenic exposure may result in the depletion of precursors of suppressor T-lymphocytes from normal spleen and, in doing so, enhance the immune response.

More studies are needed on the action of arsenicals on human immunocompetence. Investigations involving human peripheral blood lymphocytes stimulated with PHA revealed that *in vitro* exposure to arsenic resulted in inhibition of proliferation and arrest of the cells in S and G_2 phases of the cell cycle (Baron *et al.*, 1975; Petres *et al.*, 1977). More recently, Yoshida *et al.* (1991) demonstrated that workers employed in the semiconductor industry who had been exposed to arsenic dust exhibited enhanced lymphoproliferative responses to PHA in the absence of clinical symptoms. In another study, urinary arsenic concentrations, mutagenicity, and lymphocyte proliferation kinetics were evaluated in two groups of individuals from Coahuila, Mexico, a rural area with a history of arsenic contamination in the drinking water (0.39 mg As/L; primarily pentavalent) (Ostrosky-Wegman *et al.*, 1991). In arsenic-exposed individuals, there was a 10-fold higher concentration of arsenic excreted in the urine (as methylated or demethylated organic arsenic) and a decreased proliferative response of lymphocytes compared with the control group (0.019 compared with 0.06 mg/L, respectively). There were, however, no differences in chromosome aberrations or sister chromatid exchange frequencies between the two groups of individuals (Gonsebatt *et al.*, 1994). These data, in conjunction with data reported by other investigators (Petres *et al.*, 1977; McCabe *et al.*, 1983; Wen *et al.*, 1991), strongly suggest that the carcinogenic potential of arsenic may not lie solely in the mutagenicity of arsenic, but rather in the alterations in immunocompetence induced by arsenic exposure. This is supported by the observation that the types of skin cancers reported in immunosuppressed patients with no history of arsenic exposure (Walder *et al.*, 1971) are similar to those observed in individuals chronically exposed to arsenic (Cebrian *et al.*, 1983; Cebrian, 1987).

1.6.2   *Semiconductor materials*

The semiconductor industry illustrates many of the issues surrounding occupational disease in the industrialized world. Gallium arsenide and indium arsenide are being considered separately from other arsenic-containing compounds because of the relatively recent interest in their toxicology and immunotoxicology. These intermetallic compounds have found increased use as replacements or as hybrid materials in the electronics and telecommunications industries. Because their high-speed conducting performance is superior to other compounds used for the same purposes, GaAs and InAs have recently become the materials of choice in these and other industries. Their use is expected to continue to escalate in the future.

Gallium arsenide has been demonstrated to produce dose-dependent immunosuppressive effects 14 days following intratracheal exposure of mice to concentrations between 50 and 200 mg/kg (Sikorski *et al.*, 1989). Those studies showed the IgG and IgM antibody responses (humoral immunity) were decreased 48 and 66 percent, respectively, at the 200-mg/kg exposure. The delayed hypersensitivity and mixed lymphocyte responses were also dose-dependently decreased. Analysis of peritoneal exudate cells (PEC) revealed a dose-dependent decrease in total cell numbers; a population shift, consisting of an increase in the relative percentages of monocytes from 53 to 81 percent and a decrease in that of lymphocytes from 46 to 20 percent following exposure to GaAs, was also noted. Most interestingly, in a similar battery of tests with significant overlap with the study by Sikorski *et al.* (1989), intratracheal exposure to InAs produced no significant immunotoxicity (NTP Report, in preparation). Given the known similarities of these two group III–V metals, an explanation for the significant difference in the immunotoxic potential of each is at present unknown.

Investigations designed to determine the immunosuppressive component of GaAs (Burns *et al.*, 1991) used graphite atomic absorption spectrophotometry to evaluate concentrations of gallium and arsenic in various organs of mice following a single GaAs intratracheal instillation of 200 mg/kg. In these studies, arsenic was found in the blood at 2 h after exposure and remained elevated (200 ng/ml) through 48 h. Gallium concentrations (undetected in previous studies; Webb *et al.*, 1984) were detected at 2 h as well (350 ng/ml). By 14 days, the arsenic concentrations had returned to baseline; however, gallium concentrations were significantly elevated (600 ng/ml). Similar results were noted in the spleen, where arsenic concentrations peaked at 1250 ng/ml by 24 h; gallium concentrations at this time were 750 ng/ml. Unlike the blood, arsenic concentrations in the spleen remained elevated at 800 ng/ml at 14 days while gallium concentrations continued to rise (12 000 ng/ml at 14 days). Because the concentrations of gallium and arsenic in each of the organs evaluated did not occur in a 1 : 1 ratio, it was concluded that GaAs must be dissociating in the lung into gallium and arsenic species.

To further examine the roles of each metal in the observed immuno-suppression produced following GaAs exposure in mice, a series of *in vitro* experiments using gallium nitrate (Ga(NO$_3$)$_3$), GaAs, and NaAsO$_2$ was conducted. Following *in vitro* addition, GaAs directly suppressed the *in vitro*-generated AFC response in a dose-dependent (6.25 to 75 μM) and time-dependent (addition up to 36 h after immunization) manner (Burns *et al.*, 1991). Both NaAsO$_2$ and Ga(NO$_3$)$_3$ also suppressed the AFC response dose- and time-dependently when added to the *in vitro* system. However, based on IC$_{50}$ values for each salt, the role of the gallium component in the immunosuppression appears weak. The arsenic chelator DMSA dose-dependently blocked GaAs-induced immunosuppression *in vitro*, while the gallium chelator oxalic acid had no effect. In that study, DMSA was capable of preventing suppression induced by NaAsO$_2$ but not suppression induced by Ga(NO$_3$)$_3$. Likewise, oxalic acid was capable of preventing suppression induced by Ga(NO$_3$)$_3$, but not NaAsO$_2$-induced suppression. DMSA was also able to block GaAs-induced suppression of the AFC response (*in vivo* exposure) when given subcutaneously every 4 h beginning 1 h prior to GaAs exposure. It was concluded that arsenic was the primary immunosuppressive component of GaAs following both *in vivo* and *in vitro* GaAs exposure.

Studies evaluating cellular and humoral immunity following exposure to GaAs are numerous and continue to grow in number. The effects of GaAs exposure on host resistance have been discussed above and related to the chemotherapeutic potential of arsenic. Although a role for gallium in immuno-suppression induced by GaAs has not been unequivocally eliminated, studies by Burns *et al.* (1991) convincingly demonstrated that arsenic is the primary immunosuppressive moiety of GaAs following both *in vivo* and *in vitro* exposures. This model of arsenic exposure is quite unique in that the lung appears to act as a depot for the slow release of dissociated gallium and arsenic. The result is a prolonged (at least 14 days) systemic exposure to both metals (Burns *et al.*, 1991). This model of slow *in vivo* dissociation has been demonstrated by other investigators for both GaAs and InAs (Webb *et al.*, 1984; Yamauchi *et al.*, 1992).

A dose-dependent increase in spleen and lung weight in GaAs-exposed mice has also been observed. Histopathologic evaluation of spleen and lung showed gross changes in the lung indicative of an acute inflammatory response (McCay *et al.*, 1997); there were no changes in the splenic morphology. Blood glucose levels in these mice were decreased dose-dependently while serum glutamic-pyruvic transaminase (SGPT) levels were elevated (247 percent) at the 200-mg/kg exposure. Gallium oxide, gallium nitrate, and sodium arsenite were all found to suppress the antibody response. Arsenic trioxide did not suppress the AFC response and this may be related to metabolism and excretion over the 14 days following a single exposure. These studies indicate that the primary target organs for GaAs following a single intratracheal exposure are the lung and immune system, and that GaAs (as well as its component metals, gallium and arsenic) suppresses humoral immune function.

In vivo administration of GaAs dose-dependently suppresses the *in vitro*-generated primary antibody response to the T-lymphocyte-dependent antigen SRBC and to the T-lymphocyte-independent antigen dinitrophenyl (DNP)-Ficoll (Sikorski *et al.*, 1991b). Fluorescence-activated cell sorter (FACS) analysis showed decreased total numbers of T-lymphocytes, B-lymphocytes, and macrophages in GaAs-exposed splenocytes, but percentages of these cells were unchanged when compared with vehicle-exposed splenocytes. Separation and reconstitution studies indicated that GaAs affects all cells involved in the generation of a primary antibody response (macrophage, and T- and B-lymphocytes). Investigations involving the adherent population (macrophages) demonstrated that the suppression could be enhanced by titrating GaAs-exposed macrophages into reconstituted wells containing control adherent and non-adherent cells (e.g. T- and B-lymphocytes). This suppression was not due to induction of suppressor macrophages or release of prostaglandins.

Mechanistic studies on the macrophage revealed that phagocytosis of latex covaspheres and interleukin-1 (IL-1) production were not decreased in GaAs-exposed macrophages (Sikorski *et al.*, 1991a). However, exposed macrophages were impaired in their ability to present SRBC to SRBC-primed lymph node T-lymphocytes. Presentation of soluble antigens such as keyhole limpet hemocyanin (KLH), pigeon cytochrome c, and DASP (a synthetic fragment of cytochrome c) by GaAs-exposed macrophages to T-lymphocytes primed for each antigen were not suppressed. Flow cytometric analysis showed that although the number of GaAs-exposed macrophages expressing the class II molecule Ia^k (required for proper interaction of the macrophage with the T-lymphocyte) was not different from vehicle-exposed macrophages, the total amount of Ia^k on any individual GaAs-exposed macrophage was decreased. From these data it was concluded that GaAs alters splenic macrophage function by altering a step or steps involved in the processing or presentation of particulate antigens. Studies by Lewis *et al.* (1996) and Hartmann and McCoy (1996) have extended this observation. In the investigation by Lewis *et al.* (1996), GaAs was administered as a single intraperitoneal exposure (200 mg/kg) and immune parameters examined in the spleen 1 or 5 days after exposure. The results indicated that GaAs exposure resulted in suppression of $CD4^+$ T-lymphocyte responses (IL-2 production) to the soluble antigens pigeon cytochrome c and pork insulin, but not to hen egg lysozyme or chicken ovalbumin. Peptide presentation was unaffected. These observations were apparent 5 days after exposure, but not 24 h after GaAs exposure (the time utilized previously; Sikorski *et al.*, 1991a). Together with the studies by Sikorski *et al.* (1991a) these studies indicate that GaAs exposure induces a time- and antigen-dependent defect in antigen processing which is required for $CD4^+$ T-lymphocyte stimulation by splenic macrophages.

Hartmann and McCoy (1996) utilized the same exposure regime as Lewis and evaluated antigen processing and presentation of the same antigens and their peptide fragments by peritoneal macrophages. Twenty eight percent of

the peritoneal macrophages from GaAs-exposed mice contained GaAs particles as determined by light microscopy. Those investigations demonstrated
that intraperitoneal exposure enhanced MHC class II surface expression and
processing of the antigens, but not presentation of the peptides. This enhancement was not due to non-specific actions involved in particle phagocytosis. The
authors concluded that direct exposure of peritoneal macrophages to GaAs
particles may augment processing, but not presentation, of antigens by
macrophages. In addition, these investigators plausibly propose that the
paradoxical local inflammatory reactions observed in the lung following
intratracheal exposure to GaAs (Webb *et al.*, 1986), with the simultaneous
systemic immunosuppression, may be a result of the direct exposure of alveolar macrophages to GaAs particles. This may result in alveolar macrophage
activation which could contribute to inflammation, pulmonary damage, and
enhanced susceptibility to respiratory diseases. Interestingly, the only host
resistance model to show suppression consistently (following intratracheal
GaAs exposure) was the B16F10 tumor, which lodges primarily in the lung
after intravenous administration (Sikorski *et al.*, 1989).

Exposure (24 h) to a single intratracheal administration of GaAs (200 mg/
kg) not only suppresses activities associated with antigen processing, but also
alters T-lymphocyte-mediated immunologic functions (Sikorski *et al.*, 1989).
Gallium arsenide has also been shown to exert toxic effects on events occurring early in the AFC response, which may include lymphocyte activation and
proliferation (Burns *et al.*, 1991). Extensions of these observations revealed
that GaAs (200 mg/kg) decreased the ability of whole splenocytes to proliferate in response to antigen stimulation when compared with control cultures
(Burns and Munson, 1993a). Isolated GaAs-exposed T-lymphocytes were significantly suppressed in their ability to proliferate when stimulated by ConA,
PHA, anti-CD3, and IL-2, while isolated B-lymphocytes exhibited no difference in proliferative capacity between vehicle- and GaAs-exposed cells when
introduced to various stimuli. Analysis of surface receptors revealed that
expression of CD25 (the IL-2 receptor), leukocyte function antigen-I (LFA-1),
and intercellular adhesion molecules (ICAM-1) in GaAs-exposed mice were
significantly below those of vehicle-exposed mice (36, 18, and 18 percent,
respectively) (Burns and Munson, 1993b). Although expression of these molecules was upregulated by T-lymphocyte/cell reactivity (TCR) and IL-2 stimulation, a level of expression equal to vehicle was never obtained (Burns and
Munson, 1993a). These data support the studies of Petres *et al.* (1977), Wen *et
al.* (1991), McCabe *et al.* (1983), and Ostrosky-Wegman *et al.* (1991), which
also indicate that GaAs selectively inhibits T-lymphocyte proliferation, possibly by interfering with primary and secondary signals involved in mitogenic
and antigen-driven responses.

In addition to modulating T-lymphocyte proliferative capacity, exposure to
GaAs also alters production of soluble factors, which are critical for the proper
generation of a primary immune response (Burns and Munson, 1993b,c).
Supernatants from *in vivo* and *in vitro* vehicle-exposed splenocyte cultures

reversed GaAs-induced suppression of the *in vitro*-generated primary AFC response produced by both *in vitro* (50 μM) and *in vivo* (200 mg/kg) exposure to GaAs; these effects were observed to be dose (25–100%) and time dependent (24–36 h postimmunization). The concentrating of 24-h vehicle supernatants, along with treatment with proteinases, revealed that the reversing factors were proteinaceous in nature, with molecular masses of 5000 to 50000 Da; this weight range encompasses many of the lymphokines known to be necessary for the generation of an immune response. Evaluation of antibody cultures demonstrated that GaAs exposure (50 μM or 200 mg/kg) altered production of IL-2, -4, -5, and -6. Interestingly, the alterations in lymphokine production differed between the *in vivo* and *in vitro* exposure regimes. IL-2 (6.25–50 ng/ml) was able to dose-dependently reverse GaAs-induced suppression (*in vivo* exposure) and was also dependent on the concentration of GaAs (50–200 mg/kg). It would appear from these and previous data that IL-2 and/or its receptor was a primary target for GaAs following *in vivo* exposure. This action of an arsenical on lymphokine production was seen previously by Gainer (1972), where NaAsO$_2$ and *p*-arsenilic acid decreased both the production and activity of IFN. In the GaAs studies, it is not known whether the compound was altering the production or activity of IL-2.

Two possible confounding factors in the studies involving GaAs are the contribution of pulmonary inflammation to the observed immunotoxicity (as a result of instillation of a particulate agent in the lung) and the induction of corticosterone as a result of associated respiratory distress. Corticosterone is valued in medical practice for its potent immunosuppressive activity. Instillation of tantalum (an inert particulate compound) showed clearly that the presence of a particulate in the lung does not alter spleen or thymus weight, lymphocyte subpopulations, or the AFC response. Investigations utilizing the glucocorticoid antagonist RU-486 (Mifepristone, Hoecht-Roussel) revealed that a high concentration of serum corticosterone results from GaAs exposure (50–200 mg/kg; 491–757 ng/ml corticosterone compared with 86 ng/ml in controls). This elevation in serum corticosterone is responsible for the observed decrease in splenic and thymic cellularity. However, the elevated corticosterone level does not appear to contribute to the GaAs-induced suppression of the AFC response (Burns *et al.*, 1994). These data indicate that GaAs (presumably the arsenic component) exerts a direct immunosuppressive effect that is independent of the induction of endogenous corticosteroids and which is not a result of pulmonary inflammation.

When immunotoxicity of xenobiotics has been identified, it is often of interest to determine whether this toxicity is manifested as a change in host resistance to several pathogens. This approach may aid those involved in risk characterization. Following a single intratracheal exposure to GaAs, several investigators evaluated host resistance to several infectious agents including *Streptococcus pneumoniae* and *Listeria monocytogenes*, as well as to the B16F10 melanoma cell line (Sikorski *et al.*, 1989; Burns *et al.*, 1993). Investigations by Sikorski *et al.* (1989) demonstrated that 14 days after intratracheal exposure to GaAs (50 to 200 mg/kg) there was an increased incidence and growth of the B16F10 tumor in the lungs of these exposed animals. In contrast, there was no change in susceptibility to *S. pneumoniae* and protection against *L. monocytogenes* infection. Burns *et al.* (1993) examining host resistance 24 h after GaAs exposure (50–200 mg/kg) noted similar results, with the exception that resistance to *S. pneumoniae* was enhanced. These investigators had previously demonstrated the presence of detectable concentrations of dissociated gallium and arsenic in the blood, lungs, and spleen of mice 2 to 24 h after exposure to GaAs (200 mg/kg) (Burns *et al.*, 1991), and that arsenic was the primary immunosuppressive component.

Serial dilutions of GaAs (0.039 to 5 mg/ml) in brain–heart infusion (BHI) broth inoculated with either *S. pneumoniae* or *L. monocytogenes* slowed the growth of both organisms with a minimal inhibitory concentration (MIC) of 0.625 mg/ml (Burns *et al.*, 1993). Furthermore, sera from mice collected at various time intervals after exposure to GaAs (200 mg/kg) were also capable of retarding the growth of both organisms with the maximal inhibition noted 24 h after exposure. However, sera from GaAs-exposed mice (24 h postexposure) were incapable of slowing the growth of the B16F10 melanoma. These *in vitro* observations are consistent with the *in vivo* observations noted above. Direct addition of DMSA to sera obtained from mice exposed to GaAs followed by *in vitro* inoculation of that sera with *L. monocytogenes* resulted in growth of this organism, which was comparable with growth observed in vehicle cultures. These studies demonstrate that arsenic in the sera of GaAs-exposed mice may exert chemotherapeutic effects on *S. pneumoniae* and *L. monocytogenes*. These results are important to immunotoxicology because they are among the first to demonstrate directly the intricate interplay between the host, the pathogen, and exposure to a xenobiotic. While certain microbes appear to be sensitive to growth inhibition by arsenic, the B16F10 tumor was not. The reason for this contradiction is not known but may lie in the metabolic difference between normal and transformed cells. This explanation may be applicable to the studies of Schrauzer and Ishmael (1974), as well as to those of Kerkvliet *et al.* (1980).

1.6.3 *Biological indicators of arsenic exposure*

There are no specific biochemical parameters which reflect inorganic arsenic toxicity, and evaluation should be done with knowledge of the individual's known and potential exposure history. Urine, blood, and hair represent possible biological indicators of arsenic exposure. The rapid elimination of arsenic from the blood makes it only a reasonable indicator of exposure when examined within hours to a few days after an acute exposure. Hair or nails may also provide some aid in assessing exposure history; but interpretation is clouded by differentiating external contamination. By far, urine arsenic is the best indicator of recent arsenic exposure. It should be noted, however, that some

marine organisms contain significant quantities of arsenic which may be eliminated unchanged in the urine from the mammalian system.

To assess potential arsenic gas exposure, hemolysis and erythroid regeneration may be good indicators (Klimecki and Carter, 1995). Both blood and
urinalysis should assay for the presence of free hemoglobin and for other signs
of hemolysis. It has been suggested that δ-aminolevulinic acid dehydratase
(ALAD) may be a good indicator of erythrocyte maturity in rodents and may
therefore represent a potential biomarker for human monitoring.

The semiconductor industry represents a potential exposure of persons not
only to indium, gallium, and arsenic, but to binary and ternary compounds of
these elements (Conner *et al.*, 1993). Regarding these semiconductor materials, investigations into biological indicators of exposure have revealed that
heme oxygenase, an enzyme in the catabolism of heme, is inducible following
InAs exposure as are several other stress proteins. In addition, urinary
porphyrin excretion patterns have also been shown to be altered by arsenic,
indium, and InAs (Woods and Fowler, 1978; Conner *et al.*, 1995) as well as by
GaAs (Webb *et al.*, 1984). With further investigation, the porphyrins or enzymes in the heme pathway may possess significant utility as biomarkers for
exposure and/or toxicity to these compounds. More studies involving occupational exposure of human subjects are needed in order to determine the actual
risk these workers assume in industries that manufacture or utilize arsenicals.
Those future studies, in addition to those presented herein, will also aid in re-
evaluating RELs for the various forms of arsenic, including GaAs and InAs, as
well as for the various routes of exposure.

Acknowledgments

The author wishes to acknowledge Dr James M. McKim, Jr and Dr David J. Neun
(Dow Corning Corporation) for critical review of the manuscript prior to submission
for publication.

References

APOSHIAN, H.V. (1989). Biochemical toxicology of arsenic. *Rev. Biochem. Toxicol.*
10, 265–299.
ARANYI, C., BRADOF, J.N., and O'SHEA, W.J. (1985). Effects of arsenic trioxide
inhalation exposure on pulmonary antibacterial defenses in mice. *J. Toxicol.
Environ. Health* **15**, 163–172.
BARON, D., KUNICK, I., FRISCHMUTH, I., and PETRES, J. (1975). Further *in vitro*
studies on the biochemistry of the inhibition of nucleic acid and protein synthesis
induced by arsenic. *Arch. Dermatol. Res.* **253**, 15–22.
BLAIR, P.C., THOMPSON, M.B., MORRISSEY, R.E., MOORMAN, M.P., SLOANE, R.A.,
and FOWLER, B.A. (1990). Comparative toxicity of arsine gas in B$_6$C$_3$F$_1$ mice,
Fischer 344 rats, and Syrian golden hamsters: system organ studies and

comparison of clinical indices of exposure. *Fundam. Appl. Toxicol.* **14**, 776–787.

BLAKELY, B.R., SISODIA, C.S., and MUKKUR, T.K. (1980). The effect of methylmercury, tetraethyl lead, and sodium arsenite on the humoral immune response in mice. *Toxicol. Appl. Pharmacol.* **52**, 245–254.

BRIGGS, T.M. and OWENS, T.W. (1980). *Industrial hygiene characterization of the photovoltaic cell industry.* NIOSH Technical Report, DHEW (NIOSH) publication pp. 80–112, Cincinnati, OH: US Department of Health, Education and Welfare.

BROWN, J.L. and KITCHIN, K.T. (1996). Arsenite, but not cadmium, induces ornithine decarboxylase and heme oxygenase in rat liver: relevance to arsenic carcinogenesis. *Cancer Lett.* **98**, 227–231.

BURNS, L.A. and MUNSON, A.E. (1993a). Reversal of gallium arsenide-induced suppression of the AFC response by vehicle supernatants. II. Nature and identification of reversing factors. *J. Pharmacol. Exp. Ther.* **265**, 150–158.

(1993b). Gallium arsenide selectively inhibits T-cell proliferation and alters expression of CD25 (IL-2R/p55). *J. Pharmacol. Exp. Ther.* **265**, 178–186.

(1993c). Reversal of gallium arsenide-induced suppression of the AFC response by vehicle supernatants. I. Pharmacokinetics following *in vitro and in vivo* exposure. *J. Pharmacol. Exp. Ther.* **265**, 144–149.

BURNS, L.A., SIKORSKI, E.E., SAADY, J., and MUNSON, A.E. (1991). Evidence for arsenic as the primary immunosuppressive component of gallium arsenide. *Toxicol. Appl. Pharmacol.* **110**, 157–169.

BURNS, L.A., McCAY, J.A., and MUNSON, A.E. (1993). Arsenic in the sera of gallium arsenide-exposed mice inhibits bacterial growth and increases host resistance. *J. Pharmacol. Exp. Ther.* **265**, 795–800.

BURNS, L.A., SPRIGGS, T.L., and MUNSON, A.E. (1994). Gallium arsenide-induced increase in serum corticosterone is not responsible for suppression of the IgM antibody response. *J. Pharmacol. Exp. Ther.* **268**, 740–746.

BURNS, L.A., MEADE, B.J., and MUNSON, A.E. (1995). Toxic responses of the immune system, in Klaassen, C.D. (Ed.) *Casarett and Doull's Toxicology: The Basic Science of Poisons*, 5th Edn, pp. 355–402, New York: McGraw-Hill.

CASTRANOVA, V., BOWMAN, L., WRIGHT, J.R., COLBY, H., and MILES, P.R. (1984). Toxicity of metallic ions in the lung: effect of alveolar macrophages and alveolar type II cells. *J. Toxicol. Environ. Health* **13**, 845–856.

CEBRIAN, M.E. (1987). Some potential problems in assessing the effects of chronic arsenic exposure in north Mexico. Presented at the Americal Chemical Society, Division of Environmental Chemistry, 194th Meeting.

CEBRIAN, M.E., ALBORES, A., AGUILAR, M., and BLAKELEY, E. (1983). Chronic arsenic exposure in the north of Mexico. *Human Toxicol.* **2**, 121–133.

CONNER, E.A., YAMAUCHI, H., FOWLER, B.A., and AKKERMAN, M. (1993). Biological indicators for monitoring exposure/toxicity from III–V semiconductors. *J. Exposure Anal. Environ. Epidem.* **3**, 431–440.

CONNER, E.A., YAMAUCHI, H., and FOWLER, B.A. (1995). Alterations in the heme biosynthetic pathway from the III–V semiconductor metal, indium arsenide (InAs). *Chem.-Biol. Interact.* **96**, 273–285.

CRECELIUS, E.A. (1977). Changes in the chemical speciation of arsenic following ingestion by man. *Environ. Health Perspect.* **19**, 147–150.

DICKERSON, O.B. (1994). Antimony, arsenic, and their compounds, in Zenz, C., Dickerson, O.B., and Horvath, E.P., Jr (Eds) *Occupational Medicine*, 3rd Edn, pp. 468–472. St Louis: Mosby.

EPA (Environmental Protection Agency) (1987). *Special Report on Ingested Inorganic Arsenic: Skin Cancer and Nutritional Essentiality*. Risk Assessment Forum, Environmental Protection Agency, Washington, DC.

FISHER, G.G., McNIELL, K.L., and DEMOCKO, C.J. (1986). Trace element interactions affecting pulmonary macrophage cytotoxicity. *Am. J. Physiol.* **252**, C677–C683.

FOWLER, B.A., ISHINISHI, N., TSUCHIYA, K., and VAHTER, M. (1979). Arsenic, in Friberg, L. (Ed.) *Handbook on the Toxicology of Metals*, pp. 293–319, New York: Elsevier.

GAINER, J.H. (1972). Effects of arsenicals on interferon formation and action. *Am. J. Vet. Res.* **33**, 2579–2586.

GAINER, J.H. and PRY, T.W. (1972). Effects of arsenicals on viral infections in mice. *Am. J. Vet. Res.* **33**, 2299–2307.

GOERING, P.L., MARONPOT, R.R., and FOWLER, B.A. (1988). Effect of intratracheal gallium arsenide administration on δ-aminolevulinic acid dehydratase in rats: relationship to urinary excretion of aminolevulinic acid. *Toxicol. Appl. Pharmacol.* **92**, 179–193.

GONSEBATT, M.E., VEGA, L., MONTERO, R., GARCIA-VARGAS, G., DEL RAZO, L.M., ALBORES, A., CEBRIAN, M.E., and OSTROSKY-WEGMAN, P. (1994). Lymphocyte replicating ability in individuals exposed to arsenic via drinking water. *Mutat. Res.* **313**, 293–299.

GOYER, R.A. (1996). Toxic effects of metals, in Klaassen, C.D. (Ed.) *Casarett and Doull's Toxicology: The Basic Science of Poisons*, pp. 691–698, New York: McGraw-Hill.

HARRISON, R.J. (1986). Gallium arsenide. *State of the Art Reviews: Occup. Med.* **1**, 49–58.

HARTMANN, C.B. and McCOY, K.L. (1996). Gallium arsenide augments antigen processing by peritoneal macrophages for CD4$^+$ helper T-cell stimulation. *Toxicol. Appl. Pharmacol.* **141**, 365–372.

HINDMARSH, J.T. and McCURDY, R.F. (1986). Clinical and environmental aspects of arsenic toxicity. *CRC Crit. Rev. Clin. Lab. Sci.* **23**, 315–347.

HONG, H.L., FOWLER, B.A., and BOORMAN, G.A. (1989). Hematopoietic effects in mice exposed to arsine gas. *Toxicol. Appl. Pharmacol.* **97**, 173–182.

JELINEK, C.F. and COMELIUSSEN, P.E. (1977). Levels of arsenic in the United States food supply. *Environ. Health Perspect.* **19**, 83–87.

JENKINS, G.C., IND, J.E., KRAZANTZIS, G., and OWEN, R. (1965). Arsine poisoning: massive haemolysis with minimal impairment of renal function. *Br. Med. J.* **2**, 78–80.

KERKVLIET, N.I., STEPPAN, L.B., KOLLER, L.D., and EXON, J.H. (1980). Immunotoxicology studies of sodium arsenate effects of exposure on tumor growth and cell-mediated tumor immunity. *J. Environ. Pathol. Toxicol.* **4**, 65–79.

KLIMECKI, W.T. and CARTER, D.E. (1995). Arsine toxicity: chemical and mechanistic implications. *J. Toxicol. Environ. Health.* **46**, 399–409.

KOLLER, L.D. (1980). Immunotoxicology of heavy metals. *Int. J. Immunopharmacol.* **2**, 269–279.

LABEDZKA, M., GULYAS, H., SCHMIDT, N., and GERCKEN, G. (1989). Toxicity of metallic ions and oxides to rabbit alveolar macrophages. *Environ. Res.* **48**, 255–274.

LANTZ, R.C., PARLIMAN, G., CHEN, G.J., and CARTER, D.E. (1994). Effect of arsenic exposure on alveolar macrophage function. I. Effect of soluble As(III) and As(V). *Environ. Res.* **67**, 183–195.

LEWIS, T.A., MUNSON, A.E., and McCOY, K.L. (1996). Gallium arsenide selectively suppresses antigen processing by splenic macrophages for CD4[+] T-cell activation. *J. Pharmacol. Exp. Ther.* **278**, 1244–1251.

LISIEWICZ, J. (1993). Immunotoxic and hematotoxic effects of occupational exposures. *Folia Med. Cracov.* **34**, 29–47.

McCABE, M., MAGUIRE, D., and NOWAK, M. (1983). The effects of arsenic compounds on human and bovine lymphocyte mitogenesis *in vitro*. *Environ. Res.* **31**, 323–331.

McCAY, J.A., SIKORSKI, E.E., WHITE, K.L., and MUNSON, A.E. (1997). Immune modulation produced by intratracheal instillation of gallium arsenide. *Fundam. Appl. Toxicol.* (submitted).

MUNRO, I.C., CHARBONNEAU, S.M., SANDI, E., SPENCER, K., BRUCE, F., and GRICE, H.C. (1974). In *Proceedings of the 13th Annual Meeting of the Society of Toxicology*, Washington, DC, pp. 1–9.

NAS (National Academy of Sciences) (1977). *Arsenic. Medical and Biological Effects of Environmental Pollutants.* Washington, DC: National Research Council.

NIOSH (National Institute for Occupational Safety and Health) (1982). NIOSH Testimony to US Department of Labor: Comments at the OSHA Arsenic Hearing, July 14, 1982. NIOSH Policy Statement. Cincinnati, OH. US Department of Health and Human Services, PHS, CDC, NIOSH.

OSTROSKY-WEGMAN, P., GONSEBATT, M.E., MONTERO, R., VEGA, L., BARBA, H., ESPINOSA, J., PALAO, A., CORTINAS, C., GARCIA-VARGAS, G., DEL RAZO, L.M., and CEBRIAN, M. (1991). Lymphocyte proliferation kinetics and genotoxic findings in a pilot study on individuals chronically exposed to arsenic in Mexico. *Mutat. Res.* **250**, 477–482.

PETRES, J., BARON, D., and HAGEDORN, M. (1977). Effects of arsenic on cell metabolism and cell proliferation: cytogenetic and biochemical studies. *Environ. Health Perspect.* **19**, 223–227.

PIERSON, B., WAGENEN, S.V., NEBESNY, K.W., FERNANDO, Q., SCOTT, N., and CARTER, D.E. (1989). Dissolution of crystalline gallium arsenide in aqueous solutions containing complexing agents. *Am. Ind. Hyg. Assoc. J.* **50**, 455–459.

ROSENTHAL, G.J., FORT, M.M., GERMOLEC, D.R., ACKERMANN, M.F., LAMM, K.R., BLAIR, P.C., FOWLER, B.A., LUSTER, M.I., and THOMAS, P.T. (1989). Effect of subchronic arsine inhalation on immune function and host resistance. *Inhal. Toxicol.* **1**, 113–127.

ROSNER, M.H. and CARTER, D.E. (1987). Metabolism and excretion of gallium arsenide and arsenic oxides by hamsters following intratracheal instillation. *Fundam. Appl. Toxicol.* **9**, 730–737.

SCHRAUZER, G.N. and ISHMAEL, D. (1974). Effects of selenium and of arsenic on the genesis of spontaneous mammary tumors in inbred C_3H mice. *Ann. Clin. Lab. Sci.* **4**, 441–446.

SCHROEDER, H.A. and BALASSA, J.J. (1966). Abnormal trace metals in man: Arsenic. *J. Chron. Dis.* **19**, 85–106.

SIKORSKI, E.E., McCAY, J.A., WHITE, K.L., BRADLEY, S.G., and MUNSON, A.E. (1989). Immunotoxicity of the semiconductor gallium arsenide in female B$_6$C$_3$F$_1$ mice. *Fundam. Appl. Toxicol.* **13**, 843–858.

SIKORSKI, E.E., BURNS, L.A., McCOY, K.L., STERN, M., and MUNSON, A.E. (1991a). Suppression of splenic accessory cell function in mice exposed to gallium arsenide. *Toxicol. Appl. Pharmacol.* **110**, 143–156.

(1991b). Splenic cell targets in gallium arsenide-induced suppression of the primary antibody response. *Toxicol. Appl. Pharmacol.* **110**, 129–142.

SNOW, E.T. (1992). Metal carcinogenesis: mechanistic implications. *Pharmacol. Ther.* **53**, 31–65.

SQUIBB, K.S. and FOWLER, B.A. (1983). The toxicity of arsenic and its compounds, in Fowler, B.A. (Ed.) *Biological and Environmental Effects of Arsenic*, pp. 233–269, Amsterdam: Elsiever.

VAHTER, M. (1981). Biotransformation of trivalent and pentavalent inorganic arsenic in mice and rats. *Environ. Res.* **25**, 286–293.

VOS, J.G. (1977). Immune suppression as related to toxicology. *CRC Crit. Rev. Toxicol.* **5**, 67–101.

WALDER, B.K., ROBERTSON, M.R., and KREMY, J. (1971). Skin cancer and immunosuppression. *Lancet* **ii**, 1282–1283.

WEBB, D.R., SIPES, I.G., and CARTER, D.E. (1984). *In vitro* solubility and *in vivo* toxicity of gallium arsenide. *Toxicol. Appl. Pharmacol.* **76**, 96–104.

WEBB, D.R., WILSON, S.E., and CARTER, D.E. (1986). Comparative pulmonary toxicity of gallium arsenide, gallium (III) oxide, or arsenic (III) oxide intratracheally instilled into rats. *Toxicol. Appl. Pharmacol.* **82**, 405–416.

WEBB, J.L. (1966). Arsenicals, in Webb, J.L. (Ed.) *Enzyme and Metabolic Inhibitors*, Vol. 3 pp. 595–793, New York: Academic Press.

WEN, W.N., LIEU, T.L., CHANG, H.J., WON, S.W., YAU, M.L., and JAN, K.Y. (1991). Baseline and sodium arsenite-induced sister chromatid exchanges in cultured lymphocytes from patients with Blackfoot disease and healthy persons. *Human Genet.* **59**, 201–203.

WHO (World Health Organization) (1981). Environmental Health Criteria No. 18 *Arsenic*, World Health Organization, Geneva.

WOODS, J.S. and FOWLER, B.A. (1978). Altered regulation of mammalian hepatic heme biosynthesis and urinary porphyrin excretion during prolonged exposure to sodium arsenate. *Toxicol. Appl. Pharmacol.* **43**, 361.

WU, M.M., KUO, T.B., HWANG, Y., and CHEN, C.J. (1987). Dose-response relation between arsenic concentration in well water and mortality from cancers and vascular diseases. *Am J. Epidemiol.* **130**, 1123–1132.

YAMAMOTO, S., KONISHI, Y., MATSUDA, T., MURAI, T., SHIBATA, M.A., MATSUI-YUASA, I., OTANI, S., KURODA, K., ENDO, G., and FUKUSHIMA, S. (1995). Cancer induction by an organic arsenic compound, dimethylarsenic acid (cacodylic acid) in F344/DuCrj rats after pretreatment with five carcinogens. *Cancer Res.* **55**, 1271–1276.

YAMAUCHI, H., TAKAHASHI, K., and YAMAMURA, Y. (1986). Metabolism and excretion of orally and intraperitoneally administered gallium arsenide in the hamster. *Toxicology* **40**, 237–246.

YAMAUCHI, H., TAKAHASHI, K., YAMAMURA, Y., and FOWLER, B.A. (1992). Metabolism of subcutaneous administered indium arsenide in the hamster. *Toxicol. Appl. Pharmacol.* **116**, 66–70.

YOSHIDA, T., SHIMAMURA, T., and SHIGETA, S. (1986). Immunological effects of arsenic compound on mouse spleen cells *in vitro*. *Tokai J. Exp. Clin. Med.* **11**, 353–359.

YOSHIDA, T., SHIMAMURA, T., KITAGAWA, H., and SHIGETA, S. (1987). The enhancement of the proliferative response of peripheral blood lymphocytes of workers in semiconductor plant. *Ind. Health* **25**, 29–33.

YOSHIDA, T., SHIMAMURA, T., and SHIGETA, S. (1991). Immunotoxicity of arsenic: immunological changes observed in the workers contacting with arsenic and in mice exposed to it, and their possible mechanisms. *Int. J. Immunopharmacol.* **13**, 772.

2

Beryllium

LEE S. NEWMAN

Division of Environmental and Occupational Health Sciences, National Jewish Medical and Research Center; and Department of Medicine and Department of Preventive Medicine and Biometrics, University of Colorado School of Medicine, Denver, CO 80206, USA

2.1 History

Beryllium (Be) is a unique steel-gray metal valued for its unusual metallic properties. It is the lightest solid, chemically stable element. Because of its light weight, excellent electrical and thermal conductivity, and its ability to resist corrosion and oxidation, beryllium has found wide application in industry.

The history of beryllium's role as a cause of occupational and environmental disease began in the 1930s in Europe with the recognition of lung disease. While beryllium has since been shown to produce skin, upper and lower respiratory, and systemic health effects, the predominant form of disease seen today is chronic beryllium disease (CBD) also known as berylliosis. Originally described in the United States by Dr Harriet Hardy in a group of fluorescent lamp manufacturing workers in Salem, Massachusetts, this chronic granulomatous lung disease was originally coined 'Salem sarcoid' because of its resemblance to granulomatous disease of unknown etiology, sarcoidosis. This disease continues to occur today both in workers exposed in industry as well as in family members exposed as bystanders to beryllium dust. As discussed below, beryllium's unique immunologic effects appear to play a central role in the pathogenesis of this disease.

2.2 Occurrence

Occupational and non-occupational exposures to beryllium occur in and around the mining, extraction, refining, recycling, and alloy manufacturing industries as well as in specialized industrial settings, such as in manufacture of electronics, computers, fiberoptics, automobile, aircraft, aerospace, and defense industries. Pure beryllium, beryllium–metal-oxide composite materials, and beryllium–aluminum and beryllium–copper alloys have found widespread use. While the typical alloys contain less than 2 percent beryllium, disease can result from exposures to alloy dust. Environmental exposure occurs at minute levels through fossil fuel combustion and through the small amounts detectable in food, tobacco, and drinking water.

The Occupational Safety and Health Administration (OSHA) permissible exposure limit and the American Conference of Governmental Industrial Hygienist (ACGIH) threshold limit value for beryllium are set at $2\,\mu g$ Be/m^3 as an 8-h time-weighted average. The short-term (30-min) exposure limit is $25\,\mu g$ Be/m^3. The National Institute for Occupational Safety and Health (NIOSH) recommended a time-weighted average exposure limit of $0.5\,\mu g$ Be/m^3. Recent studies suggest that adherence to this $2\,\mu g$ Be standard is not sufficiently protective (Cullen *et al.*, 1987; Kreiss *et al.*, 1996). Despite improvements in industrial hygiene practices in the beryllium industry, CBD continues to occur, even at relatively low levels of exposure (Kreiss *et al.*, 1989, 1993a,b, 1996). There does appear to be an exposure–response and a work task–response relationship in CBD. Attack rates average from 2 to 5 percent, but have been found to be as high as 16 percent among workers performing certain specific tasks such as machining of metal or its oxide (Kreiss *et al.*, 1989, 1993a,b, 1996). Industry bystanders, such as security guards, secretaries, and supervisors can also contract the disease, but at lower rates. It is not known at this time how low an exposure is sufficient to produce beryllium sensitization and CBD. However, the evidence suggests that adherence to current standards is not sufficiently protective and that reduction of exposure to the lowest levels achievable is advisable.

2.3 Essentiality

Beryllium is a non-essential element, despite being the 35th most abundant element in the Earth's crust. While beryllium has a number of interesting specific biochemical effects, the biological and clinical significance of some of these is poorly understood. For example, beryllium fluoride (BeF$_2$) has been shown to bind to actin, producing structural changes (Muhlrad *et al.*, 1994). Beryllium also inhibits a number of enzyme systems including myosin ATPase and myo-inositol monophosphatase, to name but a few (see Table 2.1) (Phan and Reisler, 1992; Faraci *et al.*, 1993). As discussed below, the interaction between beryllium's non-specific biological effects and its antigen-specific effects may hold the key to understanding why this unique material is so immunogenic and pathogenic.

Table 2.1   Selected data on general toxicology of beryllium

Toxicokinetics	
Absorption	Primarily lung, poor oral and dermal absorption
Distribution	Primarily lung > bone > liver, kidney
	Injected beryllium salts penetrate placenta in mice
Metabolism	Soluble salts partially convert to insoluble forms in lung
Excretion	Following inhalation, slow clearance of oxides via mucociliary escalator and urinary elimination
	Following ingestion, most passes through gut unabsorbed, excreted in feces, urine
Toxicity	
Human	Chronic beryllium disease (granulomatous disease), lungs and systemic
	Tracheobronchitis
	Acute pneumonitis
	Dermatitis, skin ulceration, skin nodules
	Nephrolithiasis, hyperuricemia
	Lung cancer
Animal	Chemical pneumonitis, multiple species
	Granulomatous pneumonitis, self-limited, multiple species
	Rickets (due to beryllium binding phosphate in gut)
	Cancer, multiple sites, multiple species, mainly via inhalational exposure
	Increased acid phosphatase, alkaline phosphatase, 5′-ribonucleotide phosphohydrolase activity
	Limited data regarding reproductive toxicity
Bacteria and cell lines	Gene mutations, multiple bacteria, cell lines
	Chromosomal abberations, human lymphocytes, yeast, cell lines
	Sister chromatid exchange, human lymphocytes, cell lines

2.4   General Toxicology

Soluble beryllium salts rapidly complex with plasma and other endogenous proteins. In biological systems these salts both augment and disrupt the actions of a variety of proteins. For example, BeF_2 in solution assumes a tetrahedral shape that can mimic a phosphate group. It has the capacity to bind G-proteins. What impact this may have on signal transduction has not been investigated to date, but certainly raises one alternative mechanism by which beryllium may exert its immunologic effects. Table 2.1 summarizes some of the general toxicologic effects of beryllium. Toxic effects on immune effector cells, including alveolar macrophages, lymphocytes, and on fibroblast proliferation have also been demonstrated, as discussed below.

It is probable that beryllium has direct cellular and genetic effects that are not related to antigen specificity. Earlier studies have demonstrated beryllium uptake in the cell nucleus (Firket, 1953) and specific binding to nuclear proteins (Parker and Stevens, 1979). Beryllium alters cellular enzyme induction and gene transcription (Witschi, 1970; Witschi and Marchand, 1971), and selectively regulates gene expression (Perry *et al.*, 1982). Unfortunately, follow-up studies of these early observations have not been published. We do not know, for example, which nuclear proteins preferentially bind beryllium or if this binding produces physiologically relevant effects at the gene level. At this time, it is reasonable to speculate that beryllium injures and primes the lung, skin, or other organs through one or more direct toxic mechanisms. This priming, when it occurs in the presence of beryllium-specific cellular immunity, may explain how the immune response is amplified, resulting in disease. The interaction of non-specific mechanisms with the antigen-specific mechanisms remains to be worked out through future research.

2.5 Immunotoxicology

As early as 1949, investigators suggested the possibility of an 'immune mechanism' underlying a new occupational lung disease that was emerging among workers exposed to beryllium (DeNardi *et al.*, 1949; Sterner and Eisenbud, 1951). In one of the first well-described outbreaks of beryllium-related lung disease, residents of a community surrounding a beryllium extraction plant were found to be developing lung disease at a rate similar to that occurring inside the beryllium plant but at much lower levels of exposure (Sterner and Eisenbud, 1951), fueling this hypothesis.

At our present state of knowledge, it appears that CBD results from the combined non-specific inflammatory effects of beryllium plus antigen-specific, acquired immunity. Installation of beryllium sulfate in mice induces an adjuvant-like effect, inducing Ia expression on macrophages (Behbehani *et al.*, 1985). These macrophages phagocytose and vacuolize the beryllium (Sanders *et al.*, 1975), followed later by a decrease in phagocytosis (Hart *et al.*, 1984). Beryllium has other adjuvant effects, directly increasing antigen presentation (Salvaggio *et al.*, 1965; Unanue *et al.*, 1969), mediated probably in part through its stimulation of the release of interleukin 1 (IL-1) (Unanue *et al.*, 1976) and tumor necrosis factor-α (TNF-α) (Behbehani *et al.*, 1985). Non-specific toxic effects on lymphocytes may also be of importance in the pathogenesis of beryllium-related diseases. Beryllium has been shown to interact directly with lymphocyte membrane proteins (Skilleter *et al.*, 1983; Price and Skilleter, 1985; Hall, 1988). In sheep, beryllium salts induce a polyclonal B-lymphocyte response *in vivo* (Hall, 1984, 1988; Denham and Hall, 1988), and is mitogenic for mouse B-lymphocytes, but not T-lymphocytes, *in vitro* (Newman and Campbell, 1987). Pathologic alterations following installation or inhalation of beryllium salts, oxides, and ore have been observed in a wide range of animal models including rats, hamsters, mice, guinea pigs, rabbits, cats, dogs, and non-human primates. These pathologic alterations range from acute injury to the lung resulting in organ dysfunction to a pattern of lymphocytic infiltration and granuloma formation analogous to CBD in humans (Stokinger *et al.*, 1950; Vorwald *et al.*, 1966; Robinson *et al.*, 1968; Wagner *et al.*, 1969; Sanders *et al.*, 1975; Barna *et al.*, 1981, 1984a,b; Hart *et al.*, 1984; Haley *et al.*, 1989, 1991; Finch *et al.*, 1996).

2.5.1   *Effects on humoral immunity*

Several early investigations demonstrated that beryllium induces B-lymphocyte activation and polyclonal production of immunoglobulin in patients with CBD (Deodhar *et al.*, 1973). In the 1990s, Clarke and colleagues detected beryllium-specific antibodies in individuals with beryllium exposure (Clarke, 1991; Clarke *et al.*, 1993, 1995). Measures of beryllium-specific antibodies in beryllium-exposed individuals may prove useful as a monitor of internal exposure to beryllium, based on some preliminary case studies (Clarke *et al.*, 1995). Interestingly, there appears to be no direct correlation between the presence of these beryllium-specific antibodies and the development of beryllium-specific T-lymphocytes and CBD in that there is no greater chance of detecting elevated beryllium antibody levels in the blood of workers with CBD compared with other plant employees.

2.5.2   *Effects on cell-mediated immunity*

The notion that CBD is the result of a hypersensitivity response to beryllium originated first with epidemiologic evidence that after beryllium exposure CBD did not occur in a classic dose–response manner. Some individuals with seemingly low exposures developed disease (DeNardi *et al.*, 1949; Sterner and Eisenbud, 1951). The first medical evidence supporting a role for beryllium-specific cell-mediated immunity emerged in 1951 when Curtis demonstrated that patients with CBD mounted a delayed-typed hypersensitivity response to beryllium salts on skin patch testing. Biopsies of patch test sites revealed sarcoid-like non-caseating granulomas (see Figure 2.1) (Curtis, 1951, 1959; Sneddon, 1955). Recent studies in the dermatology literature confirm that individuals who develop contact dermatitis from exposure to beryllium dental alloys also mount a beryllium-specific response on patch testing (Haberman *et al.*, 1993). The impracticality of patch-testing protocols, as well as the theoretical concern that patch testing itself might induce sensitization, led to the development of *in vitro* assays that measure beryllium-specific cellular immunity. These assays have helped inform our current understanding of the immunopathogenesis of CBD and aided in diagnosis and medical surveillance.

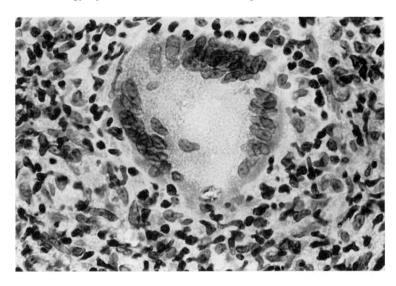

Figure 2.1 Photomicrograph (40×) of the central portion of a beryllium granuloma showing the typical aggregation of multinucleated giant cells, epithelioid cells, macrophages, and lymphocytes. This example was produced by application of beryllium sulfate as a patch test on the skin of a patient with chronic beryllium disease. The skin and lung granulomas are virtually indistinguishable.

Early *in vitro* cellular assays included the macrophage inhibitory factor assay and the morphologic scoring of lymphocyte 'blast' formation in culture (Price *et al.*, 1976; Williams and Jones Williams, 1982a). In 1970, Hanifin and colleagues (1970) demonstrated convincingly that beryllium salts affect both blood monocytes and lymphocytes isolated from patients with CBD. They described a beryllium concentration-dependent, *in vitro*, lymphocyte proliferative response to beryllium oxide and beryllium sulfate in patients who had known reactivity to patch tests. They also described phenotypic changes in blood monocytes from beryllium-sensitized individuals, noting that monocytes were larger, showed pseudopod formation, and had enhanced cytoplasmic granularity. Human blood macrophages that they preincubated with beryllium oxides showed an enhanced ability to phagocytose beryllium. Interestingly, if these macrophages were incubated along with autologous lymphocytes, the lymphocytes showed enhanced thymidine incorporation, even in the absence of any further addition of beryllium salts. These data suggested that macrophages which phagocytose beryllium have the capacity to present it as an antigen *in vitro*. Many investigators have verified this *in vitro* cellular immune response to beryllium salts using both blood and lung lymphocytes, the latter being obtained by bronchoalveolar lavage (van Ganse *et al.*, 1972; Price *et al.*, 1976; Epstein *et al.*, 1982; Williams *et al.*, 1982a,b; Jones Williams and Williams, 1983; Bargon *et al.*, 1986; Cullen *et al.*, 1987; Rossman *et al.*, 1988; Kreiss *et al.*, 1989, 1993a, 1996; Newman *et al.*, 1989, 1994; Saltini *et al.*,

1989; Mroz *et al.*, 1991). The beryllium lymphocyte proliferation assay has become a routine clinical test, used both in the screening of beryllium-exposed individuals for beryllium sensitization and early disease, as well as in diagnosis of chronic beryllium disease (Epstein *et al.*, 1982; Mroz *et al.*, 1991; Kreiss *et al.*, 1989, 1993a,b, 1996; Newman *et al.*, 1989, 1994). The lymphocyte response to beryllium salts in CBD is specific to each salt.

The demonstration of lymphocyte proliferation to beryllium salts has served as the point of departure for several interesting lines of investigation examining how metals like beryllium induce cellular immunity and produce a granulomatous response in humans. The major lymphocyte populations in-volved in this response are T-helper cells (CD4$^+$) as suggested by both *in vitro* lymphocyte studies and by the high proportion of lymphocytes found on bronchoalveolar lavage of patients with CBD (Rossman *et al.*, 1988; Newman *et al.*, 1989, 1994; Saltini *et al.*, 1989). As is typical in the lung, the great majority of these CD4$^+$ T-lymphocytes are 'memory' T-lymphocytes (CD45RO$^+$) (Saltini *et al.*, 1990). Most lines of evidence suggest that the blood and lung lymphocytes found in patients with CBD obey the same basic immunologic principles observed in other diseases in which cell-mediated immune re-sponses predominate. For example, beryllium-reactive T-lymphocytes re-spond to beryllium in a major histocompatibility complex (MHC) class II-dependent manner (Saltini *et al.*, 1989). Preliminary studies from our laboratory and others indicate that only a limited number of T-lymphocytes recognize beryllium as an antigen, based on T-lymphocyte antigen receptor utilization found in the lungs compared with blood of patients with CBD. We conclude that there is limited oligoclonality of T-lymphocytes in the lungs of these patients based on studies of the T-lymphocyte receptor V_β and V_α subunits (Rossman *et al.*, 1992; Comment *et al.*, 1994; Fontenot *et al.*, 1997). It seems likely that only a limited family of T-lymphocytes in the lungs of patients with CBD are able to recognize beryllium as an antigen. The true nature of the antigen remains unknown, although it is suspected to be a beryllium–protein conjugate. It seems unlikely that the beryllium moiety alone (molecular weight 9.01) can act as the antigen. More likely, beryllium–peptide complexes that form in physiologic solutions (Reeves, 1983; Hall, 1984) serve as the antigen. It is also possible that beryllium binds to other endogenous molecules that may affect the antigenicity of beryllium *in vivo*. Candidates that bind beryllium may include peptides, sulfated proteoglycans, and polysaccharides, such as heparin and condroitin sulfate (Hall, 1984). Future research will be required to elucidate the actual form of the antigen.

Antigen recognition by the beryllium-reactive T-lymphocyte plus macrophage activation induce a cascade of events leading to the amplification of the cellular response in the lung. With T-lymphocyte activation and prolif-eration comes release of cytokines involved in local immune and inflammatory responses by immune effector cells in the lungs. Specifically, in CBD, T-lymphocytes produce a so-called T-helper-1 pattern of cytokines: IL-2, and interferon-γ (IFN-γ), but not IL-4 (Saltini *et al.*, 1989; Tinkle *et al.*, 1996, 1997). Interestingly, beryllium stimulation of lung T-lymphocytes induces an unusually high and sustained production of IFN-γ that is regulated at the messenger RNA level (Tinkle *et al.*, 1997). IL-2 message and protein production also rise, but follow a much more limited time course in response to beryllium. Concomitantly, the macrophages in the lungs of patients with CBD produce an abundance of TNF-α and IL-6 (Bost *et al.*, 1994). This profile of lymphocyte and macrophage-derived cytokine production in the lungs of patients with CBD led us to postulate that a feedback amplification loop involving these cytokines within the pulmonary compartment results in the growth and maintenance of granulomas (see Figure 2.2). The extent to which beryllium exerts a direct modulatory effect on cytokine gene regulation in immune effector cells is an area of active investigation at this time.

Several lines of evidence suggest that genetic susceptibility contributes to the risk of beryllium sensitization and subsequent disease. In animal models,

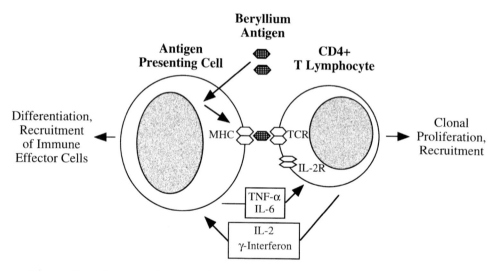

Figure 2.2 Schematic of proposed mechanism of beryllium-specific and non-specific effects leading to beryllium sensitization and chronic beryllium disease. Beryllium is phagocytosed, processed, and in some form, re-expressed on the surface of antigen-presenting cells complexed with the major histocompatibility complex (MHC) class II molecules. This antigen-presenting cell presents the antigen moiety to T-lymphocytes via the T-cell antigen receptor (TCR). At the same time, immune effector cells, such as macrophages, mast cells, and additional lymphocytes are recruited, proliferate, and/or differentiate at the site of inflammation, leading to granuloma formation. Antigen recognition by beryllium-specific T-cell clones triggers transient production of interleukin-2 (IL-2), upregulation of the IL-2 receptor (IL-2R), and sustained production of γ-interferon (IFN-γ) by lymphocytes, as well as tumor necrosis factor-α (TNF-α) and interleukin-6 (IL-6) production, presumably by activated macrophages.

MHC strain differences affect beryllium sensitization (Barna *et al.*, 1984a,b; Huang *et al.*, 1992). We and others have observed familial cases of CBD in which parents and offspring both worked with beryllium or in which a family member carried the dust home, leading to secondhand exposure and disease. CBD has been described in identical twins who both worked with beryllium (McConnochie *et al.*, 1988). Elegant work by Richeldi and colleagues demonstrated differences in the frequency of *HLA-DPβ1* gene alleles in patients with CBD compared with control subjects (Newman, 1993; Richeldi *et al.*, 1993). Similar differences in *HLA-DR* gene allelic frequencies have recently been reported by Stubbs and colleagues (Stubbs *et al.*, 1996). These data suggest that *HLA-DP* and probably *HLA-DR* serve as markers for susceptibility to CBD and beryllium sensitization among exposed individuals. In particular, the work by Richeldi suggests that the substitution of a glutamine in position 69 of the molecule encoded by *HLA-DPβ1* may be particularly influential in identifying who among beryllium-exposed workers are most likely to become diseased. Studies are underway at this time to determine if either the HLA-DP molecule or the HLA-DR molecule is involved directly in beryllium antigen presentation to CD4$^+$ T-lymphocytes. Stubbs and colleagues have presented preliminary data suggesting that when neutralizing antibodies directed against HLA-DR are added to beryllium-reactive T-lymphocytes and monocytes in culture, T-lymphocyte proliferation is blocked (Stubbs *et al.*, 1996). These lines of research on the immunogenetics suggest that the genetic susceptibility to beryllium is a multigene event. Ultimately, sensitization and disease only occur after the individual has been exposed to beryllium. Further investigations of how different chemical forms of beryllium exposure vary in their immunogenicity may ultimately lead to a clearer understanding of the interaction between genetic determinants of the immune response and environmental triggers of immune hyperreactivity.

Acknowledgments

I wish to thank my colleagues in the Division of Environmental and Occupational Health Sciences at the National Jewish Medical and Research Center for their collaboration and encouragement in our beryllium research. I thank Nina Rice for her expert secretarial assistance. I thank my patients for their willingness and enthusiasm for improving our understanding of this disease. Research supported by National Institutes of Health Grants R29 ES-04843, SCOR HL-27353, CDC/NIOSH CCU 812221, and General Clinical Research Center Grant M01 RR-00051.

References

BARGON, J., KRONENBERGER, H., BERGMANN, L., BUHL, R., MEIER-SYDOW, J., and MITROU, P. (1986). Lymphocyte transformation test in a group of foundry workers exposed to beryllium and non-exposed controls. *Eur. J. Respir. Dis.* **69**(Suppl. 136), 211–215.

BARNA, B.P., CHIANG, T., PILLARISETTI, S.G., and DEODHAR, S.D. (1981). Immuno-logic studies of experimental beryllium lung disease in the guinea pig. *Clin. Immunol. Immunopathol.* **20**, 402–411.

BARNA, B.P., DEODHAR, S.D., CHIANG, T., GAUTAM, S., and EDINGER, M. (1984a). Experimental beryllium-induced lung disease: I. Differences in immunologic response to beryllium compounds in strains 2 and 13 guinea pigs. *Int. Arch. Allergy Appl. Immunol.* **73**, 42–48.

BARNA, B.P., DEODHAR, S.D., GAUTAM, S., EDINGER, M., CHIANG, T., and McMAHON, J.T. (1984b). Experimental beryllium-induced lung disease: II. Analyses of bronchial lavage cells in strains 2 and 13 guinea pigs. *Int. Arch. Allergy Appl. Immunol.* **73**, 49–55.

BEHBEHANI, K., BELLER, D.I., and UNANUE, E.R. (1985). The effects of beryllium and other adjuvants on Ia expression by macrophages. *J. Immunol.* **134**, 2047–2049.

BOST, T.W., RICHES, D.W., SCHUMACHER, B., CARRÉ, P.C., KHAN, T.Z., MARTINEZ, J.A., and NEWMAN, L.S. (1994). Alveolar macrophages from patients with beryllium disease and sarcoidosis express increased levels of mRNA for TNF-α and IL-6 but not IL-1β. *Am. J. Respir. Cell. Mol. Biol.* **10**, 506–513.

CLARKE, S.M. (1991). A novel enzyme-linked immunosorbent assay (ELISA) for the detection of beryllium antibodies. *J. Immunol. Methods* **137**, 65–72.

CLARKE, S.M., THURLOW, S.M., and HILMAS, D.E. (1993). Occupational beryllium exposure and recovery monitored by beryllium antibody assay: a six year case study. *J. Occup. Med. Toxicol.* **2**, 277–283.

(1995). Application of beryllium antibodies in risk assessment and health surveillance: two case studies. *Toxicol. Ind. Health* **11**, 399–411.

COMMENT, C.E., KOTZIN B.L., SCHUMACHER B.A., and NEWMAN L.S. (1994). Preferential use of T-cell antigen receptors in beryllium disease. *Am. J. Respir. Crit. Care Med.* **149**, A264.

CULLEN, M.R., KOMINSKY, J.R., ROSSMAN, M.D., CHERNIACK, M.G., RANKIN, J.A., BALMES, J.R., KERN, J., DANIELE, R.P., PALMER, L., NAEGEL, G.P., McMANUS, K., and CRUZ, R. (1987). Chronic beryllium disease in a precious metal refinery: clinical, epidemiologic, and immunologic evidence for continuing risk from exposure to low level beryllium fume. *Am. Rev. Respir. Dis.* **135**, 201–208.

CURTIS, G.H. (1951). Cutaneous hypersensitivity due to beryllium: a study of thirteen cases. *AMA Arch. Dermatol. Syph.* **64**, 470–482.

(1959). The diagnosis of beryllium disease with special reference to the patch test. *AMA Arch. Ind. Health* **19**, 150–153.

DENARDI, J.M., VAN ORDSTRAND, H.S., and CARMODY, M.G. (1949). Acute dermatitis and pneumonitis in beryllium workers: review of 406 cases in 8-year period with follow-up on recoveries. *Ohio State Med. J.* **45**, 567–575.

DENHAM, S. and HALL, J.G. (1988). Studies on the adjuvant action of beryllium: III. The activity in the plasma of lymph efferent from nodes stimulated with beryllium. *Immunology* **64**, 341–344.

DEODHAR, S.D., BARNA, B., and VAN ORDSTRAND, H.S. (1973). A study of the immunologic aspects of chronic berylliosis. *Chest* **63**, 309–313.

EPSTEIN, P.E., DAUBER, J.H., ROSSMAN, M.D., and DANIELE, R.P. (1982). Bronchoalveolar lavage in a patient with chronic berylliosis: evidence for hypersensitivity pneumonitis. *Ann. Intern. Med.* **97**, 213–216.

FARACI, W.S., ZORN, S.H., BAKKER, A.V., JACKSON, E., and PRATT, K. (1993). Beryllium competitively inhibits brain myo-inositol monophosphatase, but unlike lithium docs not enhance agonist-induced inositol phosphatase accumulation. *Biochem. J.* **291**, 369–374.

FINCH, G.L., HOOVER, M.D., HAHN, F.F., NIKULA, K.J., BELINSKY, S.A., HALEY, P.J., and GRIFFITH, W.C. (1996). Animal models of beryllium-induced lung disease. *Environ. Health Perspect.* **104**, 973–979.

FIRKET, H. (1953). Mise en évidence histochimique du béryllium dans des cellules cult. *R. Soc. Biol. (Paris)* **147**, 167.

FONTENOT, A.P., KOTZIN, B.L., COMMENT, C.E., and NEWMAN L.S. (1997). Expansion of T-cell receptor variable regions in chronic beryllium disease (CBD). *Am. J. Respir. Cell. Mol. Biol*, in press.

HABERMAN, A.L., PRATT, M., and STORRS, F.J. (1993). Contact dermatitis from beryllium in dental alloys. *Contact Dermatitis* **28**, 157–162.

HALEY, P.J., FINCH, G.L., MEWHINNEY, J.A., HARMSEN, A.G., HAHN, F.F., MOOVER, M.D., and BICE, D.E. (1989). A canine model of beryllium-induced granulomatous lung disease. *Lab. Invest.* **61**, 219–227.

HALEY, P.J., FINCH, G.L., and HOOVER, M.D. (1991). The comparative pulmonary toxicity of beryllium metal and oxide in nonhuman primates. *Am. Rev. Respir. Dis.* **143**, A219.

HALL, J.G. (1984). Studies on the adjuvant action of beryllium: effect on individual lymph nodes. *Immunology* **53**, 105–120.

   (1988). Studies on the adjuvant action of beryllium: IV. The preparation of beryllium containing macromolecules that induce immunoblast responses *in vivo*. *Immunology* **64**, 345–351.

HANIFIN, J.M., EPSTEIN, W.L., and CLINE, M.J. (1970). *In vitro* studies of granulomatous hypersensitivity to beryllium. *J. Invest. Dermatol.* **55**, 284–288.

HART, B.A., HARMSEN, A.G., LOW, R.B., and EMERSON, R. (1984). Biochemical, cytological, and histological alterations in rat lung following acute beryllium aerosol exposure. *Toxicol. Appl. Pharmacol.* **75**, 454–465.

HUANG, H., MEYER, K.C., KUBAI, L., and AUERBACH, R. (1992). An immune model of beryllium-induced pulmonary granulomata in mice: histopathology, immune reactivity, and flow-cytometric analysis of bronchoalveolar lavage-derived cells. *Lab. Invest.* **67**, 138–146.

JONES WILLIAMS, W. and WILLIAMS, W.R. (1983). Value of beryllium lymphocyte transformation tests in chronic beryllium disease and in potentially exposed workers. *Thorax* **38**, 41–44.

KREISS, K., NEWMAN, L.S., MROZ, M.M., and CAMPBELL, P.A. (1989). Screening blood test identifies subclinical beryllium disease. *J. Occup. Med.* **31**, 603–608.

KREISS, K., MROZ, M.M., ZHEN, B., MARTYNY, J., and NEWMAN, L.S. (1993a). Epidemiology of beryllium sensitization and disease in nuclear workers. *Am. Rev. Respir. Dis.* **148**, 985–991.

KREISS, K., WASSERMAN, S., MROZ, M.M., and NEWMAN, L.S. (1993b). Beryllium disease screening in the ceramics industry: blood test performance and exposure–disease relations. *J. Occup. Med.* **35**, 267–274.

KREISS, K., MROZ, M.M., NEWMAN, L.S., MARTYNY, J., and ZHEN, B. (1996). Machining risk of beryllium disease and sensitization with median exposures below 2 µg/m^3. *Am. J. Ind. Med.* **30**, 16–25.

McConnochie, K., Williams, W.R., Kilpatrick, G.S., and Jones Williams, W. (1988). Chronic beryllium disease in identical twins. *Br. J. Dis. Chest* **82**, 431–435.

Mroz, M.M., Kreiss, K., Lezotte, D.C., Campbell, P.A., and Newman, L.S. (1991). Re-examination of the blood lymphocyte transformation test in the diagnosis of chronic beryllium disease. *J. Allergy Clin. Immunol.* **88**, 54–60.

Muhlrad, A., Cheung, P., Phan, B.C., Miller, C., and Reisler, E. (1994). Dynamic properties of actin. Structural changes induced by beryllium fluoride. *J. Biol. Chem.* **269**, 11852–11858.

Newman, L.S. (1993). To Be^{2+} or not to Be^{2+}: relating immunogenetics to occupational exposure. *Science* **262**, 197–198.

Newman, L.S. and Campbell, P.A. (1987). Mitogenic effect of beryllium sulfate on mouse B-lymphocytes but not T-lymphocytes *in vitro*. *Int. Arch. Allergy Appl. Immunol.* **84**, 223–227.

Newman, L.S., Kreiss, K., King, T.E., Jr, Seay, S., and Campbell, P.A. (1989). Pathologic and immunologic alterations in early stages of beryllium disease: re-examination of disease definition and natural history. *Am. Rev. Respir. Dis.* **139**, 1479–1486.

Newman, L.S., Bobka, C., Schumacher, B., Daniloff, E., Zhen, B., Mroz, M.M., and King, T.E., Jr (1994). Compartmentalized immune response reflects clinical severity of beryllium disease. *Am. J. Respir. Crit. Care Med.* **150**, 135–142.

Parker, V.H. and Stevens, C. (1979). Binding of beryllium to nuclear acidic proteins. *Chem.-Biol. Interact.* **26**, 167–177.

Perry, S.T., Kulkarni, S.B., Lee, K.L., and Kenney, F.T. (1982). Selective effect of the metallocarcinogen beryllium on hormonal regulation of gene expression in cultured cells. *Cancer Res.* **42**, 473–476.

Phan, B. and Reisler, E. (1992). Inhibition of myosin ATPase by beryllium fluoride. *Biochemistry* **31**, 4787–4793.

Price, C.D., Pugh, A., Pioli, E.M., and Jones Williams, W. (1976). Beryllium macrophage migration inhibition test. *Ann N.Y. Acad. Sci.* **278**, 204–211.

Price, R.J. and Skilleter, D.N. (1985). Stimulatory and cytotoxic effects of beryllium on proliferation of mouse spleen lymphocytes *in vitro*. *Arch. Toxicol.* **56**, 207–211.

Reeves, A.L. (1983). The immunotoxicity of beryllium, in Gibson, G.G., Hubbard, R., and Parke, D.V. (Eds) *Immunotoxicology*, pp. 261–282, London: Academic Press.

Richeldi, L., Sorrentino, R., and Saltini, C. (1993). HLA-DPβ1 glutamate 69: a genetic marker of beryllium disease. *Science* **262**, 242–244.

Robinson, F.R., Schaffner, F., and Trachtenberg, E. (1968). Ultrastructure of the lungs of dogs exposed to beryllium-containing dusts. *Arch. Environ. Health* **17**, 193–203.

Rossman, M.D., Kern, J.A., Elias, J.A., Cullen, M.R., Epstein, P.E., Preuss, O.P., Markham, T.N., and Daniele, R.P. (1988). Proliferative response of bronchoalveolar lymphocytes to beryllium: a test for chronic beryllium disease. *Ann. Intern. Med.* **108**, 687–693.

Rossman, M.D., Yan, H., Williams, W.V., and Weiner, D.B. (1992). Chronic beryllium disease: an immune response by restricted subfamilies of T-cells. *Am. Rev. Respir. Dis.* **145**, A415.

Saltini, C., Winestock, K., Kirby, M., Pinkston, P., and Crystal, R.G. (1989). Maintenance of alveolitis in patients with chronic beryllium disease by beryllium-specific helper T-cells. *N. Eng. J. Med.* **320**, 1103–1109.

SALTINI, C., KIRBY, M., TRAPNELL, B.C., TAMURA, N., and CRYSTAL, R.B. (1990). Biased accumulation of T-lymphocytes with 'memory'-type CD45 leukocyte common antigen gene expression on the epithelial surface of the human lung. *J. Exp. Med.* **171**, 1123–1140.

SALVAGGIO, J.E., FLAX, M.H., and LESKOWITZ, S. (1965). Studies in immunization: III. The use of beryllium as a granuloma-producing agent in Freund's adjuvant. *J. Immunol.* **95**, 846–854.

SANDERS, C.L., CANNON, W.C., POWERS, G.J., ADEE, R.R., and MEIER, D.M. (1975). Toxicology of high-fired beryllium oxide inhaled by rodents. 1. Metabolism and early effects. *Arch. Environ. Health* **30**, 546–551.

SKILLETER, D.N., PRICE, R.J., and LEGG, R.F. (1983). Specific G_1–S phase cell cycle block by beryllium as demonstrated by cytofluorometric analysis. *Biochem. J.* **216**, 773–776.

SNEDDON, I.B. (1955). Berylliosis: a case report. *Br. Med. J.* **i**, 1448–1450.

STERNER, J.H. and EISENBUD, M. (1951). Epidemiology of beryllium intoxication. *Arch. Ind. Hyg. Occup. Med.* **4**, 123–151.

STOKINGER, H.E., SPRAGUE, G.F., HALL, R.H., ASHENBURG, N.J., SCOTT, J.K., and STEADMAN, L.T. (1950). Acute inhalation toxicity of beryllium. *Arch. Ind. Hyg. Occup. Med.* **1**, 379–397.

STUBBS, J., ARGYRIS, E., LEE, C.W., MONOS, D., and ROSSMANN, M.D. (1996). Genetic markers in beryllium hypersensitivity. *Chest* **109**, 45S.

TINKLE, S.S., SCHWITTERS, P.W., and NEWMAN, L.S. (1996). Cytokine production by bronchoalveolar lavage cells in chronic beryllium disease. *Environ. Health Perspect.* **104**, 969–971.

TINKLE, S.S., KITTLE, L.A., SCHUMACHER, B.A., and NEWMAN, L.S. (1997). Beryllium induces IL-2 and IFN-γ, not IL-4 in berylliosis. *J. Immunol.* **158**, 518–526.

UNANUE, E.R., ASKONAS, B.A., and ALLISON, A.C. (1969). A role of macrophages in the stimulation of immune responses by adjuvants. *J. Immunol.* **103**, 71–78.

UNANUE, E.R., KIELY, J.M., and CALDERON, J. (1976). The modulation of lymphocyte functions by molecules secreted by macrophages: II. Conditions leading to increased secretion. *J. Exp. Med.* **144**, 155–166.

VAN GANSE, W.F., OLEFFE, J., VAN HOVE, W., and GROETENBRIE, J.C. (1972). Lymphocyte transformation in chronic berylliosis. *Lancet* **i**, 1023.

VORWALD, A.J., REEVES, A.L., and URBAN, E.J. (1966). Experimental beryllium toxicology, in Stokinger H.E. (Ed.) *Beryllium: Its Industrial Hygiene Aspects*, pp. 201–234, New York: Academic Press.

WAGNER, W.D., GROTH, D.H., HOLTZ, J.L., MADDEN, G.E., and STOKINGER, H.E. (1969). Comparative chronic inhalation toxicity of beryllium ores, bertrandite, and beryl, with production of pulmonary tumors by beryl. *Toxicol. Appl. Pharmacol.* **15**, 10–29.

WILLIAMS, W.R. and JONES WILLIAMS, W. (1982a). Comparison of lymphocyte transformation and macrophage migration inhibition tests in the detection of beryllium hypersensitivity. *J. Clin. Pathol.* **35**, 684–687.

(1982b). Development of beryllium lymphocyte transformation tests in chronic beryllium disease. *Int. Arch. Allergy Appl. Immunol.* **67**, 175–180.

WITSCHI, H.P. (1970). Effects of beryllium on deoxyribonucleic acid-synthesizing enzymes in regenerating rat liver. *Biochem. J.* **120**, 623–634.

WITSCHI, H.P. and MARCHAND, P. (1971). Interference of beryllium with enzyme induction in rat liver. *Toxicol. Appl. Pharmacol.* **20**, 565–572.

3

Cadmium

LOREN D. KOLLER

College of Veterinary Medicine, Oregon State University, Corvallis, OR 97331-4801, USA

3.1 Introduction

Cadmium, a silver-white metal that is not usually found in the environment in its pure form, occurs naturally in the Earth's crust in association with zinc ores. Cadmium forms a number of salts (chloride, oxide, sulfide, carbonate, selenide, sulfate) and is usually combined with these elements, with cadmium sulfate being the most prevalent form. Small amounts of cadmium are present in the air, water, and all soils and rocks. Thus, cadmium is present in meats, grains, dairy products, and vegetables.

Cadmium is used in the manufacturing of many consumer products such as batteries, stabilizers, alloys, pigments, metal coatings, and plastics. The largest source of environmental cadmium is from the disposal of rechargeable nickel–cadmium batteries. Cadmium enters the air primarily from the burning of coal and household waste, as well as from metal mining and refining processes. Cadmium can enter water via waste effluents, fertilizer, and spills or leaks from hazardous waste sites.

The largest potential sources of cadmium exposure for humans are food and cigarette smoke. Foods in the United States contain cadmium concentrations ranging from 2 to 40 ppb, while those in cigarette smoke range from 1000 to 3000 ppb. The average person consumes in food about 30 μg of cadmium daily. Air levels of cadmium in US cities range from 5 to 40 ng Cd/m^3, while most drinking water contains less than 1 ppb. Airborne cadmium is usually in the form of relatively insoluble cadmium oxide, while oral exposures are usually with highly soluble salts of cadmium, such as cadmium chloride ($CdCl_2$) and cadmium carbonate ($CdCO_3$). Occupational exposure of workers generally occurs by inhalation of cadmium fumes or dusts during smelting, battery manufacturing, soldering, and pigment production. Because the biological

41

half-life of cadmium is estimated to be 10 to 25 years, the metal tends to accumulate in the body. Approximately 5 percent of ingested cadmium is absorbed while retention of inhaled cadmium is much greater.

3.2   Toxicity

The target organs of concern for acute toxicity to cadmium are the lungs, while the kidneys and bones are the main target organs for chronic exposures. The physicochemical properties of the given cadmium compounds, like most of the inorganic metals, are important determinants of toxicity; for example, the solubilized forms are generally the most toxic. The rank order of solubility is: $CdCl_2$ > cadmium sulfate ($CdSO_4$) > cadmium oxide (CdO) > cadmium sulfide (CdS). These compounds can be absorbed through the skin, intestinal tract, or respiratory tract and transported in the blood throughout the body.

Cadmium is an irritant to the respiratory tree at high dosages. In animals, inhalation of cadmium results in pneumonia, fibrosis, and emphysema. Emphysema is known to occur in occupationally exposed workers.

The kidney is the most sensitive target organ for cadmium toxicity following chronic low-level exposure. Occupationally exposed workers develop a high incidence of abnormal renal function expressed by proteinuria and a decreased glomerular filtration rate.

Cadmium affects calcium metabolism which leads to calcium loss from bone, thereby decreasing the rate of bone formation. Likewise, many factors that affect calcium metabolism, such as renal disease, dietary deficiencies, vitamin D deficiency, and estrogen status, can contribute to the severity of cadmium toxicity to bone.

Cadmium has induced developmental toxicity in rodents following inhalation exposures. Delayed ossification, decreased locomotor activity, and impaired reflexes have occurred following exposures to low levels of cadmium. Oral exposures to cadmium can reduce both fetal and pup weights and, at higher doses, induce skeletal malformations.

3.3   Carcinogenicity

The Agency for Toxic Substances and Disease Registry in 1993 considered cadmium as a 'strong' carcinogen in animals, but with 'limited' carcinogenicity in humans. The Environmental Protection Agency classified cadmium as a probable human carcinogen (group B1) by inhalation in 1990. However, in 1993, the International Agency for Research in Cancer concluded that cadmium and its compounds were human carcinogens. This designation was based upon epidemiologic studies of occupationally exposed workers, animal studies, and the genotoxic effects in a variety of eucaryotic cell types, including human cells. The permissible exposure limit in the workplace in the United States is $5\,\mu g\,Cd/m^3$.

3.4 Metallothionein

Certain metals, such as zinc, cadmium, and mercury, enhance the production of the intracellular low-molecular-weight protein metallothionein that binds the metal and protects organs against toxicity. Small doses of these metals, when administered as a pretreatment, can induce metallothionein and thereby prevent toxic effects of subsequent larger doses of the same metal. Metallothionein binding within tissues renders the metal(s) non-toxic until the levels of metal(s) exceed a critical concentration. Metallothionein is present in most organs including the lung. Zinc can compete with cadmium for enzyme-binding sites and, thus, reduce cadmium induced renal toxicity.

3.5 Biomarkers

Biomarkers are indicators signaling events in biological systems. Biomarkers of cadmium exposure include analytical measurements in blood, urine, feces, liver, kidney, hair, and other tissues. Blood levels reflect recent exposures rather than body burden, while urinary cadmium does not correlate with recent exposure but, rather, reflects total body burden. Hair-associated cadmium is a useful measure of recent cadmium exposure; however, the hair itself can be contaminated by the environment. The liver and kidneys bioaccumulate cadmium, with impaired kidney function being a primary biomarker of effect in response to exposure.

3.6 Immunotoxicity

3.6.1 Immune system

The immune system can be a sensitive target organ for xenobiotic toxicity. The system is extremely complex and sophisticated, consisting of several immune pathways driven by numerous messengers to generate optimum responses. The ideal design is to assess the immunotoxic potential of xenobiotics in animal models. Experiments should emulate natural routes of exposure and composition(s) of the given chemical to be tested. The ensuing immune responses should be assessed using validated methodology that evaluates immune function.

Immune function is defined as the normal, special, or proper action of immunocytes and their secretory products (Koller, 1990). Action may be manifest by: macrophages phagocytosing foreign debris; B-lymphocytes producing antibody that neutralizes foreign antigen; T-lymphocytes, natural killer cells, or lymphokine-activated killer cells killing neoplastic cells; or a number of cytokines catalyzing certain segments of immune networks. Mitogen assays, on the other hand, do not provide a measure of immune function: (i) mitogens, such as concanavalin A (ConA) or phytohemagglutinin (PHA), stimulate cells to divide but do not permit the ability of the divided cells to perform their immune functions to be estimated; (ii) a specific subpopulation (helper, suppressor, cytotoxic, etc.) of cells may be affected while others are not; and (iii) mitogen responses frequently do not correlate with the responses of other reliable procedures that are representative of immune function. Likewise, antigens on the surface of cells are usually specific for identifying subpopulations of immunocytes but do not test their functional capacity. Should a positive correlation be shown to exist between non-functional and functional parameters, repetition of the correlation will permit the non-functional procedure to serve as a biological marker of effect for the functional assay.

In vitro immunotoxicology cannot duplicate the complexity of an intact immune response, which not only involves the sophisticated immune system but also frequently engages the endocrine and central nervous systems. Nevertheless, *in vitro* procedures can be used initially to screen xenobiotics for immunotoxic effects in addition to being extremely useful in dissecting and manipulating immune responses to obtain a better understanding of the underlying mechanisms of immunomodulation. Furthermore, human cells and tissues can be evaluated indirectly (*in vitro* exposures) and compared with animal responses. However, the underlying application of immunotoxicology is to interpret and relate data from animals to humans in order to correlate immune competence with disease resistance. Cadmium has also been tested in animal systems and by using human immunocytes *in vitro*.

3.6.2 *Effects of cadmium on host resistance*

Host resistance/susceptibility procedures evaluate the host's ability to combat disease successfully as a result of exposure to pathogenic microorganisms. These *in vivo* procedures utilize *in toto* an intact immune system that defends the host against invasion by infectious and neoplastic agents. Numerous bacteria, virus, parasite, and tumor models are available to test this phenomenon in rodent models. Several studies have been conducted using various host resistance models in animals that have been exposed to cadmium compounds.

Bacterial models used to assess host resistance in cadmium-exposed animals include: *Salmonella enteritidis*, *Salmonella gallinarium*, *Streptococcus* spp., *Escherichia coli*, *Klebsiella pneumoniae*, *Pseudomonas aeruginosa*, *Listeria monocytogenes*, *Pasteurella gallinarium*, and *Mycobacterium* spp. The chemical forms of cadmium included cadmium acetate ($Cd(CH_3COO^-)_2$),

CdCl$_2$, CdSO$_4$, and CdO, while the routes of administration were inhalation, oral, intraperitoneal, and intravenous. The animal species selected was primarily the mouse, with one study each in the rat and chicken.

Viruses used in host susceptibility studies include encephalomyocarditis virus (EMCV), influenza A2/Taiwan virus, coxsackievirus B3, and herpes simplex type 1 (HIV-1) and 2 (HIV-2). The chemical forms of cadmium tested were CdCl$_2$, CdSO$_4$, Cd(CH$_3$COO$^-$)$_2$, and CdO, while the routes of exposure were inhalation, ingestion, and intraperitoneal. These experiments were all conducted in laboratory mice.

The parasitic study used *Hexamita muris* in mice treated orally with CdCl$_2$. The tumor studies used a Moloney sarcoma virus (MSB) and MOPC-104E tumor model in mice that were administered CdCl$_2$ orally.

Cadmium chloride is the most soluble and toxic of the cadmium compounds. In mice treated intraperitoneally with CdCl$_2$, the effects varied from decreased to increased host resistance. For example, infection with *L. monocytogenes*, but not *K. pneumoniae* or *P. aeruginosa*, showed an increase in virulence in cadmium-treated Ham/ICR Swiss or C$_3$H mice (Berche *et al.*, 1980). These mice had been injected intraperitoneally with 10 µg CdCl$_2$ three times per week for 4 weeks. In another study (Bozelka and Burkholder, 1979), CdCl$_2$ injected intraperitoneally reduced host susceptibility to infection with mycobacteria. In contrast, C57BL mice injected intraperitoneally three times per week for 4 weeks with 0.5 mg CdCl$_2$/kg and infected with *L. monocytogenes* had enhanced susceptibility to infection (Simonet *et al.*, 1984). Similarly, an increase in mortality occurred in mice exposed to an aerosol of CdCl$_2$ prior to the inhalation of aerosolized streptococci (Gardner *et al.*, 1977). In B$_6$C$_3$F$_1$ mice treated with 0, 10, 50, or 250 ppm CdCl$_2$ in water for 90 days and exposed to an LD$_{27}$ of *L. monocytogenes*, none of the cadmium treatments significantly altered the infectivity of *L. monocytogenes* (Thomas *et al.*, 1985).

Swiss–Webster mice treated orally for 10 weeks with 0, 3, 30, or 300 ppm CdCl$_2$ had a dose-dependent (albeit non-significant) increase in mortality when challenged with an LD$_{35}$ of EMCV (Exon *et al.*, 1986). However, matched pairs challenged with an LD$_{75}$ EMCV had a dose-dependent decrease in mortality with the 300 ppm group reaching statistical significance. Exposure of B$_6$C$_3$F$_1$ mice to 10, 50, or 250 ppm CdCl$_2$ orally for 90 days had no statistically significant effect on mortality and mean survival time following primary or secondary challenge with influenza A2/Taiwan, HIV-1, or HIV-2 viruses (Thomas *et al.*, 1985). In another study when mice were injected intraperitoneally with CdCl$_2$, SLC-ICR mice had a significant increase in mortality rates and decreased time to death when injected with 0.09 mg CdCl$_2$ once and simultaneously inoculated with Japanese encephalitis virus (JEV) (Suzuki *et al.*, 1981). However, if mice were pretreated with CdCl$_2$ at the same dose at 10 days, or 3, 5, or 7 weeks prior to virus challenge, or if JEV was given intracerebrally, no effect of cadmium was observed. Cadmium administered at 2 mM orally for 10 weeks did not influence mortality due to coxsackievirus B3 infection (Ilback *et al.*, 1994).

When Swiss–Webster mice were treated with 3, 30, or 300 ppm $CdCl_2$ orally for 2 to 3 weeks and infected with the protozoan *Hexamita muris*, the animals had increased mortality rates (Exon *et al.*, 1975). Inbred C57BL mice given 3, 30, or 300 ppm $CdCl_2$ orally for 10 weeks and challenged with Moloney sarcoma virus-infected cells developed fewer tumors and had greater tumor regression rates than non-treated mice (Kerkvliet *et al.*, 1979). BALB/c mice treated with low levels (0.01 ppm) of $CdCl_2$ orally for 5 weeks had significantly less MOPC-104E tumor-induced mortality rates than controls (Gray *et al.*, 1982). However, in the same study, tumor-induced mortality in mice treated with higher levels (0.1 or 1.0 ppm) of $CdCl_2$ was no different from controls.

Oral exposure to $CdCl_2$ either has no effect on host resistance or may actually increase it. When injected intraperitoneally, the effects of $CdCl_2$ range from decreased to increased host susceptibility or, in some cases, no effect.

Fewer studies have been conducted with $CdSO_4$. Dietary exposure to $CdSO_4$ at levels of 20 to 400 ppm increased the resistance in chickens challenged with *S. gallinarium* but did not affect the mortality rates from *E. coli* (Hill, 1979). There was a significantly enhanced protection in Swiss Webster mice challenged with an LD_{35} of EMCV after either 3 or 30 ppm $CdSO_4$ exposures, and with an LD_{75} challenge of EMCV after the 300 ppm dose (Exon *et al.*, 1986). In two other studies (Gainer, 1977a,b), resistance to EMCV was decreased in CD-1 mice chronically given $CdSO_4$ orally at doses of 2 or 5 mM. Overall, exposure to $CdSO_4$ tends to increase host resistance to challenge with infectious agents.

Cadmium acetate has been used in several host resistance studies. In two particular studies (Exon *et al.*, 1979, 1986), Swiss–Webster mice exposed to 3, 30, or 300 ppm $Cd(CH_3COO^-)_2$ orally for 10 weeks had reduced host susceptibility to EMCV challenge. However, Charles River rats were markedly more susceptible to the lethal effects of *S. enteritidis* endotoxin or *E. coli* after intravenous injection of one dose of 0.6 mg $Cd(CH_3COO^-)_2$ given 3 days prior to challenge (Cook *et al.*, 1974, 1975). These contrasting effects could be due to different routes of exposure (oral or intravenous) or length of exposure (1 day or 10 weeks).

Cadmium oxide exposures are primarily by inhalation. Groups of mice free from specific pathogens were exposed to $10 mg CdO/m^3$ for a single 15-min period and then separate groups were challenged with either bacteria (*Pasteurella multocida*) or a virus (orthomyxovirus influenza A). The bacterially exposed mice that received CdO had an increased death rate, while the virus-infected cadmium-treated group had a decreased death rate compared with controls. Mortality actually decreased in mice that were exposed to CdO by inhalation and then tested with influenza virus (Chaumard *et al.*, 1983).

It is interesting to note that, in general, mice treated with cadmium (any form) and challenged with bacteria had decreased resistance to the challenge while cadmium-exposed mice had increased resistance to viral challenge.

3.6.3 *Humoral immunity*

Humoral immunity is conferred by B-lymphocytes, but for optimum response requires interaction with macrophages and T-lymphocytes and their messengers. The B-lymphocytes differentiate into plasma cells which combat extracellular pathogens and their products by releasing antibody, a molecule which specifically recognizes and binds to a particular target molecule (antigen). Many types of assays are available to assess humoral immune responses.

Assays that have been used to evaluate cadmium-induced dysfunction of humoral immunity in animal models include lymphoproliferation in response to mitogens, plaque-forming cell (PFC) response, antibody titers, migration inhibition factor, B-lymphocyte memory, and receptor activity. Most of the studies have been conducted in mice, with one each in the rabbit and guinea pig. Exposures have been acute, subacute, or chronic with treatment being either *in vivo* or *in vitro*. Cadmium chloride is the primary form that has been used in these studies.

The mitogen lipopolysaccharide (LPS) stimulates proliferation and division of B-lymphocytes. Although the cellular responses to mitogens are not an indication of the overall immune function of the lymphocytes, they serve as a tool to evaluate the ability of lymphocytes to proliferate. Mice administered 160 ppm $CdCl_2$ orally for 30 days had enhanced B-lymphocyte proliferative responses to LPS (Gaworski and Sharma, 1978). Similar increases occurred in an experiment where Swiss–Webster mice were given 300 ppm $CdCl_2$ orally for 10 weeks (Koller and Roan, 1980a). In another study, evaluating the affects of $CdCl_2$ exposure in different species of mice, the LPS proliferative responses of B-lymphocytes were significantly decreased in C_3H/He mice that were injected subcutaneously for 5 consecutive days with 0.5 and 1.0 mg $CdCl_2$/kg (Ohsawa *et al.*, 1986). Only the 1.0 mg $CdCl_2$/kg dose resulted in a significant reduction in the DBA/2 mice, while the LPS response in BALB/c mice was unaffected. In another study, LPS resulted in a non-significant increase in lymphoproliferation in BALB/c mice treated orally with 2 mM $CdCl_2$ (Ilback *et al.*, 1994). BDF_1 mice exposed orally to $CdCl_2$ at concentrations up to 50 μg $CdCl_2$/ml for 3 weeks showed a dose-dependent enhancement of B-lymphocyte activity when the lymphocytes were cultured with LPS (Blakley, 1985). No effect in LPS stimulation of splenic lymphocytes was seen in mice exposed to 30 or 300 ppm $CdCl_2$ for 10 weeks, but a dose-dependent (albeit non-significant) increase was noted (Koller *et al.*, 1979). C57Bl/6 mice exposed once to 0.88 mg $CdCl_2$/m^3 via inhalation had a significant decrease in the lymphoproliferative response to LPS (Krzystyniak *et al.*, 1987). In *in vitro* exposures, $CdCl_2$ at 10^{-5} M was blastogenic to BALB/c mice spleen cells but did not alter the LPS-induced response (Gallagher *et al.*, 1979). Cadmium chloride has been shown to enhance LPS-induced blastogenesis of BALB/c mice splenocytes *in vitro* at levels of 10^{-7} and 10^{-6} M, but this response was inhibited at higher concentrations (Shenker *et al.*, 1977).

Antibody response has been measured by titers and/or plaque-forming cell (PFC) techniques. The PFC response (IgG) to sheep red blood cell (SRBC) challenge was significantly decreased in Swiss–Webster mice given 3, 30, or 300 ppm $CdCl_2$ orally for 10 weeks and then monitored for up to 42 days thereafter (Koller *et al.*, 1975). Although the IgM response was also suppressed, it appeared to recover by day 14 following cessation of the exposure. In contrast, the primary PFC response of BALB/c mice treated with 0.01, 0.05, 0.10, 1.0, or 5.0 ppm $CdCl_2$ orally for 5 weeks was significantly enhanced at all dosage levels (Cay, 1981). Mice (C57BL/6) exposed long-term to 50, 200 or 300 ppm $CdCl_2$ orally had a moderate increase in IgM PFC formation at the 50 and 200 ppm doses, but a decrease at the 300 ppm dose (Malave and DeRuffino, 1984). In another study using spleen cells cultured from BALB/c mice, the primary antibody response to SRBC was augmented when the spleen cells were incubated for 4 to 5 days with 4 or 8 μM $CdCl_2$, but were suppressed in cultures receiving 20 or 40 μM $CdCl_2$ (Fujimaki *et al.*, 1982; Fujimaki, 1985). Reconstitution studies of both the adherent and non-adherent cells and/or T- and B-lymphocytes suggested that the effects of cadmium on B-lymphocytes were mainly responsible for the enhancement seen at the lower dosages. No effects were observed in these studies on the secondary antibody response with *in vitro* treatment of cadmium. When BDF_1 mice were exposed to $CdCl_2$ orally at concentrations ranging from 0 to 50 μg/ml for 3 weeks, the PFC response to SRBC was suppressed in a dose-dependent manner (Blakley, 1985). In a study in which CD1 mice were exposed orally to 50 ppm $CdCl_2$ for 3 weeks, a reduction in splenic PFC levels in response to the SRBC challenge was observed (Borgman *et al.*, 1987); this response recovered over a 6-week period following discontinuation of the cadmium exposure. In a more recent study, BALB/c mice administered 1, 10, or 100 ppm $CdCl_2$ orally for 7, 14, 21, 28, 60, or 90 days, and then immunized with dinitrophenyl-bovine serum albumin at 14-day intervals had serum antibody titers similar to those of unexposed controls (Schulte *et al.*, 1994). However, there was marked splenic and thymic atrophy noted in the 100 ppm $CdCl_2$ group. Chronic (10 week) exposure of rabbits to 300 ppm $CdCl_2$ reduced serum neutralizing antibody titers to a pseudorobies virus challenge (Koller, 1973).

The effects of $CdCl_2$ have also been extensively studied following exposures via inhalation or injection. In an inhalation experiment, CD-1 mice exposed to 190 μg $CdCl_2/m^3$ for 2 h had reduced ability to produce IgM antibody (Graham *et al.*, 1978). However, if the $CdCl_2$ was administered by intramuscular injection in doses ranging from 0.48 to 11.81 μg $CdCl_2/kg$, no effect was observed. C57B1/6 mice exposed to 0.88 mg $CdCl_2/m^3$ for 60 min demonstrated an inhibited IgM response to SRBC which was associated with a marked decrease in spleen cell viability 5 to 8 days following exposure (Krzystyniak *et al.*, 1987). Exposure to 9.02 mg CdO/m^3 for 15 min did not affect antibody titers in Swiss–Webster mice challenged with influenza virus (Chaumard *et al.*, 1991). The antibody response in mice exposed to a single dose of 0.15 mg $CdCl_2$ by different routes revealed that the IgM and IgG responses were increased when

CdCl$_2$ was injected intraperitoneally, but were suppressed when it was given orally (Koller *et al.*, 1976). Both the primary and secondary PFC responses were impaired in B10A2R mice chronically exposed to 2 ppm CdCl$_2$ by daily intraperitoneal injections over 40 days (Bozelka *et al.*, 1978). No effect was observed on antibody titers in SLC-ICR mice injected subcutaneously with a single dose of 0.09 mg CdCl$_2$ given either simultaneously or before challenge with JEV (Suzuki *et al.*, 1981).

The B-lymphocyte memory response was impaired in Swiss–Webster mice given 300 ppm CdCl$_2$ orally for 10 weeks, but no effect was observed at the 3 and 30 ppm CdCl$_2$ doses (Koller and Roan, 1980a). The levels of B-lymphocytes were significantly reduced in myocardial inflammatory lesions in mice treated for 10 weeks with CdCl$_2$ and then challenged with coxsackievirus (Ilback *et al.*, 1994). *In vitro* exposure of CBA/J mice spleen cells to 10^{-4} or 10^{-7} M CdCl$_2$ in culture for 5 days significantly suppressed the IgM PFC response (Lawrence, 1981); primary PFC responses were not significantly affected at the 10^{-5} or 10^{-6} M CdCl$_2$ levels.

When B-lymphocytes were collected from CBA/J mice exposed to 3, 30, or 300 ppm CdCl$_2$ for 10 weeks, there was a significant suppression of erythrocyte–antibody–complement (EAC) rosette formation by cells from mice in the 30 and 300 ppm dosage groups (Koller and Brauner, 1977). This procedure would suggest that cadmium may impair the ability of complement components to bind with C3 receptors on B-lymphocytes, thereby reducing the effectiveness of the antibody to destroy or inactivate bacterial antigens. Furthermore, *in vitro* exposure of B-lymphocytes to CdCl$_2$ at 30 μM inhibited cellular RNA and DNA synthesis by 50 percent (Daum *et al.*, 1993). The B-lymphocytes were apparently in arrest throughout the cell cycle, and IgG secretion was inhibited (with a rank order of inhibition of IgG$_3$ > IgG and IgG$_{2b}$ > IgG$_{2a}$ and IgM).

Animals have also been exposed to other forms of cadmium. NMR-1 mice treated with 30, 300, or 600 ppm Cd(CH$_3$COO$^-$)$_2$ orally for 10 weeks had normal primary and secondary PFC responses (Muller *et al.*, 1979). Sprague–Dawley rats injected with 0.6 mg CdSO$_4$ for 5 consecutive days beginning 7 days before antigenic challenge with human γ-globulin had decreased serum antibody levels, but if the injections were started 14 days prior to antigenic challenge, antibody levels were enhanced (Jones *et al.*, 1971).

The mitogenic response of B-lymphocytes tends to be enhanced while antibody synthesis tends to be suppressed by *in vivo* exposure to CdCl$_2$. However, opposite effects and no effects have also been reported after exposure to CdCl$_2$.

3.6.4 *Cell-mediated immunity*

T-lymphocytes, which act via cell-to-cell contact, are the principal cells responsible for cell-mediated immunity. Classic cell-mediated reactions are graft-versus-host (rejection), cytotoxicity (killing), and delayed-type hypersensitivity reactions (contact dermatitis). However, T-lymphocytes also produce and secrete numerous lymphokines which serve as messengers/ catalysts of immune reactions.

Delayed-type hypersensitivity and cytotoxic procedures have been used most frequently to evaluate xenobiotics for their effects on cell-mediated immunity. Other techniques such as graft rejection, blastogenesis, surface antigen/receptor recognition, and cytokine production/activity also have been used. Responses have been measured after acute, subacute, and chronic exposures and by oral, intraperitoneal injection, or *in vitro* administration of cadmium. The mouse has been the animal model of choice for these investigations.

Mitogens commonly used to stimulate T-lymphocyte division are ConA and PHA. The mixed lymphocyte reactions (MLR) are also measures of lymphoproliferation, but in response to allogeneic cell stimulation rather than to mitogens. Early studies indicated that blastogenic responses of T-lymphocytes were reported to be stimulated at low doses of cadmium but inhibited at the higher dosages (Shenker *et al.*, 1977; Gallagher *et al.*, 1979). In contrast, the blastogenic response of spleen cells obtained from CBA mice exposed to $CdCl_2$ for 10 weeks, when stimulated with purified protein derivative (PPD), was enhanced at 300 ppm $CdCl_2$ but suppressed at 3 ppm $CdCl_2$ (Koller *et al.*, 1979). However, neither dose of $CdCl_2$ affected ConA-induced lymphoproliferation or MLR activities (Koller and Roan, 1980b). In a similar study where mice were exposed to $CdCl_2$ orally at 30, 300, or 600 ppm for 10 weeks, responses to ConA and PHA were increased (Muller *et al.*, 1979). A similar increase in response to ConA and PHA occurred in C57BL/6 mice exposed to 50, 200, or 300 ppm $CdCl_2$ orally for up to 11 weeks (Malave and DeRuffino, 1984). Mice exposed to 160 ppm $CdCl_2$ orally for 30 days resulted in inhibition of spleen cell proliferation when exposed to PHA but not to ConA (Gaworski and Sharma, 1978). In another study where mice were exposed to 10, 50, or 250 ppm $CdCl_2$ orally for 90 days, the 10 and 250 ppm doses significantly suppressed ConA- and PHA-induced blastogenesis, while the 50 ppm dose had no effect on lymphoproliferation (Thomas *et al.*, 1985). ConA-induced T-lymphocyte responses were not affected in mice exposed to $CdCl_2$ orally for 3 weeks at concentrations up to 50 μg $CdCl_2$/ml (Blakley, 1985). C57B1/6 mice exposed to a single 60 min inhalation of 0.88 mg $CdCl_2$/m^3 demonstrated a significant inhibition of lymphoproliferative responses to allogenic antigens and PHA; these changes were correlated with a marked decrease in spleen cell viability (Krzystyniak *et al.*, 1987). In another study, C_3H mice were shown to be more susceptible to the suppressive effects of $CdCl_2$ on the proliferative responses to ConA and PHA than were DBA and BALB strains of mice (Ohsawa *et al.*, 1986). In this study, the mice were injected subcutaneously daily with 0.5 or 1.0 mg $CdCl_2$/kg for 5 days. While both concentrations significantly suppressed the PHA responses in C_3H mice, only the higher dose impaired the ConA response. The MLR was significantly

decreased at both concentrations in C_3H mice; however, the lower dose significantly increased both the ConA and MLR responses in DBA mice. Cadmium
chloride also depressed the ConA-induced lymphoproliferative response *in*
vitro when added at a concentration of 10 μM to cells (Otsuka and Ohsawa,
1991).

The cytotoxic response of splenocytes against tumor cells was increased in
C57Bl mice given 3, 30, or 300 ppm $CdCl_2$ orally for 10 weeks and challenged
with Moloney sarcoma virus-infected cells (Kerkvliet *et al.*, 1979). However,
this response was inversely correlated to cadmium dose. In another study
analyzing cytotoxicity, rejection of allografts was accelerated and time to
isograft acceptance increased in C57Bl mice treated with 0.01, 0.1, 1.0, or
10 ppm $CdCl_2$ orally for 4 weeks prior to grafting (Balter *et al.*, 1982).

The delayed-type hypersensitivity (DTH) response was significantly reduced in NMR-1 mice treated with 30, 300, or 600 ppm $CdCl_2$ orally for 10
weeks (Muller *et al.*, 1979). These data were supported by an impaired DTH
reaction to ovalbumin in mice given a single intraperitoneal injection of $CdCl_2$
(0.75 or 6 mg $CdCl_2$/kg) within 2 days of immunization (Kojima and Tamuar,
1981). This cadmium-induced suppression was observed during both the induction and progression phase of the delayed response; memory- and suppres-
sor-T-lymphocyte induction was also impaired and thymic weights decreased
in this study. However, antibody titers to ovalbumin were not affected as
measured by passive hemagglutination. In ICC mice gavaged daily for 14 days
with either 0.65, 32.6, or 65.2 mg $CdCl_2$/kg, the highest dose significantly suppressed the DTH response in male and female mice while the lowest dose
decreased the reaction in females only (Barnes and Munson, 1978). After 90
days of treatment, the delayed response was decreased by all doses administered to the females, but only by the high dose in the males. The cutaneous
hypersensitivity reaction, as measured by skin thickness, was significantly
suppressed in goats orally exposed to 10 mg $CdCl_2$/kg for 42 days (Haneef
et al., 1995). In an *in vitro* study where guinea pig lymphocytes were exposed
to 10^{-3} to 10^{-6} M $CdCl_2$, a lower negative surface charge on the lymphocytes
was proposed to interfere with the interaction of antigen on the cell surface
membrane (Kiremidjian-Schumacher *et al.*, 1981a,b).

From the cell-mediated investigations, cadmium appears to suppress DTH
responses and, possibly, accelerates allograft rejection. Mitogen-induced
lymphoproliferative responses following exposure to $CdCl_2$ range widely to
each end of the spectrum (increased → no effect → decreased). Furthermore,
the inconsistent results with mitogen-induced proliferation suggest that this
procedure has little, if any, relevance for assessing immune function.

3.6.5   *Non-specific immunity*

For the purpose of this chapter, non-specific immunity will include those
compartments of the immune system that are not listed in the previous

humoral and cell-mediated immune sections. Cytokines (lymphokines and monokines) are grouped together in this section, although many are important both in humoral and cell-mediated immune responses. Some non-specific responses include those involving macrophages, natural killer cells, lymphokine-activated killer cells, and cytokine production/activity.

Macrophages, which are composed of both effector and regulator subpopulations, phagocytose, process, and transfer antigenic information to other immunocytes, particularly T-lymphocytes. Macrophages not only orchestrate most immune networks but are also cytotocidal, bactericidal, and possess antitumor activity.

Early studies showed that intravascular clearance of lipid emulsions was increased in rats given a single intravenous dose of 0.6 mg CdCl$_2$/kg (Cook *et al.*, 1974). Similar effects were observed *in vitro* where phagocytosis of SRBC was stimulated and acid phosphatase levels in macrophages were increased in Swiss–Webster mice treated with 300 ppm CdCl$_2$ orally for 10 weeks (Koller and Roan, 1977). ICC mice gavaged with 0.65, 32.6, or 62.5 mg CdCl$_2$/kg daily for 90 days displayed a significant reduction in the phagocytosis of *L. monocytogenes* by liver cells (Barnes and Munson, 1978). Exposure to 10 to 300 ppm CdCl$_2$ orally for 3 months impaired the ability of reticuloendothelial cells to bind and catabolize immune complexes in mice (Knutson *et al.*, 1980). This effect was compared with that of cadmium levels in liver tissue and was thought to be due to cadmium perturbation of the Fc or complement receptors on Kupffer cells. Oral cadmium administration to rabbits impaired rosette formation by alveolar macrophages, suggesting that the Fc receptor for immunoglobulins may be altered on those cells (Hadley *et al.*, 1977). Reduced phagocytic activity of alveolar macrophages has been observed in rabbits that were exposed via inhalation to 2.5 µg CdCl$_2$/ml (Graham *et al.*, 1975) .

In vitro studies have shown depressed phagocytic activity and microbial killing by macrophages of peritoneal or alveolar origin treated *in vitro* with 8×10^{-1} to 8×10^{-3} mEq/L of CdCl$_2$ or Cd(CH$_3$COO$^-$)$_2$ for 30 min (Loose *et al.*, 1977, 1978a), although cadmium had a direct cytotoxic effect on these same cells (Loose *et al.*, 1978b). Similar results were reported in a study by Hilbertz *et al.* (1986) in which mouse peritoneal macrophages exposed to 10^{-1} to 10^{-8} M cadmium for 1 h had significant reductions in their ability to phagocytose zymogen granules. However, after 20 h of cadmium exposure, phagocytic activity had returned to normal levels. Nevertheless, cell viability was significantly affected at all levels of cadmium exposure. In another study, phagocytosis by guinea pig alveolar macrophages was unaffected when incubated with 1, 3, or 10 µg CdCl$_2$/L for 45 min (Kramer *et al.*, 1990).

The oxidative burst of rat alveolar macrophages during phagocytosis was depressed when the macrophages were cultured with CdCl$_2$ or Cd(CH$_3$COO$^-$)$_2$ at a dose of 10^{-3} M for 15 min (Castranova *et al.*, 1980). Cadmium inhibited oxygen consumption, glucose metabolism, and the release of active oxygen species, which was postulated to inhibit the antibacterial activity of the alveolar macrophages. However, the oxidative metabolism (chemiluminescence) of mouse peritoneal macrophages was increased when cultured with $CdCl_2$ at 10^{-4} to 10^{-8} M for 1 h (Hilbertz *et al.*, 1986), but no effect was observed on superoxide production in alveolar macrophages incubated with up to 10 μg $CdCl_2$/L for 45 min (Kramer *et al.*, 1990). Exposure to cadmium has also been shown to alter macrophage mobility, responsiveness to lymphokines, and cytotoxic activity (Kiremidjian *et al.*, 1981a,b). Guinea pig peritoneal macrophages and splenic lymphocytes exposed *in vitro* to doses of $CdCl_2$ ranging from 10^{-4} to 10^{-6} M had reduced migration properties and impaired capacities to react to migration inhibition factor. *In vitro* exposure of 10^{-5} or 10^{-6} M $CdCl_2$ also inhibited macrophage cytotoxicity of sarcoma I tumor cells (Nelson *et al.*, 1982). Cadmium chloride has also been reported to affect the immune system of rainbow trout (Zelikoff *et al.*, 1995). Trout exposed in the water to 2 ppb cadmium for 8 days had a significant increase in the phagocytic activity of peritoneal macrophages; the response returned to control levels by day 17 of treatment. However, in this study, the production of reactive oxygen intermediates by the peritoneal macrophages was reduced in a time-dependent manner.

Natural killer (NK) cells do not require sensitization and therefore manifest spontaneous cytotoxic properties. Mice (CD-1 and C_3H) injected intramuscularly with a single 6.25 mg dose of $CdCl_2$ had a significant decrease in both murine cytomegalovirus-augmented and spontaneous cytotoxicity of NK cells (Daniels *et al.*, 1987). In this same study, a dose of 3.13 mg $CdCl_2$/kg had no effect on NK activity. Sprague–Dawley rats exposed to $CdCl_2$ had a significant suppression of NK and killer cell activities. The suppression occurred only when cadmium was administered intraperitoneally (4.0 mg $CdCl_2$/kg, single injection) but not by gavage (250 μg $CdCl_2$/kg for 6 weeks) (Stacey *et al.*, 1988). Wistar rats exposed orally to 200 or 400 ppm $CdCl_2$ orally for 170 days had much lower NK activity than control rats for the first 30 days of treatment; the exposed rats also had a noticeable, but non-significant, increase in NK activity from day 40 until the end of the experiment on day 170 (Cifone *et al.*, 1989a). The NK activity was unaffected in BALB/c mice exposed to 2 mM of cadmium orally for 10 weeks (Ilback *et al.*, 1994).

Cytokines serve as messengers that catalyze immune circuits. Normal production, secretion, and activity of these immunohormones are necessary for optimum responses in most, if not all, immune pathways. Interferon activity was unaffected in C_3H and CD-1 mice injected intramuscularly with up to 6.25 mg $CdCl_2$/kg (Daniels *et al.*, 1987). Failure of interleukin-2 (IL-2) secretion and a reduction in IL-2 receptor expression has been reported in cadmium-treated cells (Cifone *et al.*, 1989b; Payette *et al.*, 1995). It was proposed that cadmium interferes with lymphocyte signal transduction and activation. Interleukin-6 (IL-6) mRNA transcripts and IL-6 secretion were measured in the kidney of $B_6C_3F_1$ mice following exposure of LPS-primed mice injected subcutaneously with 0.5 mg $CdCl_2$/kg three times per week for 14 weeks (Kayama *et al.*, 1995). When slices of the kidneys were cultured, the IL-6 was found to be secreted by the renal mesangial cells. Results of other cytokine investigations are reported in the human immunoresponsiveness section of this chapter.

Cadmium can also affect lymphocyte populations. A cytologic shift in lymphocyte populations, consisting of a significant increase in the numbers of large lymphocytes accompanied by a decrease in the numbers of small lymphocytes, occurred in Sprague–Dawley rats and ICR mice exposed to a 0.1 mg CdO/m^3 aerosol for 4 weeks or injected subcutaneously at a dose of 2 mg CdO/kg (Ohsawa and Kawai, 1981). Splenomegaly, anemia, neutropenia, and lymphopenia were also observed in these animals. Thymic atrophy and splenomegaly have been reported to develop in mice after a single intra-peritoneal injection of 1.8 mg CdCl$_2$/kg (Yamada *et al.*, 1981). *In vitro* exposure of murine lymphocytes to cadmium has also inhibited the formation of RNA (Gallagher and Gray, 1982).

Cadmium has been suggested to cause autoimmune-like diseases in experi-mental animal models. In one study, anti-laminin antibodies were detected in the serum of Sprague–Dawley rats treated orally with 20 or 100 ppm CdCl$_2$ for 13 months or injected intraperitoneally with 1 mg CdCl$_2$/kg five times per week for 4 months (Bernard *et al.*, 1984). The antibodies were not associated with concomitant immunoglobulin deposits in the kidneys. Renal immune complex disease developed in rats exposed orally to 200 ppm CdCl$_2$ for 30 weeks (Joshi *et al.*, 1981) or when a cadmium wire was implanted into Sprague–Dawley rats (Greenburg, 1980).

3.6.6   *Effects of cadmium on the human immune response*

Only a few studies have been conducted to evaluate the immunotoxic effects of cadmium on human immunity, and almost all of these have examined cells exposed in culture. A mitogen study in the early 1980s demonstrated that exposure of human lymphocytes *in vitro* to 1.6 or 3.3 M CdCl$_2$ significantly inhibited PHA-induced blastogenesis and MLR activity (Kastelan *et al.*, 1981). These results were confirmed in a second more recent study by Borella *et al.* (1990).

Cadmium chloride concentrations varying from 1×10^{-4} to 3.3×10^{-6} M inhibited the PHA-induced production of IL-1 and tumor necrosis factor-α in *in vitro* activated human peripheral blood mononuclear cells (Theocharis *et al.*, 1994). This effect was accompanied by decreased levels of mRNA, which indicated that cadmium suppressed the production of these cytokines at the transcriptional level.

An early study found that workers industrially exposed to cadmium had up to 900 mg of antibody light-chain molecules in urine samples compared with a mean of 5 mg in unexposed controls (Vigliani, 1969). Antibody light chains are the major component of proteinuria seen after chronic exposure to cadmium. Cadmium appears to inhibit the catabolism of light chains, probably via a general inhibition of zinc-dependent aminopeptidase and carboxypeptidase. In another epidemiological study, serum IgM, IgG, and IgA levels (measured by radial immunodiffusion) were analyzed in groups of workers exposed to cadmium while employed in zinc/cadmium smelters (Karakaya *et al.*, 1994). No difference in the serum immunoglobulin levels were noted between the exposed workers and unexposed controls.

Immune cells, mostly lymphokine-activated killer cells generated by incubation in tissue culture with human recombinant IL-2, exhibited reduced cytotoxic activity in the presence of cadmium (Herberman *et al.*, 1987). Cadmium chloride has been shown to inhibit NK cell cytotoxicity and the antibody-dependent cellular toxicity of cultured human peripheral blood lymphocytes (Cifone *et al.*, 1990). This suppression could only be partially prevented by increasing external calcium concentrations or by adding zinc to the culture medium (Cifone *et al.*, 1991). Cadmium chloride at 10^{-4} M induced human peripheral blood mononuclear cells to produce large amounts of bioactive IL-8, with an increase in IL-8 mRNA levels within 30 min following addition of $CdCl_2$. In this study, IL-8 levels reached a maximum after 2 h and decreased thereafter (Horiguchi *et al.*, 1993). In contrast, 0.06, 0.08, and 0.1 mM $CdCl_2$ significantly reduced the levels of both IL-6 and its mRNA in a human monocytic cell line (Funkhouser *et al.*, 1994). Cadmium chloride (4 µM) stimulated apoptosis in a human T-lymphocyte line, CEM-C12, while higher doses (8–10 µM) resulted in necrosis of the cells (Azzouri *et al.*, 1994). Further studies are needed to delineate the immunotoxic effects of cadmium on the human immune system.

3.6.7 *Biological markers of cadmium immunotoxicity*

Biological markers can serve as surrogates to identify early stages of disease and to foster an understanding of the basic mechanisms of biological responses resulting from exposure to environmental substances (Koller, 1993). Immunological markers of effect can be any detectable change within the immune system and changes in other tissues resulting from immune-mediated dysfunction (Koller, 1996). Biomarkers of immunity must be able to recognize exposure, predict species susceptibility to exposure, and/or identify biological effects or disease as a result of exposure to a xenobiotic.

There do not appear to be many consistent reproducible immune effects following either *in vivo* or *in vitro* cadmium exposures that can serve as immunological markers. Cadmium has a wide range of effects on mitogen-induced lymphoproliferation, and tends to suppress antibody synthesis and DTH responses; while macrophage-mediated immune functions are extremely sensitive to the effects of cadmium, there is no definitive effect upon non-specific immunity although the tendency is towards suppression. Assessment of the effects of cadmium on human immune cells in tissue culture also are inconsistent, but the general pattern is toward suppression.

3.6.8 Summary

The effect of cadmium on the immune system has been studied extensively in animals, particularly rodents. Although results of many of these investigations appear to be contradictory, it is clear that cadmium can modulate certain immune responses. Host resistance to infectious agents or tumors may either be enhanced or suppressed, depending on the target used and/or species of animal tested. Humoral immune responses appear to be enhanced or unaffected by low-dose acute exposure to cadmium while, conversely, antibody production is usually suppressed in animals chronically exposed to moderate or high levels of cadmium. The cell-mediated immune responses have been both enhanced and depressed by cadmium, depending on the index measured. There is also considerable variation within non-specific immune procedures following exposure to cadmium. Human data are inconsistent and could be improved by additional epidemiological studies. None the less, the overall tendency is for cadmium to induce various degrees of immunosuppression in human and animal immune systems.

References

AZZOURI, B., TSANGARIS, G.T., PELLEGRIN, O., MANUEL, Y., BENVENISTE, J., and THOMAS, Y. (1994). Cadmium induces apoptosis in a human T-cell line. *Toxicology* **88**, 127–139.

BALTER, N., KAWEKI, M.E., GINGOLD, B., and GRAY, I. (1982). Modification of skin-graft rejection and acceptance by low concentrations of cadmium in drinking water of mice. *J. Toxicol. Environ. Health*. **10**, 433–439.

BARNES, D. and MUNSON, A.E. (1978). Cadmium-induced suppression of cellular immunity in mice. *Toxicol. Appl. Pharmacol.* **45**, 350–361.

BERCHE, P., SIMONET, M., THEVENIN, M., FAUCHERE, J.L., PRAT, J.J., and VERON, M. (1980). Susceptibility of mice to bacterial infections after chronic exposure to cadmium. *Ann. Microbiol.* **131B**, 145–151.

BERNARD, A., LAUWERYS, R., GENGOUX, P., MAHIEU, P., FOIDART, J.M., DRUET, P., and WEENING, J.J. (1984). Anti-laminin antibodies in Sprague–Dawley and Brown Norway rats chronically exposed to cadmium. *Toxicology* **31**, 307–313.

BLAKLEY, B.R. (1985). The effect of cadmium chloride on the immune response in mice. *Can. J. Comp. Med.* **49**, 104–108.

BORELLA, P., MANNI, S., and GIARDINO, A. (1990). Cadmium, nickel, chromium and lead accumulate in human lymphocytes and interfere with PHA-induced proliferation. *J. Trace Elem. Electrolyte Health. Dis.* **4**, 87–95.

BORGMAN, R.F., AU, B., and CHANDRA, R.K. (1987). Recovery of immune response and renal ultrastructure following cadmium administration in mice. *Nutr. Res.* **7**, 35–41.

BOZELKA, B.E. and BURKHOLDER, P.M. (1979). Increased mortality of cadmium-intoxicated mice infected with the BCG strain of *Mycobacterium bovis*. *J. Reticuloendothel. Soc.* **26**, 229–347.

BOZELKA, B.E., BURKHOLDER, P.M., and CHANG, L.W. (1978). Cadmium, a metallic inhibitor of antibody-mediated immunity in mice. *Environ. Res.* **17**, 390–402.

CASTRANOVA, V., BOWMAN, L., REASON, M.J., and MILES, P.R. (1980). Effects of heavy metal ions on selected oxidative metabolic processes in rat alveolar macrophages. *Toxicol. Appl. Pharmacol.* **53**, 14–23.

CAY, B.S. (1981). The effect of cadmium on lymphocyte transformation. *J. Am. Vet. Med. Assoc.* **36**, 143–147.

CHAUMARD, C., QUERO, A.M., BOULEY, G., GIRARD, F., BOUDERE, C., and GERMAN, A. (1983). Influence of inhaled cadmium microparticles on mouse influenza pneumonia. *Environ. Res.* **31**, 428–439.

CHAUMARD, C., FORESTIER, F., and QUERO, A.M. (1991). Influence of inhaled cadmium on the immune response to influenza virus. *Arch. Environ. Health* **41**, 50–56.

CIFONE, M.G., ALLESSE, E., DIEUGENIO, R., NAPOLITANO, T., MORRONE, S., PAOLINI, R., SANTONI, G., and SANTONI, A. (1989a). *In vivo* cadmium treatment alters natural killer activity and large granular lymphocyte numbers in the rat. *Immunopharmacology* **18**, 149–156.

CIFONE, M.G., ALESSE, E., PROCOPIO, A., PAOLINI, R., SANTONI, G., and SANTONI, A. (1989b). Effects of cadmium on lymphocyte activation. *Biochim. Biophys. Acta* **1011**, 25–32.

CIFONE, M.G., PROCOPIO, A., NAPOLITANO, T., ALESSE, E., SANTONI, G., and SANTONI, A. (1990). Cadmium inhibits spontaneous (NK), antibody-mediated (ADCC) and IL-2-stimulated cytotoxic functions of natural killer cells. *Immunopharmacology* **20**, 73–80.

CIFONE, M.G., NAPOLITANO, T., FESTUCAIA, C., CANTALINI, M.G., DeNUNTIIS, G., SANTONI, G., MARINELLI, G., and SANTONI, A. (1991). Effects of cadmium on cytotoxic functions of human natural killer cells. *Toxicol. in Vitro.* **5**, 525–528.

COOK, J.A., MARCONI, E.A., and DILUZIO, N.R. (1974). Lead, cadmium, endotoxin interaction: effect on morality and hepatic function. *Toxicol. Appl. Pharmacol.* **28**, 292–302.

COOK, J.A., HOFFMAN, E.O., and DILUZIO, N.R. (1975). Influence of lead and cadmium on the susceptibility of rats to bacterial challenge. *Proc. Soc. Exp. Biol. Med.* **150**, 741–747.

DANIELS, M.J., MENACHE, M.G., BURLESON, G.R., GRAHAM, J.A., and SELGRADE, M.J. (1987). Effects of $NiCl_2$ on susceptibility to murine cytomegalovirus and virus-augmented natural killer cell and interferon responses. *Fundam. Appl. Toxicol.* **8**, 443–453.

DAUM, J.R., SHEPHERD, D.M, and NOELLE, R.J. (1993). Immunotoxicology of cadmium and mercury on B-lymphocytes. I: Effects on lymphocyte function. *Int. J. Immunopharmacol.* **15**, 383–394.

EXON, J.H., PATTON, N.M., and KOLLER, L.D. (1975). Hexamitiasis in cadmium-exposed mice. *Arch. Environ. Health* **31**, 463–464.

EXON, J.H., KOLLER, L.D., and KERKVLIET, N.I. (1979). Lead–cadmium interaction: effect on viral-induced mortality and tissue residues in mice. *Arch. Environ. Health* **34**, 469–475.

EXON, J.H., KOLLER, L.D., and KERKVLIET, N.I. (1986). Tissue residues, pathology and viral-induced mortality in mice chronically exposed to different cadmium salts. *J. Environ. Pathol. Toxicol.* **7**, 109–114.

FUJIMAKI, H. (1985). *In vitro* effect of cadmium on primary antibody response to T-cell independent antigen (DNP-Ficoll). *Toxicol. Lett.* **24**, 21–24.

FUJIMAKI, H., MURAKAMI, M., and KUBOTA, K. (1982). *In vitro* evaluation of cadmium-induced alteration of the antibody response. *Toxicol. Appl. Pharmacol.* **62**, 288–293.

FUNKHOUSER, S.W., MARTINEZ-MAZA, O., and VREDEVOE, D. (1994). Cadmium inhibits IL-6 production and IL-6 mRNA expression in a human monocytiocell line, THP-1. *Environ. Res.* **66**, 77–86.

GAINER, J.H. (1977a). Effects of heavy metals and of deficiency of zinc on mortality rate in mice infected with encephalomyocarditis virus. *Am. J. Vet. Res.* **38**, 869–872. (1977b). Effects on interferon of heavy metal excess and zinc deficiency. *Am. J. Vet. Res.* **38**, 863–867.

GALLAGHER, K.E. and GRAY, I. (1982). Cadmium inhibition of RNA metabolism in murine lymphocytes. *J. Immunopharmacol.* **3**, 339–361.

GALLAGHER, K., MATTARAZZO, W.J., and GRAY, I. (1979). Trace metal modification of immunocompetence: II. Effect of Pb^{2+}, Cd^{2+}, and Cr^{3+} on RNA turnover, hexokinase activity, and blastogenesis during B-lymphocyte transformation *in vitro*. *Clin. Immunol. Immunopathol.* **13**, 369–377.

GARDNER, D.E., MILLER, F.J., ILLING, J.W., and KIRTZ, J.M. (1977). Alterations in bacterial defense mechanisms of the lung induced by inhalation of cadmium. *Bull. Eur. Physiopathol. Respir.* **13**, 157–174.

GAWORSKI, C.L. and SHARMA, R.P. (1978). The effects of heavy metals ^{3}H-thymidine uptake in lymphocytes. *Toxicol. Appl. Pharmacol.* **46**, 305–313.

GRAHAM, J.A., GARDNER, D.E., WATERS, M.D., and COFFIN, D.C. (1975). Effect of trace metals on phagocytosis by alveolar macrophages. *Infect. Immun.* **11**, 1278–1283.

GRAHAM, J.A., MILLER, F.J., DANIEL, M.J., PAYNE, E.A., and GARDNER, D.E. (1978). Influence of cadmium nickel, and chromium on primary immunity in mice. *Environ. Res.* **16**, 77–87.

GRAY, I., ARRIEH, M., and BALTER, N.J. (1982). Cadmium modification of the response in Balb/c mice to MOPC-104E tumor. *Arch. Environ. Health* **37**, 342–345.

GREENBERG, S.R. (1980). Fluorescent studies on the potential existence of vascular metallic immune complexes. *Arch. Environ. Health* **35**, 148–151.

HADLEY, J.G., GARDNER, D.E., COFFIN, D.L., and MENZEL, D.B. (1977). Inhibition of antibody-mediated rosette formation by alveolar macrophages: a sensitive assay for metal toxicity. *J. Reticuleondothel. Soc.* **22**, 417–426.

HANEEF, S.S., SWARUP, D., KALICHARAN, and DWIVEDI, S.K. (1995). The effect of concurrent lead and cadmium exposure on the cell-mediated immune response in goats. *Vet. Human Toxicol.* **37**, 428–429.

HERBERMAN, R.B., HISERODT, J., VUJANOVIC, N., BALCH, N., BOLHUIS, R., GOLUB, S., LANIER, L.L., PHIULLIPS, J.H., RICCARDI, C., SANTONI, A., SCHMIDT, R.E., and UCHDA, A. (1987). Lymphokine-activated killer cell activity. Characteristics of effector cells and their progenitors in blood and spleen. *Immunol. Today* **8**, 178–185.

HILBERTZ, U., KRAMER, U., RUITER, N.D., and BAGINSKI, B. (1986). Effects of cadmium and lead on oxidative metabolism and phagocytosis by mouse peritoneal macrophages. *Toxicology* **39**, 47–57.

HILL, C.H. (1979). Dietary influences on resistance to *Salmonella* infection in chicks. *Fed. Proc.* **38**, 2129–2133.

HORIGUCHI, H., MUKAIDA, N., OKAMOTO, S., TERANISHI, H., KASUYA, M., and MATSUSHIMA, K. (1993). Cadmium induces interleukin-8 production in human peripheral blood mononuclear cells with the concomitant generation of superoxide radicals. *Lymphokine Cytokine Res.* **12**, 421–428.

ILBACK, N.G., FOHLMAN, J., FRIMAN, G., and EHRNST, A. (1994). Immune responses and resistance to viral-induced myocarditis in mice exposed to cadmium. *Chemosphere* **29**, 1145–1154.

JONES, R.H., WILLIAMS, R.L., and JONES, A.M. (1971). Effects of heavy metals on the immune response. Preliminary findings for cadmium in rats. *Proc. Soc. Exp. Biol. Med.* **137**, 1231–1236.

JOSHI, B.C., DWIVEDI, C., POWELL, A., and HOLSCHER, M. (1981). Immune complex nephritis in rats induced by long term oral exposure to cadmium. *J. Comp. Pathol.* **91**, 11–15.

KARAKAYA, A., YULESOY, B., and SARDAS, O.S. (1994). An immunological study on workers occupationally exposed to cadmium. *Human Exp. Toxicol.* **13**, 73–75.

KASTELAN, M., GERENCER, M., KASTELAN, A., and GAMULIN, S. (1981). Inhibition of mitogen- and specific antigen-induced human proliferation by cadmium. *Exp. Cell Biol.* **49**, 15–19.

KAYAMA, F., YOSHIDA, T., ELWELL, M.R., and LUSTER, M.I. (1995). Cadmium-induced renal damage and proinflammatory cytokines: possible role of IL-6 in tubular epithelial cell recoognition. *Toxicol. Appl. Pharmacol.* **134**, 26–34.

KERKVLIET, N.I., KOLLER, L.D., BAECHER, L.G., and BRAUNER, J.A. (1979). Effect of cadmium exposure on primary tumor growth and cell-mediated cytotoxicity in mice bearing MSB-6 sarcomas. *J. Natl Cancer Inst.* **63**, 479–485.

KIREMIDJIAN-SCHUMACHER, L., STOTZKY, G., DICKSTEIN, R.A., and SCHWARTZ, J. (1981a). Influence of cadmium, lead, and zinc on the ability of guinea pig macrophages to interact with macrophage inhibitory factor. *Environ. Res.* **24**, 106–116.

KIREMIDJIAN-SCHUMACHER, L., STOTZKY, G., LIKHITE, V., SCHWARTS, J., and DICKSTEIN, R.A. (1981b). Influence of cadmium, lead and zinc on the ability of sensitized guinea pig lymphoctyes to interact with specific antigen and to produce lymphokines. *Environ. Res.* **24**, 96–105.

KNUTSON, D.W., VREDEVOE, D.L., AOKI, K.R., HAVS, E.J., and LEVY, L. (1980). Cadmium and the reticuloendothelial system (RES): a specific defect in blood clearance of soluble aggregates of IgG by the liver in mice given cadmium. *Immunology* **40**, 17–26.

KOJIMA, A. and TAMUAR, S.I. (1981). Acute effects of cadmium on delayed-type hypersensitivity in mice. *Jpn J. Med. Sci. Biol.* **34**, 281–291.

KOLLER, L.D. (1973). Immunosuppression produced by lead, cadmium and mercury. *Am. J. Vet. Res.* **34**, 1457–1458.

(1990). Immunotoxicology: functional changes, in Dayan, A.D., Hertel, R.F., Haseltine, E., Dazantzis, G., Smith, E.M., and van der Venne, M.T. (Eds), *Immunotoxicity of Metals and Immunotoxicology*, pp. 233–239, New York: Plenum Press.

(1993). Biomarkers of immunotoxicology, in Travis, C.C. (Ed.), *Use of Biomarkers in Assessing Health and Environmental Impacts of Chemical Pollutants*, pp. 201–207, New York: Plenum Press.

(1996). Profiling immunotoxicology: past, present and future, in Stolen, J.S., Fletcher, T.C., and Bayne, C.J. (Eds), *Modulators of Immune Responses, the Evolutionary Trail*, pp. 301–310, Fair Haven, NJ: SOS Publications.

KOLLER, L.D. and BRAUNER, J.A. (1977). Decreased B cell response after exposure to lead and cadmium. *Toxicol. Appl. Pharmacol.* **42**, 621–624.

KOLLER, L.D. and ROAN, J.G. (1977). Effects of lead and cadmium on mouse peritoneal macrophages. *J. Reticuloendothel. Soc.* **21**, 7–12.

(1980a). Effects of lead, cadmium and methylmercury on immunological memory. *J. Environ. Pathol. Toxicol.* **4**, 47–52.

(1980b). Response of lymphocytes from lead, cadmium and methylmercury on immunological memory. *J. Environ. Pathol. Toxicol.* **4**, 393–398.

KOLLER, L.D., EXON, J.H., and ROAN, J.G. (1975). Antibody suppression by cadmium. *Arch. Environ. Health* **30**, 598–601.

KOLLER, L.D., ROAN, J.G., and EXON, J.H. (1976). Humoral antibody response in mice after exposure to lead or cadmium. *Proc. Soc. Exp. Biol. Med.* **151**, 339–342.

KOLLER, L.D., ROAN, J.G., and KERKVLIET, N.I. (1979). Mitogen stimulation of lymphocytes in CBA mice exposed to lead and cadmium. *Environ. Res.* **19**, 177–188.

KRAMER, C.M., COLES, R.B., and CARCHMAN, R.A. (1990). *In vitro* effects of cadmium chloride on calcium metabolism in guinea pig alveolar macrophages: lack of correlation with superoxide anion release or phagocytosis. *In Vitro Toxicol.* **3**, 153–160.

KRZYSTYNIAK, K., FOURNIER, M., TROTTER, B., NADIEU, and CHEVALIER, G. (1987). Immunosuppression in mice after inhalation of cadmium aerosol. *Toxicol. Lett.* **38**, 1–12.

LAWRENCE, D.A. (1981). Heavy metal modulation of lymphocyte activities. I. *In vitro* effects of heavy metals on primary humoral immune responses. *Toxicol. Appl. Pharmacol.* **57**, 439–451.

LOOSE, L.D., SILKWORTH, J.B., and WARRINGTON, D. (1977). Cadmium-induced depression of the respiratory burst in mouse pulmonary alveolar macrophages, peritoneal macrophages and polymorphonuclear neutrophils. *Biochem. Biophys. Res. Commun.* **79**, 326–332.

LOOSE, L.D., SILKWORTH, J.B., and SIMPSON, D.W. (1978a). Influence of cadmium on the phagocytic and microbial activity of murine peritoneal macrophages, pulmonary alveolar macrophages, and polymorphonuclear neutrophils. *Infect. Immun.* **22**, 378–381.

LOOSE, L.D., SILKWORTH, J.B., and WARRINGTON, D. (1978b). Cadmium-induced phagocyte cytotoxicity. *Bull. Environ. Contam. Toxicol.* **20**, 582–588.

MALAVE, I. and DERUFFINO, D.T. (1984). Altered immune response during cadmium administration in mice. *Toxicol. Appl. Pharmacol.* **74**, 46–56.

MULLER, S., GILBERT, K.-E., KRAUSE, C.H., JAUTZKE, G., GROSS, U., and DIAMANTSTEIN, T. (1979). Effects of cadmium on the immune system of mice. *Experentia* **35**, 909–910.

NELSON, D.J., KIREMIDJIAN-SCHUMACHER, L., and STOTZKY, G. (1982). Effects of cadmium, lead and zinc on macrophage-mediated cytotoxicity toward tumor cells. *Environ. Res.* **28**, 154–163.

OHSAWA, M. and KAWAI, K. (1981). Cytological shift in lymphocytes induced by cadmium in mice and rats. *Environ. Res.* **24**, 192–200.

OHSAWA, M., MASUKO-SATO, K., TAKAHASHI, K., and OTSUKA, F. (1986). Strain difference in cadmium-mediated suppression of lymphocyte proliferation in mice. *Toxicol. Appl. Pharmacol.* **84**, 379–388.

OTSUKA, F. and OHSAWA, M. (1991). Differential susceptibility of T- and B-lymphocyte proliferation to cadmium: relevance to zinc requirement in T-lymphocyte proliferation. *Chem.-Biol. Interact.* **78**, 193–205.

PAYETTE, Y., LACHAPELLE, M., DANIEL, C., BERNIER, J., FOURNIER, M., and KRZYSTYNIAK, K. (1995). Decreased interleukin-2 receptor and cell cycle changes in murine lymphocytes exposed *in vitro* to low doses of cadmium chloride. *Int. J. Immunopharmacol.* **17**, 235–246.

SCHULTE, S., MENGEL, K., GATKE, U., and FRIEDBERG, K.D. (1994). No influence of cadmium on the production of specific antibodies in mice. *Toxicology* **93**, 263–268.

SHENKER, B.J., MATARAZZO, W.J., HIRSCH, R.L., and GRAY, I. (1977). Trace metal modification of immunocompetence. I. Effect of trace metals in cultures on *in vitro* transformation of B-lymphocytes. *Cell. Immunol.* **34**, 19–24.

SIMONET, M., BERCHE, P., FAUCHERE, J.L., and VERNON, M. (1984). Impaired resistance to *Listeria monocytogenes* in mice chronically exposed to cadmium. *Immunology* **53**, 155–163.

STACEY, N.H., CRAIG, G., and MULLER, L. (1988). Effects of cadmium on natural killer cell functions. *Environ. Res.* **45**, 71–77.

SUZUKI, M., SIMIZU, B., YABE, S., OYA, A., and SETA, H. (1981). Effect of Japanese encephalitis virus infection in mice. I. Acute and single dose exposure experiments. *Toxicol. Lett.* **9**, 231–235.

THEOCHARIS, S.E., SOULIOTIS, T., and PANAYIOTIDIS, P. (1994). Suppression of interleukin-1β and tumor necrosis factor-α biosynthesis by cadmium in *in vitro* activated human peripheral blood mononuclear cells. *Arch. Toxicol.* **69**, 132–136.

THOMAS, P.T., RATAJCZAK, H.V., ARANYI, C., GIBBONS, R., and FENTERS, J.D. (1985). Evaluation of host resistance and immune function in cadmium-exposed mice. *Toxicol. Appl. Pharmacol.* **80**, 446–456.

VIGLIANI, E.C. (1969). The biopathology of cadmium. *J. Am. Ind. Hyg. Assoc.* **30**, 329–340.

YAMADA, Y.K., SHIMIZU, F., KAWAMURE, R., and KUBOTA, K. (1981). Thymic atrophy in mice induced by cadmium administration. *Toxicol. Lett.* **8**, 49–55.

ZELIKOFF, J.T., BOWSER, D., SQUIBB, K.S., and FRENKEL, K. (1995). Immunotoxicity of low level cadmium exposure in fish: an alternative animal model for immunotoxicological studies. *J. Toxicol. Environ. Health* **45**, 235–248.

4

Chromium

DARRYL P. ARFSTEN, LESA L. AYLWARD, AND NATHAN J. KARCH

Karch & Associates Inc., 1701K Street N.W., Suite 1000, Washington, DC 20006, USA

4.1 History

Chromium (Cr) was discovered in Russia in 1765 by P.S. Pallas, but the element was not isolated until 1797 by the French chemist L.N. Vauquelin. By 1820, chromium had found industrial application in the form of potassium dichromate ($K_2Cr_2O_7$), which was used as a pigment in textile manufacture. By the mid-1800s, leather tanning with chromic acid had become a common practice. The ability of products manufactured from chromium to resist high temperatures was also exploited early on and by 1879 chromite ore was routinely used in the manufacture of high-temperature refractory furnaces. As the industrial revolution progressed into the 21st century, the use of chromium for metal finishing, manufacturing, and alloy production became commonplace. Following World War I, chrome metal became an important raw material in the manufacture of metal alloys and also found extensive application in growing markets for automobiles and household appliances (Burrows and Adams, 1990).

Chromium is used extensively in the metallurgical industry for manufacturing stainless steel, alloy cast irons, non-ferrous alloys, and other metal-based materials. Chromium is also employed in the manufacture of magnesite-chrome bricks and 'granular chromite', which are used for manufacturing high-temperature industrial furnaces. Both chromium(III) and chromium(VI) are used in the chemical industry for the manufacture of pigments, paints, and dyes. Chromium(III) has been used in the past to tan shoe leather and other products. Chromium(VI) has been used in the preservation of wood products and as a rust inhibitor that is incorporated into radiator and cooling tower fluids. Small amounts of chromium are used in the manufacture of magnetic tapes, textile pigments, toner for copying machines, cement, and joint prostheses.

63

The commercial source of chromium is chromite ore ($FeO \cdot Cr_2O_3$), which is composed of iron and chromium oxides. Chromite ore was mined in the United States prior to 1961 but now all chromium ores are imported from South Africa, Turkey, and Zimbabwe. A number of countries belonging to the former Soviet Union, as well as Finland, Albania, India, and the Philippines contain significant deposits of chromite ore (Stern, 1982). United States manufacturers produce chromium metal by reducing chromite ore with carbon, aluminum, or silicon, after which the product undergoes subsequent purification steps (ATSDR, 1993). Sodium chromate and dichromate are formed by the roasting of chromite ore with soda ash. Sodium chromate and dichromate are commonly converted to lead chromate and chromic sulfate, which have a number of industrial applications.

4.2 Occurrence

Chromium is a relatively common mineral that can be found in soil, air, and water in trace amounts. It is estimated that the total chromium in conterminous US soils may range from 1 to 2000 mg Cr/kg (estimated mean of 37.0 mg Cr/kg). Continental dust flux is the main natural source of chromium in the atmosphere while volcanic dust may also account for a minor portion (Fishbein, 1981). The arithmetic mean concentration of total chromium in US urban, suburban, and rural ambient air for 1979 to 1984 ranged from a low of 0.005 µg Cr/m^3 to a high of 0.525 µg Cr/m^3. Typical total chromium concentrations for rural air are usually less than 0.01 µg Cr/m^3, while urban air usually contains 0.01 to 0.03 µg Cr/m^3 (Fishbein, 1984). Snow collected from two urban areas, Toronto and Montreal, Canada, contained 100 to 3500 mg Cr/kg, indicating that precipitation may play an important role in chromium deposition, particularly in urban areas (Landsberger *et al.*, 1983).

Chromium concentrations in US river waters range from less than 1 to 30 µg Cr/L (EPA, 1984). The median value of total chromium in US river waters is 10 µg Cr/L, while the median value for US lakes rarely exceeds 5 µg Cr/L (ATSDR, 1993). The EPA (1984) reports that sediment and suspended-material chromium concentrations typically range from 1 to 500 mg Cr/kg. Measurements of chromium in US tap water taken in 1974 to 1975 were between 0.4 and 8.0 µg Cr/L (mean of 1.8 µg Cr/L; Greathouse and Craun, 1978). This study is thought to overestimate chromium concentrations because of the high prevalence of stainless steel piping at that time and the inadequate flushing period used before sample collection (EPA, 1984).

Chromium is also commonly found in fresh foods. Typical chromium concentrations in vegetables range between 20 and 50 µg Cr/kg, while the mean concentration in fruit is 20 µg Cr/kg (EPA, 1984). The mean concentration of chromium found in grains and cereals is 40 µg Cr/kg (EPA, 1984). Seafood is also a potential source of chromium; fish, oysters, clams, and mollusks typically contain chromium concentrations of 0.1 to 6.8 mg Cr/kg (ATSDR, 1993).

Stationary point sources are the predominant source of airborne chromium. The major point sources of airborne chromium are industrial, commercial, and residential sources that burn oil, coal, or natural gas. Airborne emissions from metal industries are major sources of anthropogenic chromium. Recent studies have found that the emissions released from cooling towers treated with chromate rust inhibitors are also major contributors of anthropogenic chromium (ATSDR, 1993). Contemporary estimates place the US annual release of airborne chromium emissions at 2700 to 2900 metric tons (EPA, 1990). Non-point sources of airborne chromium include road dusts, the gradual wearing of asbestos brake-linings, emissions from catalytic converters containing chromium, cement manufacturing plants, and from the incineration of municipal refuse.

The most significant anthropogenic sources for the deposition of chromium into US waterways are electroplating operations, leather tanning industries, and textile manufacturing (ATSDR, 1993). Very few data are available to estimate the amount of chromium that is deposited into US soils annually. According to Nriagu and Pacyna (1988), the disposal of commercial products containing chromium is the largest source of anthropogenic chromium in soil (55 percent) followed by the deposition of coal fly ash/bottom fly ash from electric utilities and other industries (33.1 percent), agricultural and food wastes (5.3 percent), animal wastes containing chromium (3.9 percent), atmospheric fallout (2.4 percent), and solid wastes derived from metal manufacturing activities (less than 0.2 percent).

Occupational sources of chromium exposure are as many and varied as the industries that utilize chromium as a raw material in their manufacturing processes. Potential sources of chromium exposure in the workplace include the use of cleaning compounds, cement manufacture and use, paint manufacture and use, lubricants and greases, and exposure to anticorrosive agents containing chromium compounds. Exposure to chromium(VI) and chromium(III), present in both soluble and insoluble forms, is common. Dermal exposure is the most common route of exposure in industry, but inhalation of particulates or fumes containing chromium can occur among welders, chromite miners, and workers involved in ore processing and purification. The National Occupational Exposure Survey (NOES) of 1981 to 1983 conducted by NIOSH estimated that about 304 829 US workers were exposed to Cr(VI) daily (NIOSH, 1989; ATSDR, 1993). Typical airborne concentrations of Cr(VI) associated with various industrial processes are as follows: chromate production, 100 to 500 μg Cr/m^3; stainless steel welding, 50 to 400 μg Cr/m^3; chromium plating, 5 to 25 μg Cr/m^3; ferrochrome alloys, 10 to 140 μg Cr/m^3; and chrome pigment, 60 to 600 μg Cr/m^3 (Stern, 1982). Airborne exposure to Cr(III) may range from 10.0 to 50.0 μg Cr/m^3 in industries using a one-bath tanning process.

Blood chromium concentrations are a viable measure of relative exposure. Average blood chromium levels for workers involved in the manufacture of lead chromate range from 3 to 216 µg Cr/L, while urine chromium concentrations were found to range from a low of 1.8 to a high of 575 µg Cr/g creatinine (McAughey *et al.*, 1988). A control population of non-occupationally exposed individuals contained less than 1.0 µg Cr/L in whole blood and more than 0.5 µg Cr/g creatine in urine. Blood and urine chromium concentrations present in stainless steel welders working for an average of 13 ± 6 years (±SD) ranged from 0.26 to 5.2 µg Cr/L and 19.3 to 67.2 µg Cr/g in whole blood and urine, respectively (Rahkonen *et al.*, 1983). Chromium concentrations for control subjects were below the detection limits of 0.26 µg Cr/L in whole blood and 0.9 µg Cr/g creatine in urine. In a similar study, Bonde and Christensen (1991) report a post-shift mean urinary chromium concentration of 2.1 nmol Cr/mmol creatine for stainless steel welders as compared with 0.7 nmol Cr/mmol in unexposed workers. Wood treaters (*n* = 83) using arsenic-based wood preservatives containing chromium had urinary concentrations of 41.0 µg Cr/L as compared with 63.0 µg Cr/L in 232 individuals of a nearby town (Takahashi *et al.*, 1983). The study noted that the urinary chromium levels were not significantly different from those of the controls despite the fact that chromium air concentrations in the plant (2 to 9 µg Cr/m^3) exceeded the NIOSH-recommended TLV (threshold limit value) of 1 µg Cr/m^3. Urinary chromium concentrations from two leather tanners exposed to Cr(III) in the form of chromium sulfate varied between 104 µg and 620 µg Cr/L (Aitio *et al.*, 1984); concentrations for workers not in direct contact with the wet leather hides were 1 µg Cr/L, the level of detection.

The mean concentration of chromium in blood serum for US citizens is estimated to be 0.06 µg Cr/L (range of 0.01 to 0.17 µg Cr/L; Sunderman *et al.*, 1989), while the mean concentration of chromium in urine is estimated as 0.4 µg Cr/L (range of 0.024 to 1.8 µg Cr/L; Iyengar and Woittiez, 1988). It has been noted that blood and urine chromium concentrations in the US general population have decreased significantly over time (Guthrie, 1982). This decrease can be attributed to improved methods for measuring total chromium in biological samples and the discontinued practice of using stainless steel equipment in collecting, storing, and analyzing tissue samples.

The inhalation of ambient air, ingestion of drinking water, and the consumption of foodstuffs containing chromium are potential sources of exposure to chromium. Dermal contact with certain products such as leather, textiles, cleaning materials, and wood products containing chromium-based wood preservatives are also a potential source of chromium exposure (ATSDR, 1993). Daily intake of chromium via inhalation or tap water is estimated to be more than 0.2 to 0.6 µg Cr and more than 4.0 µg Cr, respectively (Fishbein, 1984). The estimated daily chromium intake for the US general population ranges between 25 and 224 µg Cr, with an average of 76 µg Cr (ATSDR, 1993).

National regulations and guidelines pertinent to human exposure to chromium are summarized in Table 4.1.

Table 4.1 National regulations and guidelines applicable to chromium

Agency	Description	Valence	Level	Reference
Air:				
OSHA[a]	PEL TWA[b]	Cr(0) and insoluble salts	1.0 mg/m³	OSHA 1989 (29 CFR 1910)
		Cr(II)	0.5 mg/m³	
		Cr(III)	0.5 mg/m³	
	Ceiling concentration	Chromic acid and chromates	0.1 mg/m³	
ACGIH[c]	TLV-TWA[d]	Cr(0)	0.5 mg/m³	ACGIH (1995)
		Cr(II)	0.5 mg/m³	
		Cr(III)	0.5 mg/m³	
		Cr(VI)	0.05 mg/m³	
		Zinc chromates	0.01 mg/m³	
Water:				
EPA[e]	MCL[f] in drinking water	Chromium	0.1 mg/L	EPA (1991)
EPA	MCLG[g]	Chromium	0.1 mg/L	
EPA	AWQC[h]	Cr(III)	170 mg/L	EPA (1980, 1987)
		Cr(VI)	0.05 mg/L	
Biological Exposure Indices (BEI):				
ACGIH	BEI, end of workweek	Cr(VI)	30 µg/g creatinine	ACGIH (1995)
	BEI, increase during shift	Cr(VI)	10 µg/g creatinine	

[a] Department of Occupational Safety and Health.
[b] Permissible Exposure Limit, time-weighted average.
[c] American Conference of Governmental Industrial Hygienists.
[d] Threshold Limit Value, time-weighted average.
[e] United States Environmental Protection Agency.
[f] Maximum Concentration Limit.
[g] Maximum Concentration Limit Guideline.
[h] Ambient Water Quality Criteria.

4.3 Essentiality

Chromium is an essential element in human nutrition that is required for energy metabolism. Chromium(III) serves as a cofactor for insulin action by forming a tertiary complex with the insulin receptor which facilitates the attachment of insulin to these sites (Goyer, 1991). The exact mechanism of the interaction of Cr(III) with these receptor sites is not known (ATSDR, 1993). The biologically active form of chromium(III) is sometimes referred to as the 'glucose tolerance factor' (GTF). This term originated from studies that showed that rats fed torula yeast developed a progressive impairment in glucose tolerance (Mertz and Schwartz, 1955). As the feeding of brewer's yeast restored normal glucose tolerance, the unknown nutrient in the latter yeast was designated the 'glucose tolerance factor'. After systematically testing 47 different elements, it was determined that chromium was the essential constituent in the GTF (Mertz, 1969). Clinical studies found that chromium supplementation improved glucose tolerance in some hyperglycemic or diabetic patients, although other studies found no such effect in patients with similar conditions. Differences in response to chromium supplementation could be attributed to factors such as actual doses of chromium administered compared with reported doses, differences in dose regimen, and age of the patient. Genetics may also contribute to the divergent results reported. Thus far, there have been no other theories proposed as to the role of chromium in human physiology. High concentrations of Cr(III) are found in RNA, but its role has not been determined (Goyer, 1991).

The chemical form or speciation of a metal is an important factor in determining its absorption, distribution, and toxicological effects. As summarized in Table 4.2, there are six valence states of chromium, but only Cr(III) and Cr(VI) play any significant role in chromium toxicology. Valency is important in determining the ability of chromium to penetrate skin and biological membranes. Cr(VI) readily penetrates skin and cell membranes and is quickly reduced to chromium(III) by skin and intercellular proteins. In the blood, Cr(VI) is rapidly taken up by red blood cells through sulfate anion channels where it is converted to Cr(III)–hemoglobin complexes (Paustenbach *et al.*, 1996). Formation of these complexes can cause chromium concentrations in the red blood cell to remain elevated for weeks, while plasma levels return to background within days. The protein albumin, α-, β-, and γ-globulins, and fibrinogens may also bind with free chromium (Jandl and Simmons, 1957). Lactic acid readily binds with and reduces Cr(VI) to Cr(III) (Polak, 1983).

While Cr(VI) penetrates the skin and cell membranes virtually unimpeded, Cr(III) compounds penetrate human skin very poorly or not at all. The Cr(III) ion is a very reactive cation that readily binds with proteins in the stratum corneum, effectively preventing its entrance into the bloodstream. Studies have shown that there are some exceptions to this general observation (Polak, 1983). It has been shown that chromium(III) chloride can penetrate human

Table 4.2 The valence states of the chromium molecule

Valence	
Cr^0	Metal in alloys (e.g. in stainless steel) and in plating
Cr^{3+}	Salts of inorganic acids (e.g. chlorides), salts of organic acids (e.g. oxalate), and basic sulfate; hydroxides and salt complexes
Cr^{4+}	Dioxide in magnetic tapes; forms Cr^{3+} and Cr^{6+} in presence of water
Cr^{6+}	Chromates, dichromates of K, Na, Ca, NH_4; dichromate of Zn; chromate and dichromates of Pb

skin almost as well as hexavalent $K_2Cr_2O_7$ (Samitz and Katz, 1964; Samitz *et al.*, 1967). The penetrating capability of chromium(III) nitrate was marginal while chromium(III) sulfate did not penetrate human skin to any measurable extent. When the stratum corneum is 'stripped' from human skin, both Cr(III) and Cr(VI) penetrate human skin equally well, reinforcing the finding that it is the proteins within the stratum corneum that prevent Cr(III) from penetrating human skin at the same rate as Cr(VI). Penetration of the skin by Cr(III) ions is probably a function of dose and may be important during exposure to high concentrations of chromium compounds. More experimental work needs to be conducted to address this issue adequately.

4.4 General Toxicology

The health effects of chromium exposure reported in the literature range from acute dermal irritation to cancer. A number of deaths have been reported after the accidental or intentional ingestion of large doses of chromium. The ultimate causes of death most commonly reported are gastrointestinal hemorrhage and renal tubular necrosis. According to the ATSDR (1993), there are no reports in the literature of death occurring after the acute inhalation of chromium or chromium compounds.

The most commonly reported effects among workers exposed to chromium are dermal and respiratory irritation associated with working with or near chromium compounds. Breathing high levels (more than $2.0 \mu g \, Cr/m^3$) of chromium(VI) for several months to many years can cause effects ranging from irritation of the nose (runny nose, sneezing, and itching) to nosebleeds, ulcers, or perforation of the nasal septum (ATSDR, 1993). The early literature contains a large number of case reports of tanners, electroplaters, or workers involved in bichromate manufacture who developed deep ulcerations or 'chrome holes' after years of exposure. These begin as papular lesions on the hands, forearms, or feet and gradually progress to pustules which develop into 'punched out' ulcerates, usually 2 to 5 mm in diameter, with an undermined border and tenacious crust. The ulcers may penetrate deeply into the skin and include underlying bone. The lesions usually do not destroy bone tissue, but cartilage and joints may be penetrated. However, the presence of chrome holes in workers is not predictive of other consequences of chromium exposure such as chromium hypersensitivity or dermatitis. Edmundson (1951) found that only 2 of 56 workers with chrome ulcers tested positive for chromium hypersensitivity when dermally challenged with 0.5 percent $K_2Cr_2O_7$.

Chromium(VI) is an International Agency for Research on Cancer (IARC) Group I carcinogen. Long-term exposure to chromium has been associated with squamous cell carcinoma and adenocarcinoma of the lung in workers exposed to chromium air concentrations 100 to 1000 times higher than those found in the natural environment (ATSDR, 1993). Increased rates of lung cancer have been noted especially among workers in chromate production and chromate pigment industries. The high incidence of lung cancer among workers involved in the chrome plating industry, where exposure is almost exclusively to Cr(VI), supports the hypothesis that Cr(VI) is the carcinogenic chromium species. Epidemiological studies of the incidence of cancer among stainless steel welders exposed primarily to Cr(VI) and of ferrochromium workers exposed primarily to Cr(0) and Cr(III) do not demonstrate a clear increase in the incidence or death from cancer. Cancer incidence and mortality rates of leather tanners predominately exposed to Cr(III) have consistently shown no correlation between exposure and disease (ATSDR, 1993). A recent study found that machine workers had a higher relative risk for squamous cell carcinoma of the oral cavity, adjusted for alcohol and tobacco consumption, which correlated with their exposure to metal dust that presumably contained chromium, and to chromium-containing paints, lacquers, or varnishes (Tisch *et al.*, 1996). Future studies hope to address whether a gradient of cancer risk exists in industries where exposure to Cr(VI) and Cr(III) occur simultaneously and at varying concentrations (Goyer, 1991). Chromium(VI) is a carcinogen in mice and rats when administered by inhalation (ATSDR, 1993).

The molecular mechanism(s) by which chromium causes toxic effects is not well established. Chromium(VI) ions readily penetrate membranes and enter cells where they are rapidly converted to Cr(III) by interactions with cellular enzymes and other proteins (Figure 4.1). In the erythrocyte, Cr(VI) is

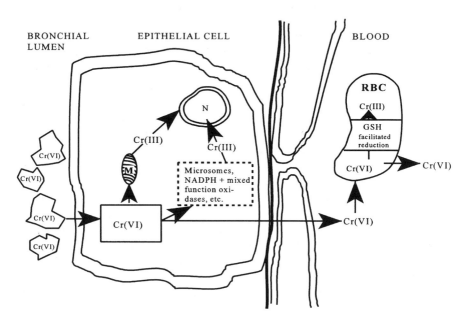

Figure 4.1 Suggested model of Cr(VI) uptake in the lungs and in the gastrointestinal tract. GSH, glutathione; M, mitochondria; N, nucleus; RBC, red blood cell. (Adapted from Langård, 1982; reproduced with permission).

converted to Cr(III) primarily by interaction with glutathione (Alpoim *et al.*, 1995). There is evidence that chromium is capable of damaging DNA in epithelial cells, but the mechanism by which this metal causes genetic damage is not clear. It is thought that the by-products formed in the cell during the reduction of chromium(VI) to chromium(III) may damage DNA by forming adducts. However, a recent study found that neither Cr(VI) nor its reduction product, Cr(V), formed DNA or RNA adducts as measured by changes in either their optical or electron paramagnetic resonance signals in a cell-free system (Molyneux and Davies, 1995). Reduction of Cr(VI) to Cr(V) by NADH/NADPH in the presence of hydrogen peroxide and nucleic acids produced hydroxyl-nucleotide adducts, suggesting that Cr(VI) may damage DNA through the formation of oxygen radicals. Alpoim *et al.* (1995) have shown that Cr(VI) reduces erythrocyte glutathione levels, lending further support to the hypothesis that Cr(VI) can cause toxicity by inducing oxidative stress. Studies conducted in bacteria have shown that certain Cr(III) complexes are mutagenic, indicating that Cr(III) may also play an active role in the mechanisms by which chromium causes toxicity.

4.5 Immunotoxicology

4.5.1 *Immunologically mediated contact dermatitis*

Chromium contact dermatitis is one of the most extensively studied immunological reactions to chromium exposure (Polak, 1983). Chromium contact dermatitis is an inflammatory skin reaction characterized by redness and vesicles resulting from contact with chromium-containing materials. The term contact dermatitis is very broad and includes all skin dermatitis reactions whether they are instigated by an allergen or an irritant. The differences between allergic and irritant contact dermatitis are significant and it is important to distinguish between the two types of reactions. Allergic dermatitis (contact allergy) is immunologically mediated, involves T-lymphocytes sensitized to allergens, and is characteristically a delayed reaction that occurs 12 to 48 h, or even later, after contact (Marzulli and Maibach, 1991). Irritant contact dermatitis is a non-immunologically mediated inflammatory response to a contact irritant characterized by redness and vesicles which are sharply demarcated with a brownish eroded or glistening surface. Compression of the vesicles between the fingers may reveal a finely wrinkled surface referred to as a 'soap effect' (Hjorth, 1991). Because some irritant reactions may take hours to develop, they are sometimes mistaken for an allergic response. Cumulative allergic irritant dermatitis is a condition that manifests itself after chronic exposure to large doses of irritant over a relatively short period in relation to skin recovery time. Cumulative allergic irritant dermatitis develops slowly over a period of days or weeks after subtle exposure to antigen and is most often classified as a delayed-type (type IV) allergic reaction. Irritative skin responses can be distinguished from allergic ones using skin patch testing techniques, but the investigator must be aware of the chemical concentration and exposure duration that produces irritation to avoid misinterpreting results from these tests (Hjorth, 1991).

Chromium contact sensitivity or 'hypersensitivity' is an adaptive immune response to chromium contact exposure in an exaggerated form. There are four types of hypersensitivity reactions (type I, II, III, and IV), which are not necessarily isolated from each other (Roitt *et al.*, 1993). Type I, II, and III reactions are antibody mediated, whereas type IV hypersensitivity is a result of an activated T-lymphocyte response to antigen. In sensitive individuals, chromium elicits a type IV response that manifests itself in four distinct phases: refractory, induction, elicitation, and persistence (Polak, 1983; Haines and Nieboer, 1988). The refractory period is demarcated by an initial contact with chromium compounds. In the refractory phase, Cr(VI) ions penetrate cell membranes and are reduced to Cr(III) by sulfhydryl group-containing cysteine or methionine moieties. The Cr(III) ions readily bind with cellular proteins to form Cr(III)–protein complexes which are then presented to T-lymphocytes by antigen-presenting cells (APC) such as macrophages or Langerhans cells. Presentation of the antigen complex to T-lymphocytes promotes the expansion and proliferation of both memory and effector T-lymphocytes specific for chromium–protein complexes. Expansion and pro-liferation of specific T-lymphocyte populations usually takes place in the re-gional draining lymph nodes. Lastly, the memory T-lymphocytes specific for Cr(III)–protein complexes then enter the general circulation and become a part of the host immune repertoire.

Following any subsequent exposure to chromium compounds, a full-blown type IV immune response takes place within 48 h after exposure. The develop-ment of this secondary response to chromium is characterized by induction and elicitation. Induction is characterized by presentation of chromium–protein complexes to the appropriate memory T-lymphocytes by the APC. Presentation of antigen induces the production of lymphokines by T-lymphocytes that promote chemotaxis, edema, inflammation, and prompt the proliferation of additional antigen-specific effector T-lymphocytes (Haines and Nieboer, 1988). Sensitivity to future exposure to chromium is maintained by continual renewal of memory T-lymphocytes specific for chromium–protein complexes.

4.5.2 *Chromium contact sensitivity in humans*

Besides nickel, chromium is the second most common skin allergen in humans (Haines and Nieboer, 1988). Although chromium hypersensitivity is a com-mon skin allergy, its prevalence among the general population is quite low, while only slightly higher rates occur among people with a history of chronic exposure to chromium. Zachariae *et al.* (1996) report that 79 of 4511 consecutive patients with eczematous skin disease (1.75 percent) tested positive in a skin patch test using 0.5 percent $K_2Cr_2O_7$. In 1973, 8 percent of 1200 patients evaluated by the North American Contact Dermatitis Group (1973) tested positive for chromium allergy. Interestingly, the prevalence rate varied widely by geographic location (2 to 20 percent) suggesting a possible link with local environment. Careful evaluation of 142 eczema cases among 1045 Dutch males and 947 Dutch females from Vlagtwedde determined that only 6 percent of these cases were positive for chromium allergy (Lantinga *et al.*, 1984). Peltonen and Fraki (1983) found that 2 percent of 410 male and 1.5 percent of the 412 female volunteers taken from the general Finnish population tested positive to chromate when administered skin patch tests. However, 10 of the 14 positive reactions occurred among a subgroup of 110 patients with previous occupational exposure to chromium. Sipos (1967) found only 5 positive reactions to chromate skin patch tests among 659 hospitalized patients with either healthy skin or with dermatoses other than eczema. Nethercott *et al.* (1994) have determined a 10 percent minimum elicitation threshold of chromium sensitivity of $0.089\,\mu g\,Cr(VI)/cm^2$ of skin based on patch testing in 54 chromium-sensitized volunteers. Unlike previous chromium contact sensitivity studies, the authors provide a reliable measure of a threshold dose for contact hypersensitivity through the use of a patch test system capable of delivering a controlled amount of allergen per surface area of skin.

It is apparent from investigations among the general population that occupational exposure to chromium compounds is a major risk factor for chromium hypersensitivity. Unfortunately, data concerning the incidence of chromium hypersensitivity among industrial workers is primarily limited to case and case series reports (Haines and Nieboer, 1988). The incidence of chromium contact sensitivity among workers employed in the building trades examined at St John's Hospital for Skin Diseases was found to be 46 percent (Cronin, 1980). Of the 119 metal workers diagnosed with contact skin allergy, 31 percent developed a positive reaction to $K_2Cr_2O_7$ patch tests (Alomar *et al.*, 1985). Contact sensitivity to chromium compounds has been reported in workers employed in other trades where contact with chromium-containing compounds is possible (Table 4.3). Chromium dermatitis is most frequently reported among cement workers, carpenters, and painters. Cement contains chromium compounds (Fregert and Gruvberger, 1972; Perone *et al.*, 1974), as do some wood preservatives. Chromium is also used as a pigment in paints. Frequent contact with cleaning fluids, coolants, or bleaches containing chromium is thought to increase the risk of chromium hypersensitivity. US household bleach does not appear to be a significant source of chromium exposure (Hostynek and Maibach, 1988); however, chromate was commonly added to bleaching and cleaning products manufactured in Italy, France, Spain, and Belgium during the 1960s and 1970s. Postage stamps, magnetic tapes, tattoos, and artificial chromium steel-made dentures are also potential sources of chromium exposure (Polak, 1983).

Table 4.3 The ten most important examples of eczematogenic exposures to chromium compounds (Polak, 1983)

Chromium-containing material or objects	Profession or places of contact	Chromium compounds responsible
Chromium ore	Industrial chromium production	Chromite
Chrome baths	Electroplating industry	Chromic acid, sodium dichromate
	Graphic trade	Chromates
	Metal industry	Chromates, zinc chromates
Chrome colors and dyes	Painters and decorators; graphic trades; textile, rubber, glass, and china industries	Chromic oxide green, chromic hydroxide green, chrome yellow (lead chromate)
Lubricating oils and greases	Metal industry	Chromic oxide
Anticorrosive in water systems	Diesel locomotive workshops and sheds, central heating and air-conditioning systems	Alkali dichromates
Wood preservation (Wolman salts)	Wood impregnation, furniture industry, carpenters, miners	Alkali dichromates
Cement, cement products, quick-hardening (e.g. Sika One)	Cement production, manufacture of cement products, building trades	Chromates
Cleaning materials (Eau de Javelle), washing and bleaching materials	Housewives, cleaners, laundry workers	Chromates
Textiles, furs	Textile and fur industries, everyday life	Chromates
Leather and artificial leather tanned with chromium	Leather and footwear industry, everyday life	Chromium sulfate and chromium alum

Hexavalent chromium is the predominant form of chromium cited in the literature as causing contact dermatitis, but a number of authors report that Cr(III) compounds may elicit a positive response in chromium-sensitive individuals (Fregert and Rorsman, 1964, 1966; Zelger, 1964; Mali *et al.*, 1966; Samitz and Shrager, 1966). Positive responses to skin patch tests were obtained using 0.187 M chromium chloride (Samitz and Shrager, 1966) and 0.5 M chromium trichloride (Fregert and Rorsman, 1964). Depending on the Cr(III) compound used, the development of positive tests required doses of Cr(III) that were 10 to 100 times higher than those needed for Cr(VI) compounds (Fregert and Rorsman, 1966). Mali *et al.* (1966) found that the concentration of Cr(III) required for a positive patch test was at least 20 times higher than those required for hexavalent chromium compounds. Although it cannot be determined from the data provided in the studies, it is possible that the positive patch test results observed for Cr(III) are not a T-lymphocyte-mediated allergic response but are the manifestations of irritant dermatitis in response to high concentrations of Cr(III).

A majority of the reports of severe chromium dermatitis are limited to the hands and lower forearms, but it is not uncommon for chromium dermatitis to develop at other sites on the body that may have frequent contact with chromium compounds. Contact dermatitis reactions have been known to occur on parts of the body that have frequent contact with jewelry containing chromium, such as the neck, ear, and ankle. Contact dermatitis of the foot has been attributed to wearing shoes made with chromium-tanned leather in certain sensitized individuals (Morris, 1958). Fregert *et al.* (1978) describes two patients who suffered repetitive contact dermatitis reactions on both the arms and thighs that were attributed to wearing military uniforms dyed with water-soluble chromate dyes. Contact hypersensitivity reactions in conjunction with stainless-steel prosthetic devices have been observed near the site of implantation (Langård and Hensten-Pettersen, 1981; Fisher, 1984), which most often involves sites located near the arm, leg, or hip.

Chromium contact dermatitis is a persistent condition that may reoccur upon challenge with chromium compounds months or years after initial recovery. Thormann *et al.* (1979) found that 38 of 48 patients (79 percent) initially diagnosed with chronic eczema had a positive reaction to dichromate patch testing 4 to 7 h after diagnosis; 72 percent of the 38 patients had a history of occupational exposure to chromate. Only 7 percent of the women and 10 percent of the men re-evaluated in a 10-year follow-up of occupational dermatitis associated with chromium exposure were clear of symptoms as compared with 16 and 24 percent of women and men, respectively, who had healed from occupational dermatitis associated with exposure to other sources (Fregert, 1975). In a study of the prognosis of industrial contact dermatitis, Burrows (1972) found that only 8 percent of workers diagnosed with cement contact dermatitis were clear of the disease 10 to 13 years later; a majority of these workers had positive patch tests to chromium.

There is some evidence that genetic factors may play a role in chromium contact dermatitis. A number of authors have demonstrated clear differences in chromium dermatitis rates among separate guinea pig strains (Polak, 1983). However, little has been done to investigate the possible role of genetic susceptibility among human populations. In an investigation of 52 Israeli Jews with chrome dermatitis, Wahba and Cohen (1979) observed that patients of Kurdish origin manifested the disease at a significantly earlier age than other non-Ashkenazi patients, even though the socio-economic conditions and occupations of the different ethnic groups were similar. The authors concluded that these findings may point to a genetic predisposition for chromium contact dermatitis, but other factors such as exposure frequency may account for the difference in the rate of chromium dermatitis among these populations.

Oral administration of chromium in chromium-sensitive individuals is capable of triggering skin eruptions associated with chromium contact dermatitis. Goitre *et al.* (1982) describe a case in which a building worker with a 20-year history of chromium dermatitis was given 7 mg $K_2Cr_2O_7$ (2.5 mg Cr). Two days after the oral dose, dermal irritation increased; administration of a double oral dose caused dyshidrotic lesions on the hands in combination with lymphangitis, auxiliary lymphadenitis, and fever 12 h after administration. Kaaber and Veien (1978) administered 2.5 mg chromium or a placebo to 31 patients with dermatitis who were positive for the chromium patch test. Contact dermatitis was 'aggravated' in 11 of the patients after ingestion of chromium. Fourteen patients had no reaction to either the chromium or placebo tablets while one patient had a reaction to administration of both chromium and placebo tablets.

Concomitant positive reactions to cobalt and/or nickel are sometimes observed in patients testing positive for chromium hypersensitivity. There is some question as to whether these reactions are independent or simultaneous, occurring as a result of cross-sensitivity to one or more metals. Double allergy to chromium and cobalt is common in patients with cement eczema (Polak, 1983). Cement contains trace amounts of both chromium and cobalt. Cronin (1980) investigated the incidence of cross-sensitivity to nickel and cobalt in 285 patients diagnosed with chromate allergy: 70 percent of the men and 57 percent of the women were allergic to chromium only; 19 percent of the men were allergic to cobalt and chromium but not nickel, while only 9 percent of the women tested exhibited this pattern; sensitivity to both chromium and nickel but not cobalt occurred in 5 percent of the men and 11 percent of the women; 6 percent of the men and 23 percent of the women were allergic to all three metals. Polak (1983) notes that patch testing of several antigens at a time can sometimes produce a condition in which false positives occur when certain allergens are applied, therefore knowledge of the exposure history of the patient is vital in identifying double allergies.

4.5.3 *Experimental chromium contact sensitivity in animals*

Early experimental research on chromium contact sensitivity focused on developing a suitable animal model sensitive to secondary challenge with chromium compounds. Guinea pigs were used in initial studies because of their tendency to react homogeneously to antigenic challenge, ease in handling, and cost-effectiveness (Klecak, 1991). Early studies found that guinea pigs could not be sensitized by dermal application of chromium salts dissolved in water (Levin *et al.*, 1959; Darabos *et al.*, 1963; Magnusson and Kligman, 1970). Successful sensitization of laboratory animals to chromium has been reported using numerous methodologies. Sensitization has been achieved via dermal application of chromium salts dissolved in sodium lauryl sulfate (Wikstrom, 1962), Triton X-100 (Schwarz-Speck and Keil, 1965), or DMSO (Heise *et al.*, 1969). Pretreatment of skin with petrol or heparin prior to dermal application has been shown to increase the success rate of chromium sensitization, presumably by enhancing skin permeability to chromium (Cohen, 1966).

Intradermal injection of chromium salts dissolved in homogenized Freund's complete adjuvant (FCA) has proven to be a dependable method for inducing chromium sensitivity in guinea pigs (van Neer, 1963; Gross *et al.*, 1968; Jansen and Berrens, 1968), but the success rate of immunization differs as a result of variation in methodology and the laboratory employing them (Skog and Wahlberg, 1970; Hicks *et al.*, 1979). Freund's complete adjuvant is a critical component for ensuring chromium sensitization. Magnusson and Kligman (1970) found that chromium salts mixed with this adjuvant were able to induce chromium sensitization in 21 of 25 animals, while the combination of Freund's incomplete adjuvant (FIA) plus chromium salts induced sensitization in 10 out of 25 animals. Sensitization was successful in only 1 of 25 animals when chromium salts were administered without adjuvant.

Chromium contact sensitivity can be delayed, curbed, or prevented in guinea pigs by giving oral doses of $K_2Cr_2O_7$ prior to inducing sensitization by injection (van Hoogstraten *et al.*, 1991, 1994). Oral doses of $10\,mg\,K_2Cr_2O_7$ per week given for 6 weeks prevented positive reactions in animals challenged with injections of $2\,mg\,K_2Cr_2O_7$/ml intradermally. The tolerance to chromium in these animals persisted for 2 years. Oral administration of chromium after confirmed sensitization caused a 'transient desensitization' characterized by flare-up reactions at previous chromium injection sites. These reactions subsided with continued feeding of chromium. In addition, contact dermatitis reactions were strongly reduced when rechallenged with chromium. The intensity of reaction to challenge with chromium returned within 2 months after cessation of oral treatment with $K_2Cr_2O_7$.

In addition to the guinea pig, both ICR and BALB/c mice have been successfully sensitized to chromium (Mor *et al.*, 1988; Vreeburg *et al.*, 1991). Mor *et al.* (1988) successfully sensitized ICR mice and an inbred strain of BALB/c mice by repeated topical applications of $K_2Cr_2O_7$ dissolved in DMSO. The covering of dermal application sites in BALB/c mice with plaster casts significantly increased the number of mice that developed sensitivity to dermally applied chromium (Vreeburg *et al.*, 1991). The authors concluded that the plaster casts prevented mice from grooming the application site, swallowing chromium particles, and inadvertently becoming tolerant to chromium. On the other hand, CBA mice were not successfully sensitized with or without the use of plaster casts to cover the application site. Intradermal injections of $K_2Cr_2O_7$ mixed with FCA did not induce sensitization to chromium in either BALB/c or CBA mice.

A number of authors have demonstrated that Cr(III) compounds can be used to sensitize guinea pigs to Cr(VI), suggesting that Cr(III) may be involved in the sensitization process. Jansen and Berrens (1968), Schwarz-Speck (1968), and Schwarz-Speck and Keil (1965) all report that Cr(III) and Cr(VI) were equally capable of sensitizing guinea pigs to Cr(VI). Schneeberger and Forch (1974) found that either Cr(III) or Cr(VI) could induce contact sensitivity in guinea pigs when dissolved in 2 percent Triton X-100, but chromium(VI) compounds were much more effective than Cr(III) in eliciting hypersensitivity when rechallenged with antigen. Similarly, Mali *et al.* (1963) were able to sensitize guinea pigs to Cr(VI) using Cr(III) compounds; however, guinea pigs sensitized initially with Cr(VI) induced sensitivity to Cr(VI) in a greater number of animals and to a greater degree than animals initially treated with Cr(III). In contrast, van Neer (1963) found that only Cr(VI) was capable of causing a positive response in guinea pigs sensitized with either Cr(VI) or Cr(III). Polak (1983) found no difference in the ability of either Cr(VI) or Cr(III) salts dissolved in FCA to sensitize guinea pigs to secondary challenge with either Cr(VI) or Cr(III) compounds. When compared on a molar basis, Cr(VI) was more effective in eliciting hypersensitivity than Cr(III) regardless of the species used in the sensitization process. Gross *et al.* (1968) found that the administration of Cr(III) dissolved in FCA consistently established chromium contact sensitivity in guinea pigs, but subsequent intradermal challenge with Cr(III) induced either a weak or negative type IV response.

The conjugation of Cr(III) compounds with proteins has been shown to enhance chromium hypersensitivity in rodents, providing support for the idea that Cr(III)–protein complexes are the ultimate haptens that mediate chromium hypersensitivity. Administration of Cr(III)–protein complexes caused hypersensitivity reactions of much greater intensity and at greater frequency than Cr(VI)–protein conjugates in guinea pigs initially sensitized with Cr(VI) (Shmunes *et al.*, 1973). The fact that anti-Cr(III) antibodies are commonly found in Cr(VI)-sensitized animals and that anti-Cr(VI) antibodies have not been isolated in these animals provides additional evidence in favor of the ultimate importance of Cr(III) compounds in chromium allergy (Cohen, 1962; Mali *et al.*, 1963; Novey *et al.*, 1983).

Genetic factors may influence the development of chromium hypersensitivity in laboratory animals. Polak *et al.* (1968) evaluated the effect of animal strain on chromium hypersensitivity in guinea pigs. Contact sensitivity to $K_2Cr_2O_7$, beryllium fluoride, and mercury chloride was measured in Hartley, Pirbright, white-spotted Himalayan, and inbred strains II and XIII guinea pigs. White-spotted Himalayan guinea pigs had the highest ratio of positive responses to $K_2Cr_2O_7$ (83 percent) followed by the Hartley (66 percent) and Pirbright guinea pigs (67 percent) strains. Eighty percent of the inbred strain II were successfully sensitized to $K_2Cr_2O_7$, while none of the XIII strain developed hypersensitivity to chromium. Reactivity to other metals also differed by strain, but as in the case with chromium, the prevalence of hypersensitivity within strains never reached 100 percent, indicating genetic variation within each strain, in addition to genetic differences, segregated by strain. The observed differences in immune response between groups may be attributed to restricted genetic inhomogeneity, gene penetrance, and/or non-immunological factors (Polak, 1983). It has been proposed that differences in reactivity among animals of the same strain may be the result of a genetically determined deficiency in the presentation of antigen to T-lymphocytes by APC (Polak, 1983). A deficiency in antigen presentation may explain differences in the time required for the appearance of the first signs of hypersensitivity among animals of the same strain. Polak (1983) found that 35 percent of the guinea pigs tested developed full-blown type VI hypersensitivity reactions to chromium within 14 days of first immunization. In contrast, 33 percent of the guinea pigs tested acquired hypersensitivity to chromium within 6 to 8 weeks. Fifteen percent of the animals tested did not develop sensitivity to chromium after repeated attempts to induce sensitization.

4.5.4 *Occupational asthma*

Asthma, a chronic illness characterized by persistent bronchial hyperactivity (Boushey and Fahy, 1995), is an immune-mediated response that has been historically linked with exposure to common allergens such as dust mites, pollen, cat dander, and cigarette smoke. Early case reports of asthma among industrial workers implicated toxic air emissions as a potential trigger of asthmatic episodes. Over 200 industrial compounds have been associated with high incidences of occupational asthma (Chan-Yeung, 1995; Koren, 1995) including nickel, cobalt, styrene, formaldehyde, and chromium.

There are two subtypes of occupational asthma: immune-mediated asthma and non-immune-mediated asthma. The onset of immune-mediated asthma occurs shortly after exposure and is triggered by the binding of antigen to mast cell-bound IgE antibodies. Antigen binding prompts the release of mast cell granules which in turn promote leukocyte chemotaxis and bronchial smooth muscle spasm. The onset of non-immune-mediated asthma is delayed and is promulgated by the expansion and proliferation of activated T-lymphocytes which in turn secrete lymphokines that promote chemotaxis, smooth-muscle contraction, edema, and mucus hypersecretion. A recent study indicates that mast cells may play as prominent a role in non-immune-mediated asthma as in immune-mediated asthmatic reactions (Boushey and Fahy, 1995).

There have been isolated reports in the literature of a reoccurring subform of occupational asthma in workers exposed to a large dose of chemical irritant (Brooks *et al.*, 1985; Chan-Yeung, 1995). A single exposure may render the worker sensitive to subsequent exposure to subacute doses of similar compounds and sensitization may persist for many years. This condition is referred to as reactive airways dysfunction syndrome (RADS). The biological mechanism(s) by which RADS invokes asthmatic response in sensitive individuals is not clear; the lung pathology of affected individuals contains features of both immune- and non-immune-mediated asthma.

Cases of occupational asthma have been reported in conjunction with occupational exposure to dichromates, ammonium bichromate, chromic acid, chromite ore, chromate pigments, and to welding fumes (Williams, 1969, 1973; Haines and Nieboer, 1988). Evidence suggesting that chromium exposure is a cause of occupational asthma is limited to a small number of case reports and case series. In most of these instances, a consistent pattern between chromium exposure and the onset of asthma is noted. In some cases, hypersensitivity to chromium is confirmed either by patch testing or by challenge of a chromium bronchodilator (Haines and Nieboer, 1988). Both immediate (Novey *et al.*, 1983) and delayed (Moller *et al.*, 1986) asthmatic reactions have been reported, suggesting that chromium-induced asthma may be both immune mediated or non-immune mediated.

Exposure to chromium has been implicated as a potential cause of occupational asthma among metal platers and polishers. Joules (1932) describes a metal plater who developed both asthma and dermatitis in conjunction with chromium exposure. Intradermal injection of this worker with $K_2Cr_2O_7$ caused an immediate local reaction followed shortly by a severe asthmatic reaction that lasted 18 h. Similarly, Card (1935) was able to precipitate an asthmatic reaction in a chrome plater/polisher who had been previously diagnosed with occupational asthma with an intradermal injection of 4 mg $K_2Cr_2O_7$. Broch (1949) found that a ferrochrome worker's history of asthma triggered by factory furnace fumes correlated well with his positive reaction to chromium in a skin prick test.

There is evidence that chromium may play a role in occupational asthma among painters as well as stainless steel welders. Exposure to chromium among workers of either occupation is common. In four cases of occupational asthma among painters, manifestation of symptoms was closely associated with exposure to spray paints containing chromium, lead, and zinc chromates (Marechal, 1957). In one of these cases, an asthmatic response could be induced through exposure to paint containing chromium with the severity of response increasing in a dose-dependent fashion. Keskinen *et al.* (1980) describe two workers who consistently developed asthma-like symptoms after welding stainless steel. In the first patient, a 51-year-old man with severe eczema attributed to chromium exposure, skin prick tests for chromium sensitivity were negative, but positive results were obtained via patch testing. Bronchial challenge with manual metal arc (MMA) stainless steel welding fumes produced a severe bronchial obstruction within 30 min of exposure. Subsequent testing with nickel compounds showed that the patient was not hypersensitive to nickel exposure, suggesting that either chromium or another compound within the welding fumes was triggering the asthmatic response. Bronchial challenge of the second patient with MMA stainless steel fumes induced a decrease in FEV_1 (forced expiratory volume) within 30 min, but no positive hypersensitivity reactions to either chromium or nickel were reported.

Voluntary inhalation of chromium sulfate induced a clear depression in FEV_1 within 15 min in an electroplater who developed asthma-like symptoms after 1 week of employment (Novey *et al.*, 1983). Challenge with nickel sulfate also induced a depression in FEV_1, but the response differed from chromium in that it was biphasic, i.e. immediate symptoms, followed by a quick recovery, with recurrence of symptoms at 3 h postexposure. Skin tests for either chromium or nickel were negative, but elevated blood levels of IgE chromium- and nickel-specific antibodies were detected, indicating an elevated immune response to these metals. Similarly, voluntary challenge of a chromium-sensitive welder with sodium chromate aerosol ($29 \mu g/m^3$) produced a delayed-type onset of asthma beginning 3.5 h postexposure and peaking with a 40 percent decrease in FEV_1 at 6 h postexposure (Moller *et al.*, 1986). The worker also tested positive for leukocyte inhibition indicating that the delayed asthmatic reaction was immunological in nature. Olaguibel and Basomba (1989) report that bronchial challenge of five asthmatic chromium workers with $K_2Cr_2O_7$ induced a clear decrease in FEV_1 in all five, but that none of the workers tested positive for chromium hypersensitivity when administered skin prick tests. The authors note that the type of asthmatic reaction induced with $K_2Cr_2O_7$ differed between individuals. One worker exhibited an immediate response to bronchial challenge, while two of the workers developed a delayed-type response that manifested itself 1 h after exposure. Two workers developed immediate decreases in FEV_1 followed by a quick recovery and then by a secondary asthmatic response at 1 h postexposure that peaked at 8 h postexposure. Sodium cromoglycate failed to prevent the onset of asthma in these workers suggesting that these reactions were not immunologically mediated and may have been irritative in nature.

There have been few studies conducted on the effects of chromium exposure on the animal lung. Johansson *et al.* (1986) found that lung macrophage counts were significantly elevated in rabbits exposed to $900 \mu g\, Cr(VI)/m^3$ for 5 days per week, 6 hours per day, for 4 to 6 weeks, but neither macrophage morphology nor metabolic activity differed significantly when compared with controls. Exposure of rabbits to $600 \mu g\, Cr(III)/m^3$ for 5 days per week, 6 hours per day, for 4 to 6 weeks had no effect on lung macrophage counts, but macrophage metabolic activity was elevated and correlated with the presence of numerous, enlarged lysosomes and intercellular inclusions rich in chromium. Exposure of Wistar rats to 25 or $50 \mu g\, Cr(VI)/m^3$ for 28 days resulted in a stimulation of both macrophage cell numbers and macrophage phagocytic activities (Glaser *et al.*, 1985). Macrophage counts were significantly decreased in rats exposed to 50 or $200 \mu g\, Cr(VI)/m^3$ for 90 days; however, macrophage activity in rats exposed to $50 \mu g/m^3$ was significantly elevated while activity in rats exposed to $200 \mu g/m^3$ was significantly depressed. The reduction of both macrophage cell count and macrophage phagocytic activity was directly related to a fourfold decrease in lung clearance of inhaled iron oxide in rats treated with $200 \mu g\, Cr(VI)/m^3$ for 90 days.

4.5.5 *Chromium-induced immunosuppression*

The suppression of the immune system by xenobiotics is a major public health concern. Suppression of immune function is thought to result in increased susceptibility to infectious disease and a reduction in immunological control of cancer, but there is much uncertainty in the extrapolation of experimental immunotoxicological data to humans, and immunosuppression by anthropogenic sources is probably a rare occurrence (Haley, 1993). However, experimental data indicate that the immune activity of both T- and B-lymphocytes is reduced in the presence of high concentrations of various chemicals, and in particular, heavy metals. Mercury, lead, nickel, cadmium, and zinc can alter immune processes in experimental animals and humans (Koller, 1980; Sharma and Reddy, 1987). Cadmium, mercury, and zinc have been shown to inhibit lymphocyte proliferation, while lead alters lymphocyte blastogenesis and reduces macrophage function and interferon production. Nickel suppresses T-dependent immune responses and also disrupts lymphocyte blast formation. Metals may also suppress or enhance macrophage nitric oxide formation, and both endpoints can have detrimental effects within living organisms (Tian and Lawrence, 1996).

Immunotoxicity data for chromium are sparse and inconsistent but indicate that chromium exposure may cause immunosuppression in certain tissues. Two studies have evaluated the modulatory effects of chromium on lymphocyte proliferation following chromium exposure. Snyder and Valle (1991) report that chromium exposure reduced mitogen-induced proliferation of rat T-lymphocytes *in vitro* at concentrations of 0.01 to 100 mg/L. Similarly, rat B-lymphocyte proliferation was also inhibited *in vitro* by chromium concentrations of more than 0.01 mg/L (range: 0.01 to 100 mg/L). Thymidine uptake by mixed rat lymphocytes in cell culture was reduced at chromium concentrations of more than 0.1 to 100 mg/L and stimulated at 0.1 mg/L. Concentrations ranging from 0.01 to less than 0.1 mg/L had no effect on thymidine uptake as compared with controls. T- and B-lymphocyte counts isolated from rats fed 200 mg Cr/L in their drinking water were reduced compared with controls, while a slight stimulation occurred in rats fed 100 mg Cr/L. Proliferation in the presence of the mitogen lipopolysaccharide (LPS) did not differ from controls in the 200 mg Cr/L dose group and was enhanced in rats fed 100 mg Cr/L. Neither chromium(III) nor chromium(VI) inhibited mitogen-induced dog lymphocyte blastogenesis over the concentration range of 0.02 to 2 μg Cr/L (Shifrine *et al.*, 1984). Glaser *et al.* (1985) found that humoral immune function directed against sheep red blood cells (SRBCs) was increased in Wistar rats treated with 25 to 100 μg Cr(VI)/m^3 for 25 or 90 days via inhalation, but humoral function was significantly depressed at Cr(VI) concentrations greater than 100 μg/m^3. No overt signs of toxicity to chromium(VI) were observed at any dose level suggesting that immunosuppression was specific for Cr(VI) exposure.

Chromium was included in a mixture study designed to evaluate the combined immunotoxic effect in $B_6C_3F_1$ mice of 25 groundwater contaminants commonly found near toxic waste sites (Germolec *et al.*, 1989). The mice were exposed to the chemical mixture via their drinking water for a period of either 14 or 90 days. None of the animals developed overt signs of toxicity related to exposure to the mixture; however, animals fed a 20 percent concentration of the mixture for 14 days or a 10 percent concentration for 90 days did show clinical signs of immunosuppression characterized by a decrease in hematopoietic stem cells and antibody activity toward SRBC. The effect of the contaminated groundwater in mouse host-resistance to *Plasmodium yoeli* infection was also evaluated to determine toxicant effects on humoral immune response. The number of red blood cells infected with *P. yoeli* was higher in mice treated with the highest concentration of groundwater when assessed on day 10, 12, or 14 after infection suggesting a toxicant induced depression in humoral immune response.

Sudhan and Michael (1995) have demonstrated the immunosuppressive effects of chrome-tannery effluents on the freshwater cichlid fish *Oreochromis mossambicus*. The authors report a significant decrease in antibody response to challenge with bovine serum albumin (BSA) in fish treated with 0.12, 0.24, or 0.48 percent effluent ($p < 0.01$) which was 2.5, 5.0, or 10.0 percent of the LC_{50} of the tannery effluent, respectively. Antibody activity was dose dependent with the highest amount of activity occurring in fish treated with 0.48 percent effluent. No other signs of toxicity were observed.

Few studies have evaluated the effects of chromium on the immune response or function of human cells in culture. There is preliminary evidence that chromium causes a decrease in human lymphocyte proliferation and total antibody production *in vitro*. Human lymphocytes treated with hexavalent chromium proliferated rapidly in a dose-related fashion over the concentration range of 10^{-8} to 10^{-6} mol Cr/L, but lymphocyte proliferation was reduced at concentrations of 1×10^{-6} to 2.5×10^{-6} mol Cr/L (Borella and Bargellini, 1993). Immunoglobulin production in cultured human lymphocytes was reduced by 80 percent compared with controls after exposure to 2.0 μmol Cr/L. Trivalent chromium (10^{-8} to 10^{-3} μmol Cr/L) had no effect on the rate of thymidine uptake, a surrogate of lymphocyte blastogenesis (Borella *et al.*, 1990). Analysis of cell chromium content showed that both suppressed proliferation and immunoglobulin production were a direct result of the amount of intracellular chromium present within the cells.

The release of bone-resorbing cytokines interleukin 1β (IL-1β) and tumor necrosis factor-α (TNF-α) was significantly enhanced while the bone-forming cytokine transforming growth factor-$β_1$ (TGF-$β_1$) was significantly reduced in human macrophages/monocytes simultaneously treated with chromium and lipopolysaccharide (Wang *et al.*, 1996). Similar results have been observed in the peripheral monocytes of patients with failed total hip replacements containing chromium (Lee *et al.*, 1997). The enhancement of bone-resorbing cytokine release, with a concomitant inhibition of bone-forming cytokine release, may be an important factor in the development of osteolysis in total joint arthroplasty.

Clinical studies also indicate that chromium may have an effect on human immune parameters. Circulating lymphocyte levels in 15 dye workers exposed to lead chromate for 16 months were significantly reduced compared with controls, but no differences in serum IgM, IgG, or IgA levels were noted (Boscolo *et al.*, 1995). Snyder *et al.* (1996) present limited but interesting field data which suggest that chromium exposure might affect human lymphocyte production of IL-6, a cytokine which stimulates B-lymphocyte growth and differentiation. Lymphocyte IL-6 levels from 46 individuals who lived and/or worked in Hudson County, New Jersey, were compared with 45 college faculty members who lived and worked outside of Hudson County. Hudson County soil contains high levels of chromium as a result of chrome ore processing activities (Burke *et al.*, 1991), and the authors theorized, based on their earlier study, that suppressed lymphocyte responses might serve as biomarkers of chromium exposure. After controlling for gender, age, and smoking status, lymphocyte responses to ConA and PWM were found not to be different between the two populations. However, IL-6 production by PWM-stimulated lymphocytes was 64 percent of that of controls ($p < 0.001$). As the number of IL-6-producing cells did not differ between groups, it was suggested that the deficit in IL-6 levels could have been caused by a chromium-induced signaling deficiency or by the blockage of IL-6 gene transcription factors. However, elevated chromium exposure was not confirmed through biological measurements (e.g. urinary Cr excretion) and the two populations differed in at least one key respect: the Hudson County participants were from a largely urban environment, while the controls were from rural or suburban areas. Numerous factors that may affect immune system parameters were not accounted for in the study design. Thus, the study results should be interpreted with caution. Because these findings were of a preliminary nature, their interpretation and evaluation should await mechanistic analyses and replication of the results in better-controlled studies.

4.6 Future Directions

Most of the existing data concerning the effects of chromium on immune function provide an adequate description of the metal's ability to induce type IV delayed hypersensitivity in laboratory animals and man; however, to date there has been no concerted effort to identify systematically and describe adequately the effects of chromium species on immune system function using the most modern experimental techniques available. A number of the heavy metals that have been studied to some degree have been shown to suppress immune function *in vivo*. Both lead and cadmium are capable of increasing the susceptibility of laboratory animals to infection with bacterial pathogens, and some data suggest that cadmium and mercury (organic and inorganic) may also possess immunosuppressive properties (Burns *et al.*, 1996). Based on the prevalence of chromium contamination among USEPA Superfund sites (Fay and Mumtaz, 1996) and the widespread use of chromium in industry, there is a definite need for basic immunotoxicology data for chromium in order to assess accurately the human health risks associated with this metal.

Guidelines for systematically evaluating the potential immunotoxicity of chemicals have been put forward by the National Toxicology Program (NTP) (Luster *et al.*, 1988) and have been implemented in the testing of over 50 commonly encountered chemicals (Luster *et al.*, 1992). These guidelines recommend using a battery of standardized immunotoxicological assays that demonstrate good reproducibility between laboratories. Agents are initially screened for immunotoxicity using a combination of simple, *in vitro* assays and pathological markers of immune effects in whole animals (tier I). Potential immunotoxicants are then subjected to a comprehensive series of tests designed to assess immunotoxic effects on lymphocyte population levels, humoral response, and host resistance to pathogens (tier II). The use of less sensitive screening tests during the initial testing phase does present the possibility that this approach may fail to identify a weak immunotoxicant. Luster *et al.* (1988) report that none of the immunotoxic compounds tested using the NTP guidelines failed to produce a measurable effect during initial screening. A set of USEPA immunotoxicity testing guidelines for chemical pesticides, similar to the current NTP guidelines, are nearing completion and will recommend many of the same immunotoxicity testing protocols adopted by the NTP (Sherl Riley, USEPA, personal communication). There are currently no immunotoxicology guidelines for metals but the testing protocols contained in the NTP guidelines provide a standardized and systematic approach to begin evaluating the possible effects of chromium compounds on immune function.

References

ACGIH (American Conference of Governmental Industrial Hygienist) (1995). *Threshold limit values (TLVs) for chemical substances and physical agents and biological exposure indices (BEIs) 1995–1996*. Cincinnati, OH.

AITIO, A., JARVISALO, J., KIILUNEN, M., TOSSAVAINEN, A., and VAITTINEN, P. (1984). Urinary excretion of chromium as an indicator of exposure to trivalent chromium sulphate in leather tanning. *Intl Arch. Occup. Health* **54**, 241–249.

ALOMAR, A., CONDE-SALAZAR, L., and ROMAGUERA, C. (1985). Occupational dermatoses from cutting oils. *Contact Dermatitis* **12**, 129–138.

ALPOIM, M.C., GERALDES, C.F., OLIVEIRA, C.R., and LIMA, M.C. (1995). Molecular mechanisms of chromium toxicity: oxidation of hemoglobin. *Biochem. Soc. Transact.* **23**, 241S.

ATSDR (Agency for Toxic Substances and Disease Registry) (1993). *Toxicological Profile for Chromium.* ATSDR PB93-182434, Atlanta, Georgia.

BONDE, J.P. and CHRISTENSEN, J.M. (1991). Chromium in biological samples from low-level exposed stainless steel and mild steel workers. *Arch. Environ. Health* **46**, 225–229.

BORELLA, P. and BARGELLINI, A. (1993). Effects of trace elements on immune system: results in cultured human lymphocytes. *J. Trace Elem. Electrolytes Health Dis.* **7**, 231–233.

BORELLA, P., MANNI, S., and GIARDINO, A. (1990). Cadmium, nickel, chromium, and lead accumulate in human lymphocytes and interfere with PHA-induced proliferation. *J. Trace Elem. Electrolytes Health Dis.* **4**, 87–95.

BOSCOLO, P., DI GIOACCHINO, M., CERVONE, M., DI GIACOMO, F., BAVAZZANO, P., and GIULIANO, G. (1995). Lymphocyte subpopulations of workers in a plant producing plastic materials (preliminary study). *G. Ital. Med. Lav.* **17**, 27–31.

BOUSHEY, H.A. and FAHY, J.V. (1995). Basic mechanisms of asthma. *Environ. Health Perspect.* **103**(Suppl. 6), 229–233.

BROCH, C. (1949). Bronchial asthma caused by chromium trioxide. *Nord. Med.* **41**, 996–997.

BROOKS, S.M., WEIS, M.A., and BERNSTEIN, I.L. (1985). Reactive airways dysfunction syndrome (RADS). *Chest* **88**, 376–384.

BURKE, T., FAGLIANO, J., GOLDOFT, M., HAZEN, R.E., IGLEWICZ, R., and McKEE, T. (1991). Chromite ore processing residue in Hudson County, New Jersey. *Environ. Health Perspect.* **92**, 131–137.

BURNS, L.A., MEADE, B.J., and MUNSON, A.E. (1996). Toxic responses of the immune system, in Klaassen, C.D. (Ed.) *Casarett and Doull's Toxicology: The Basic Science of Poisons*, 5th Edn. pp. 355–403, New York: McGraw-Hill.

BURROWS, D. (1972). Prognosis in industrial dermatitis. *Br. J. Dermatol.* **87**, 145–148.

BURROWS, D. and ADAMS, R.M. (1990). Metals, in Adams, R.M. (Ed.) *Occupational Skin Diseases*, 2nd Edn. pp. 353–386. Philadelphia: WB Saunders.

CARD, W.I. (1935). A case of asthma sensitivity to chromates. *Lancet* **229**, 1348–1349.

CHAN-YEUNG, M. (1995). Occupational asthma. *Environ. Health Perspect.* **103**(Suppl. 6), 249–252.

COHEN, H.A. (1962). Experimental production of circulating antibodies to chromium. *J. Invest. Dermatol.* **38**, 13–20.

(1966). On the pathogenesis of contact dermatitis. *Isr. J. Med. Sci.* **2**, 37.

CRONIN, E. (Ed.) (1980). *Contact Dermatitis*, 1st Edn. pp. 287–313. New York: Churchill Livingstone.

DARABOS, L., SZABO, Z., and MEGYERI, J. (1963). Beitrage zur Bichromatallergie. IV. Tierexperimentelle Prufungen unter besonderer Berucksichtigung der Arbeitsverhaltnisse in Schieferfabriken. *Hautarzt* **14**, 364.

EDMUNDSON, W.F. (1951). Chrome ulcers of the skin and nasal septum and their relation to patch testing. *J. Invest. Dermatol.* **17**, 17–19.

EPA (Environmental Protection Agency) (1980). Ambient water quality criteria for chromium, EPA 440/5-80-035. Office of Water Regulations and Standards, Criteria and Standards Division, Washington, DC.

(1984). Health assessment document for chromium, EPA 600/8-83-014F. Environmental Assessment and Criteria Office, Research Triangle Park, NC.

(1987). Quality criteria for water 1986, EPA 440/586-001. Office of Water Regulations and Standards, Washington DC.

(1990). Noncarcinogenic effects of chromium: update to health assessment document, EPA 600/8-87-/048F. Environmental Assessment and Criteria Office, Research Triangle Park, NC.

(1991). National primary drinking water regulations – synthetic organic chemicals and inorganic chemicals; monitoring for unregulated contaminants; national primary drinking water regulations implementation; national secondary drinking water regulations. Final Rule. *Fed. Regist.* **56**, 3526–3597 (As cited in ATSDR (1993)).

FAY, R.M. and MUMTAZ, M.M. (1996). Development of a priority list of chemical mixtures occurring at 1188 hazardous waste sites, using a HazDat database. *Food Chem. Toxicol.* **34**, 1163–1165.

FISHBEIN, L. (1981). Sources, transport, and alterations of metal compounds: an overview. I. Arsenic, beryllium, cadmium, chromium, and nickel. *Environ. Health Perspect.* **40**, 43–64.

(1984). Overview of the analysis of carcinogenic and/or mutagenic metals in biological and environmental samples. I. Arsenic, beryllium, cadmium, chromium, and selenium. *Int. J. Environ. Anal. Chem.* **17**, 113–170.

FISHER, A.A. (1984). The role of patch testing in the management of dermatitides caused by orthopedic metallic prostheses. *Cutis* **33**, 258–264.

FREGERT, S. (1975). Occupational dermatitis in a 10-year material. *Contact Dermatitis* **1**, 96–107.

FREGERT, S. and GRUVBERGER, B. (1972). Chemical properties of cement. *Berufsdermatosen* **20**, 238–248.

FREGERT, S. and RORSMAN, H. (1966). Allergic reactions to trivalent chromium compounds. *Arch. Dermatol.* **93**, 711–713.

(1964). Allergy to trivalent chromium. *Arch. Dermatol.* **90**, 4–6.

FREGERT, S., GRUVBERGER, B., GORANSSON, K., and NORMAN, S. (1978). Allergic contact dermatitis from chromate in military textiles. *Contact Dermatitis* **4**, 223–224.

GERMOLEC, D.R., YANG, R.S., ACKERMANN, M.F., ROSENTHAL, G.J., BOORMAN, G.A., BLAIR, P., and LUSTER, M.I. (1989). Toxicology studies of a chemical mixture of 25 groundwater contaminants. II. Immunosuppression in $B_6C_3F_1$ mice. *Fundam. Appl. Toxicol.* **13**, 377–387.

GLASER, U., HOCHRAINER, D., KLOPPEL, H., and KUHMEN, H. (1985). Low level chromium(VI) inhalation effects on alveolar macrophages and immune functions in Wistar rats. *Arch. Toxicol.* **57**, 250–256.

GOITRE, M., BEDELLO, P.G., and CANE, D. (1982). Chromium dermatitis and oral administration of the metal. *Contact Dermatitis* **3**, 203–204.

GOYER, R.A. (1991). Toxic effects of metals, in Amdur, M.O., Doull, J., and Klaassen, C.D. (Eds) *Casarett and Doull's Toxicology: The Basic Science of Poisons*, 4th Edn pp. 623–680, New York: McGraw-Hill.

GREATHOUSE, D.G. and CRAUN, G.F. (1978). Cardiovascular disease study – occurrence of inorganics in household tap water and relationships to cardiovascular mortality rates, in *Trace Substances in Environmental Health*, Vol. 12, pp. 31–39, Columbia: University of Missouri Press.

GROSS, P.R., KATZ, S.A., and SAMITZ, M.H. (1968). Sensitization of guinea pigs to chromium salts. *J. Invest. Dermatol.* **50**, 424.

GUTHRIE, B.E. (1982). The nutritional role of chromium, in Langård, S. (Ed.) *Topics in Environmental Health, Vol. 5. Biological and Environmental Aspects of Chromium* pp. 117–140, Amsterdam: Elsevier Biomedical Press.

HAINES, A.T. and NIEBOER, E. (1988). Chromium hypersensitivity, in Nriagu, J. and Nieboer, E. (Eds) *Chromium in the Natural and Human Environments* pp. 497–532, New York: John Wiley.

HALEY, P.J. (1993). Immunological responses within the lung after inhalation of airborne chemicals, in Gardner, D.E. (Ed.) *Toxicology of the Lung*, 2nd Edn, pp. 389–416, New York: Raven Press.

HEISE, H., MATTHEUS, A., and FLEGEL, H. (1969). Sensibilisierungsversuche gegen Chromverbindungen unter Mitwirkung von DMSO. *Allerg. Asthma (Leipzig)* **15**, 151.

HICKS, R., HEWITT, P.J., and LAM, H.F. (1979). An investigation of the experimental induction of hypersensitivity in the guinea pig by material containing chromium, nickel, and cobalt from arc welding fumes. *Int. Arch. Allergy Appl. Immunol.* **59**, 265.

HJORTH, N. (1991). Diagnostic patch testing, in Marzulli, F.N. and Maibach, H.I. (Eds) *Dermatotoxicology*, 4th Edn, pp. 441–452, New York: Hemisphere Publishing.

HOSTYNEK, J.J. and MAIBACH, H.I. (1988). Chromium in US household bleach. *Contact Dermatitis* **18**, 206–209.

IYENGAR, V. and WOITTIEZ, J. (1988). Trace elements in human clinical specimens: evaluation of the literature data to identify reference values. *Clin. Chem.* **34**, 474–481.

JANDL, J.H. and SIMMONS, R.L. (1957). The agglutination and sensitization of red cells by metallic cations: interaction between multivalent metals and the red cell membrane. *Br. J. Haematol.* **3**, 19.

JANSEN, L.H. and BERRENS, L. (1968). Sensitization and partial desensitization of guinea pigs to trivalent and hexavalent chromium. *Dermatologica* **137**, 65.

JOHANSSON, A., WIERNIK, A., JARSTRAND, C., and CAMNER, P. (1986). Rabbit alveolar macrophages after inhalation of hexa- and trivalent chromium. *Environ. Res.* **39**, 372–385.

JOULES, H. (1932). Asthma from sensitization from chromium. *Lancet* **223**, 182–183.

KAABER, K. and VEIEN, N.K. (1978). Chromate ingestion in chronic chromate dermatitis. *Contact Dermatitis* **4**, 119–120.

KESKINEN, H., KALLIOMAKI, P.L., and ALANKO, K. (1980). Occupational asthma due to stainless steel welding fumes. *Clin. Allergy* **10**, 151–159.

KLECAK, G. (1991). Identification of contact allergens: predictive tests in animals, in Marzulli, F.N. and Maibach, H.I. (Eds) *Dermatotoxicology*, 4th Edn, pp. 363–415, New York: Hemisphere Publishing.

KOLLER, L.D. (1980). Immunotoxicology of heavy metals. *Int. J. Immunopharmacol.* **2**, 269–279.

KOREN, H.S. (1995). Associations between criteria air pollutants and asthma. *Environ. Health Perspect.* **103**(Suppl. 6), 235–242.

LANDSBERGER, S., JERVIS, R.E., KAJRYS, G., and MONARO, S. (1983). Characterization of trace elemental pollutants in urban snow using proton induced X-ray emission and instrumental neutron activation analysis. *Int. J. Environ. Anal. Chem.* **16**, 95–130.

LANGÅRD, S. (1982). Absorption, transport, and excretion of chromium in man and animals, in Langård, S. (Ed.) *Topics in Environmental Health, Vol. 5, Biological and Environmental Aspects of Chromium* pp. 5–44, Amsterdam: Elsevier Bio-medical Press.

LANGÅRD, S. and HENSTEN-PETTERSEN, A. (1981). Chromium toxicology, in Williams, D.F. (Ed.) *Systemic Aspects of Biocompatibility*, Vol. 1 pp. 144–161, Boca Raton, FL: CRC Press.

LANTINGA, H., NATER, J.P., and COENRAADS, P.J. (1984). Prevalence, incidence and course of eczema on the hands and forearms in a sample of the general population. *Contact Dermatitis* **10**, 135–139.

LEE, S.H., BRENNAN, F.R., JACOBS, J.J., URBAN, R.M., RAGASA, D.R., and GLANT, T.T. (1997). Human monocyte/macrophage response to cobalt–chromium corrosion products and titanium particles in patients with total joint replacements. *J. Orthopaed. Res.* **15**, 40–49.

LEVIN, H.M., BRUNNER, M.J., and RATTNER, H. (1959). Lithographer's dermatitis. *J. Am. Med. Assoc.* **169**, 566–569.

LUSTER, M.I., MUNSON, A.E., THOMAS, P.T., HOLSAPPLE, M.P., FENTER, J.D., WHITE, K.L., JR, LAUR, L.D., GERMOLEC, D.R., ROSENTHAL, G.J., and DEANS, J.H. (1988). Development of a testing battery to assess chemical-induced immunotoxicity: National Toxicology Program's Guidelines for immunotoxicity evaluation in mice. *Fundam. App. Toxicol.* **10**, 2–19.

LUSTER, M.I., PORTIER, C., PAIT, G.D., WHITE, K.L., JR, GENNINGS, C., MUNSON, A.E., and ROSENTHAL, G.J. (1992). Risk assessment in immunotoxicology. I. Sensitivity and predictability of immune tests. *Fundam. Appl. Toxicol.* **18**, 200–210.

MAGNUSSON, B. and KLIGMAN, A.M. (Eds) (1970). Allergic contact dermatitis in the guinea pig, in *Identification of Contact Allergens*, Springfield: Charles C. Thomas Publishers.

MALI, J.H.W., VAN KOOTEN, W.J., and VAN NEER, F.C.J. (1963). Some aspects of behavior of chromium compounds in the skin. *J. Invest. Dermatol.* **41**, 111–122.

MALI, J.W.H., MALTEN, K., and VAN NEER, F.C.J. (1966). Allergy to chromium. *Arch. Dermatol.* **93**, 41–44.

MARECHAL, M. (1957). Irritation et allergie respiratoires au jaune de chrome dans la peninture au pisctolet. *Arch. Mal. Prof.* **18**, 284.

MARZULLI, F.N. and MAIBACH, H.I. (1991). Contact allergy: predictive testing in humans, in Marzulli, F.N. and Maibach, H.I. (Eds) *Dermatotoxicology*, 4th Edn, pp. 415–439, New York: Hemisphere Publishing.

McAUGHEY, J.J., SAMUEL, A.M., BAXTER, P.J., and SMITH, N.J. (1988). Biological monitoring of occupational exposure in the chromate pigment production industry. *Sci. Total Environ.* **71**, 317–322.

MERTZ, W. (1969). Chromium occurrence and function in biological systems. *Physiol. Rev.* **49**, 163–239.

MERTZ, W. and SCHWARTZ, K. (1955). Impaired intravenous glucose tolerance as an early sign of dietary necrotic liver damage. *Arch. Biochem. Biophys.* **58**, 504–508.

MOLLER, D.R., BROOKS, S.M., BERNSTEIN, D.I., CASSEDY, K., ENRIONE, M., and BERNSTEIN, L. (1986). Delayed anaphylactoid reaction in a worker exposed to chromium. *J. Allergy Clin. Immunol.* **77**, 451–456.

MOLYNEUX, M.J. and DAVIES, M.J. (1995). Direct evidence for hydroxyl radical-induced damage to nucleic acids by chromium(VI)-derived species: implications for chromium carcinogenesis. *Carcinogenesis* **16**, 875–882.

MOR, S., BEN-EFRAIM, S., LEIBOVICI, J., and BEN-DAVID, A. (1988). Successful contact sensitization to chromate in mice. *Int. Arch. Allergy Appl. Immunol.* **85**, 452–457.

MORRIS, G.E. (1958). Chrome dermatitis: a study of the chemistry of shoe leather with particular reference to basic chromic sulfate. *Arch. Dermatol.* **78**, 612–618.

NETHERCOTT, J., PAUSTENBACH, D., ADAMS, R., FOWLER, J., MARKS, J., MORTON, C., TAYLOR, J., HOROWITZ, S., and FINLEY, B. (1994). A study of chromium induced allergic contact dermatitis with 54 volunteers: implications for environmental risk assessment. *Occup. Environ. Med.* **51**, 371–380.

NIOSH (National Institute of Occupational Safety and Health) (1989). *National Occupational Exposure Survey Data Base.* US Dept. of Health and Human Services, Public Health Service, Centers for Disease Control, Cincinnati, OH.

North American Contact Dermatitis Group (1973). Epidemiology of contact dermatitis in North America: 1972. *Arch. Dermatol.* **108**, 537–540.

NOVEY, H.S., HABIB, M., and WELLS, I. (1983). Asthma and IgE antibodies induced by chromium and nickel salts. *J. Allergy Clin. Immunol.* **72**, 407–412.

NRIAGU, J.O. and PACYNA, J.M. (1988). Quantitative assessment of worldwide contamination of air, water, and soils by trace metals. *Nature* **333**, 134–139.

OLAGUIBEL, J.M. and BASOMBA, A. (1989). Occupational asthma induced by chromium salts. *Allerg. Immunopathol. (Madrid)* **17**, 133–136.

OSHA (Occupational Safety and Health Administration) (1989). Air contaminants. Final rule. *Fed. Reg.* **54**, 2930. US Department of Labor, Washington, DC.

PAUSTENBACH, D.J., HAYS, S.M., BRIEN, B.A., DODGE, D.G., and KERGER, B.D. (1996). Observation of steady state in blood and urine following human ingestion of hexavalent chromium in drinking water. *J. Toxicol. Environ. Health* **49**, 453–461.

PELTONEN, L. and FRAKI, J. (1983). Prevalence of dichromate sensitivity. *Contact Dermatitis* **9**, 190–194.

PERONE, V.B., MOFFITT, A.E., and POSSICK, P.A. (1974). The chromium, cobalt, and nickel contents of American cement and their relationship to cement dermatitis. *Am. Ind. Hyg. Assoc. J.* **35**, 301–306.

POLAK, L. (1983). Immunology of chromium, in Burrows, D. (Ed.) *Chromium: Metabolism and Toxicity*, pp. 51–136, Boca Raton, FL: CRC Press.

POLAK, L., BARNES, J.M., and TURK, J.L. (1968). The genetic control of contact sensitization to inorganic metal compounds in guinea pigs. *Immunology* **14**, 707.

RAHKONEN, E., JUNTTILA, M.L., KALLIOMAKI, P.L., OLKINOUORA, M., KOPONEN, M., and KALLIOMAKI, K. (1983). Evaluation of biological monitoring among stainless steel workers. *Int. Arch. Occup. Environ. Health* **52**, 243–255.

ROITT, I., BROSTOFF, J., and MALE, D. (1993). *Immunology*, 3rd Edn, pp. 19.1–19.22, Saint Louis: Mosby.

SAMITZ, M.H. and KATZ, S. (1964). A study of the chemical reactions between chromium and the skin. *J. Invest. Dermatol.* **84**, 404–409.

SAMITZ, M.H., and SHRAGER, J. (1966). Patch test reactions to hexavalent and trivalent chromium compounds. *Arch. Dermatol.* **94**, 304–306.

SAMITZ, M.H., KATZ, S., and SCHRAGER, J.D. (1967). Studies of the diffusion of chromium compounds through skin. *J. Invest. Dermatol.* **48**, 514–520.

SCHNEEBERGER, H.W. and FORCH, G. (1974). Tieresperimentelle Chromallergie. Einfluss der Valenzen und der chemischen Umgebung der Chromionen. *Arch. Dermatol. Forsch.* **249**, 71.

SCHWARZ-SPECK, M. (1968). Experimentelle Sensibilisierung mit drei- und sechswertigem Chrom, XIII. *Congressees Internationalis Dermatologiae*, Munchen, 1967, Berlin: Springer-Verlag.

SCHWARZ-SPECK, M. and KEIL, H. (1965). Experimentelles Chrom-III-Ekzem. *Dermatologica* **130**, 373.

SHARMA, R.P. and REDDY, R.V. (1987). Toxic effects of chemicals on the immune system, in Haley, T.J. and Berndt, W.O. (Eds) *Handbook of Toxicology*, pp. 555–591, New York: Hemisphere Publishing.

SHIFRINE, M., FISHER, G.L., and TAYLOR, N.J. (1984). Effect of trace elements found in coal fly ash on lymphocyte blastogenesis. *J. Environ. Pathol. Toxicol. Oncol.* **5**, 15–24.

SHMUNES, E., KATZ, S.A., and SAMITZ, M.H. (1973). Chromium–amino acid conjugates as elicitors in chromium-sensitized guinea pigs. *J. Invest. Dermatol.* **60**, 193–196.

SIPOS, K. (1967). Chemical hypersensitivity and dermatological diseases. *Dermatologica* **135**, 421–432.

SKOG, E. and WAHLBERG, J.E. (1970). Passive transfer of chromium allergy in guinea pigs. *Acta Derm. Venereol.* **50**, 189.

SNYDER, C.A. and VALLE, C.D. (1991). Immune function assays as indicators of chromate exposure. *Environ. Health Perspect.* **92**, 83–86.

SNYDER, C.A., UDASIN, I., WATERMAN, S.J., TAIOLI, E., and GOCHFELD, M. (1996). Reduced IL-6 levels among individuals in Hudson County, New Jersey, an area contaminated with chromium. *Arch. Environ. Health* **51**, 26–28.

STERN, R.M. (1982). Chromium compounds: production and occupational exposure, in Langård, S. (Ed.) *Topics in Environmental Health, Vol. 5, Biological and Environmental Aspects of Chromium*, pp. 5–44, Amsterdam: Elsevier Biomedical Press.

SUDHAN, T. and MICHAEL, R.D. (1995). Modulation of humoral immune response by tannery effluent in *Oreochromis mossambicus* (Peters). *Indian J. Exp. Biol.* **33**, 793–795.

SUNDERMAN, F.W., JR, HOPFER, S.M., SWIFT, T., REZUKE, W.N., ZIEBKA, L., HIGHMAN, P., EDWARDS, B., FOLCIK, M., and GOSSLING, H.R. (1989). Cobalt, chromium, and nickel concentrations in body fluids of patients with porous-coated knee or hip prostheses. *J. Orthopaed. Res.* **7**, 307–315.

TAKAHASHI, W., PFENNIGER, K., and WONG, L. (1983). Urinary arsenic, chromium, and copper levels in workers exposed to arsenic-based wood preservatives. *Arch. Environ. Health* **38**, 209–214.

THORMANN, J., JESPERSEN, N.B., and JOENSEN, H.D. (1979). Persistence of contact allergy to chromium. *Contact Dermatitis* **5**, 261–264.

TIAN, L. and LAWRENCE, D.A. (1996). Metal-induced modulation of nitric oxide production *in vitro* by murine macrophages: lead, nickel, and cobalt utilize different mechanisms. *Toxicol. Appl. Pharmacol.* **141**, 540–547.

TISCH, M., ENDERLE, G., ZOLLER, J., and MAIER, H. (1996). Cancer of the oral cavity in machine workers. *Laryngo-Rhino-Otologie.* **75**, 759–763.

VAN HOOGSTRATEN, I.M.W., VON BLOMBERG, B.M.E., BODEN, D., KRAAL, G., and SCHEPER, R.J. (1991). Effects of oral exposure to nickel or chromium on cutaneous sensitization. *Curr. Probl. Dermatol.* **20**, 237–241.

(1994) Non-sensitizing epicutaneous skin tests prevent subsequent induction of immune tolerance. *J. Invest. Dermatol.* **102**, 80–83.

VAN NEER, F.C.J. (1963). Sensitization of guinea pigs to chromium compounds. *Nature* **198**, 1013.

VREEBURG, K.J.J., DE GROOR, K., VAN HOOGSTRATEN, I.M.W., VON BLONBERG, B.M.E., and SCHEPER, R.J. (1991). Successful induction of allergic contact dermatitis to mercury and chromium in mice. *Int. Arch. Allergy Appl. Immunol.* **96**, 179–183.

WAHBA, A. and COHEN, T. (1979). Chrome sensitivity in Israel. *Contact Dermatitis* **5**, 101–107.

WANG, J.Y., WICKLUND, B.H., GUSTILO, R.B., and TSUKAYAMA, D.T. (1996). Titanium, chromium, and cobalt ions modulate the release of bone-associated cytokines by human monocytes/macrophages *in vitro*. *Biomaterials* **17**, 2233–2240.

WIKSTROM, K. (1962). Epidermal treatment of guinea pigs with potassium bichromate. *Acta Derm. Venereol.* **42**(Suppl. 49), 1–59.

WILLIAMS, C.D. (1969). Asthma related to chromium compounds. Report of two cases and review of the literature on chromate diseases. *N. C. Med. J.* **30**, 482–491.
    (1973). Asthma and metals. *Ann. Intern. Med.* **79**, 761–762.

ZACHARIAE, C.O., AGNER, T., and MENNE, T. (1996). Chromium allergy in consecutive patients in a country where ferrous sulfate has been added to cement since 1981. *Contact Dermatitis* **35**, 83–85.

ZELGER, J. (1964). Zur Klinik und Pathogenere des Chromatekzems. *Arch. Klin. Exp. Dermatol.* **218**, 498–542.

5

Indium

Colgate Palmolive, 909 River Road, Piscataway, NJ 08855-1343, USA

5.1 Introduction

This chapter reviews the history of indium's industrial and medical uses, its exposure, disposition, mechanism of action, and excretion. A comprehensive review of the current state of knowledge regarding the general toxicology and immunotoxicology of indium and its compounds is provided. Further discussion of indium arsenide, a semiconductor material, can be found in Chapter 1.

5.2 History

Since its discovery in 1863, indium has been used for many different applications. Indium (In) was first used commercially in the 1940s to increase the tensile strength and ductility of dental alloys (EPA, 1975; Stokinger, 1991); in addition, its oxygen scavenging properties prevented their discoloration (Stokinger, 1991). In the 1940s to 1960s, the solder and alloy industry used indium for its unique properties as an anti-corrosive and anti-fatigue agent. It was found that electroplating the surface of metals such as silver and copper or alloys containing cadmium (Cd) and lead (Pb) with indium prevented corrosion (EPA, 1975; Stokinger, 1991). When added to metal alloys used in automotive or airplane bearings, indium increased the hardness and prevented the development of metal fatigue (EPA, 1975; Venugopal and Luckey, 1978). Other uses of indium include its use in motion picture screens, mirrors, and cathode oscillographs, because it reflects all spectral wavelengths of color equally (Venugopal and Luckey, 1978).

Recently, the electronics industry has shown a growing interest in indium-containing compounds for use in photovoltaic cells, solar cells, optical communications, and intermetallic semiconductors (Smith *et al.*, 1978; Perez-Albuerne and Tyan, 1980). Compounds used to make intermetallic semiconductors include indium antimonide (InSb), indium arsenide (InAs), and indium phosphide (InP). Compared with semiconductor materials currently used, InP offers as much as a 30 percent performance improvement; these materials are also faster, lower in power consumption, handle more output power, and have good thermal characteristics (Electronic Materials and Packaging, 1994). Potential improvements in efficiency by using intermetallic compounds is likely to increase their use in the semiconductor and photovoltaic industries in the future, raising concerns about the potential occupational exposure to indium. The pattern of indium usage (estimated in 1995) was 45 percent in coatings, 35 percent in solders and alloys, 15 percent in electrical components and semiconductors, and 5 percent in research (Mineral Commodity Summaries, 1996). Thin film coatings, which are used in applications such as liquid crystal displays and electroluminescent lamps, are currently the largest end use of indium (Mineral Commodity Summaries, 1996).

While it represents only a small portion of the total amount of indium used on a yearly basis, indium isotopes (i.e. ^{111}In and ^{113}In) have been used diagnostically to locate and identify tumors in the brain, neck, colon, rectum, ovary, breast, and lung (Stern *et al.*, 1966, 1967; Thakur *et al.*, 1973). Despite the fact that indium does accumulate in tumors, studies have shown that it is not effective as a chemotherapeutic agent (Hart and Adamson, 1971; Adamson *et al.*, 1975; Smith *et al.*, 1978; Venugopal and Luckey, 1978).

5.3   Occurrence

Indium is a group 3B non-essential metal with an ionic radius of 0.081 nm and an electron configuration of $5s^2 5p^1$; it can exist in oxidation state of +1, +2, or +3, with the trivalent form being the most stable. Indium is a soft, malleable, silver-white metal that resembles tin (Sn) and ferric iron (Fe^{3+}) in its chemical and physical properties and, compared with gallium (Ga), another group 3 metal used extensively by the semiconductor industry, it is more electropositive and less amphoteric. Indium will bind to free and bound phosphate groups in biological fluids and tissues (Venugopal and Luckey, 1978).

Indium is widely distributed in nature in very small quantities, with the Earth's crust containing approximately 50 to 200 ppb In. In the environment, indium is associated with zinc (Zn) in sphalerite (ZnS), marmatite, and christophite (FeS:ZnS), and with tin in siderite (Stokinger, 1991). Indium is recovered as a byproduct of Zn smelting (Fowler, 1986) and by acid leaching of crude zinc liquors followed by phosphate precipitation (Fowler, 1988). World production of indium in 1995 was estimated to be 150 tons (Mineral Commodity Summaries, 1996).

The average daily human intake of indium is estimated to be 8 to 10 μg In/day (Smith *et al.*, 1978). Indium levels in food range from undetectable to

0.01 mg In/kg for beef and ham to 10 to 15 mg In/kg for fish and shellfish from contaminated waters (Fowler, 1988). Occupational exposure to indium-containing compounds is likely the primary source of human exposure since the dietary intake of indium is very low.

At present, there is no evidence to suggest that any ill effects in workers exposed to indium metal or its compounds could be attributed solely to the indium agents alone. In almost every instance of occupational exposure, compounds of lead and arsenic (As) are present in greater amounts in the workplace environment than indium, so it is rare for a worker to be exposed to indium without being exposed to higher concentrations of more toxic elements (Pb, As, Cd, thallium). Flue dusts from zinc smelters constitute the largest commercial source of indium (Stokinger, 1991). Atmospheric emissions are largely in the form of indium oxide (In_2O_3) (Smith *et al.*, 1978). The threshold limit value (TLV) for indium and its compounds recommended by the American Conference of Governmental Hygienists is 0.1 mg In/m^3 based on rat studies showing alveolar proteinosis after exposure to In_2O_3 dust for 224 h (4 h/day) at concentrations of 64 mg In/m^3 (Leach *et al.*, 1961). OSHA has a recommended permissible exposure limit (PEL) for indium of 0.1 mg In/m^3 of air as a time-weighted average (TWA) over an 8-h shift (Lewis, 1992).

5.4   Essentiality and Metabolism

There are no published reports dealing with dietary intake of indium by humans. The current understanding of indium absorption, distribution, metabolism, and elimination is largely based on a series of studies conducted for the Atomic Energy Commission in the late 1950s to early 1960s (Smith and Scott, 1957; Smith *et al.*, 1957, 1959; Morrow *et al.*, 1958; Downs *et al.*, 1959; Leach *et al.*, 1961). These studies found that the distribution, metabolism, and resultant toxicity of indium is dependent on the route of administration, chemical form of the compound administered, and the pH of the dosing solution.

5.4.1   *Absorption*

Gastrointestinal absorption of insoluble In_2O_3 is poor, with approximately 0.5 percent of the dose being taken up in rats (Smith *et al.*, 1960); the gastrointestinal absorption rate for water-soluble indium salts, such as indium chloride ($InCl_3$) and indium sulfate ($In_2(SO_4)_3$), is only slightly higher (Vignoli *et al.*, 1946), while indium–citrate complexes are better absorbed. Following inhalation or intratracheal instillation, indium salts are retained in the lung and rapidly absorbed, having half-lives of approximately 1 h; insoluble indium compounds, like In_2O_3, are absorbed slowly, with half-lives of approximately 2 months (Smith *et al.*, 1978). Morrow *et al.* (1958) reported that rats absorbed

3 to 6 percent of inhaled indium following a 1-h exposure to $^{114}In_2O_3$ particles. Two weeks after exposure, approximately 50 percent of an administered dose had been cleared from the lungs with the rest of the dose being retained in the tracheobronchial lymph nodes for up 2 months (Venugopal and Luckey, 1978). Similarly, Smith *et al.* (1957) found that most of an intratracheally administered dose of $^{114}In(OH)_3$ or ^{114}In–citrate was taken up by tracheobronchial lymph nodes. Following subcutaneous administration of In–citrate, 85 to 90 percent was absorbed into the blood within 2 days, whereas $InCl_3$ was slowly absorbed (Smith and Scott, 1957). Following intravenous administration, indium salts were quickly removed from the blood due to the binding of indium to transferrin and distribution to soft tissues and bone (Venugopal and Luckey, 1978). Whereas ionic indium will also bind to albumin and globulin, colloidal forms of indium will not bind to plasma proteins and are engulfed by leukocytes and removed from the liver and spleen (Venugopal and Luckey, 1978).

5.4.2 *Distribution*

An extensive review of indium distribution following parenteral administration can be found in Smith *et al.* (1978). The distribution of indium is influenced by a number of factors including the route of administration and the nature of the chemical being administered. For example, administration of $InCl_3$ has been shown to damage the renal proximal convoluted tubule (EPA, 1975) while administration of colloidal In_2O_3 targets phagocytic cells (i.e. Kupffer cells) in the liver and reticuloendothelial system (RES) (Castronovo and Wagner, 1971). Concentration of insoluble indium compounds in reticuloendothelial cells by phagocytosis may represent a mechanism for the progressive toxic effects, as exemplified by the fact that hydrated colloidal $In(OH)_3$ is 40 times as toxic as ionic indium in mice (Castronovo and Wagner, 1971). Furthermore, blocking the RES with the less toxic $Ga(OH)_3$ will decrease the RES accumulation and toxicity of $In(OH)_3$ (Evdokimoff *et al.*, 1972a).

The initial distribution of indium is dependent on the route of administration. While the skin, muscle, and bones accumulate indium following a subcutaneous dose, the highest levels of indium following an intramuscular dose are found in the kidney (Venugopal and Luckey, 1978). In general, indium administered subcutaneously is absorbed faster than doses administered intramuscularly. When administered intraperitoneally indium initially accumulates in the liver with subsequent redistribution to the spleen, kidneys, and bone (Venugopal and Luckey, 1978). This is in agreement with several acute toxicity studies that have shown the liver to be the principle site of injury following an intraperitoneal injection (EPA, 1975). Intravenously administered ionic indium initially accumulates in the kidneys and to a lesser extent the liver (Castronovo and Wagner, 1971, 1973), with subsequent redistribution to the skin, bones, and muscle (Venugopal and Luckey, 1978).

Following intratracheal instillation of In(OH)$_3$, 76 to 80 percent of the dose was present in the lungs and trachea after 12 h (Smith *et al.*, 1978); after 64 days, 8 to 14 percent of the initial dose still remained in the lungs. Using [114]In–citrate, Smith *et al.* (1960) found that more than 50 percent of a dose administered intratracheally to rats had been absorbed within the first 4 days. While the initial target tissues included the liver, kidney, spleen, and salivary glands, the largest amount of indium was found in the skeleton, muscle, and pelt. The distribution of [114]In in the rat was the same whether administered in the form of the hydroxide or the citrate (Smith *et al.*, 1957). Zheng *et al.* (1994) showed that 4 percent of 10 mg/kg InP administered intratracheally was retained in the lungs of rats after 96 h with the recovery from all other tissues being 0.36 percent; tissues with higher amounts of indium included the hair, liver, and skin.

An acute inhalation study found that rats exposed nose-only to an aerosol of less than 0.1 mg In/m^3 (as hydrolyzed InCl$_3$) for 0.5 to 3 h had 69 percent of the dose in the lower respiratory tract, 21 percent in the upper respiratory tract, and approximately 7 percent in the esophagus, stomach, and duodenum (Morrow *et al.*, 1958). A subchronic inhalation study in which rats were exposed to an average dose of 64 In$_2$O$_3$/m^3 for 3 months found that indium accumulated in the lung > tracheobronchial lymph nodes > spleen > kidney > liver and femur (Leach *et al.*, 1961). At 12 weeks postexposure, 58 percent of the indium remaining in the lungs at the end of exposure had been mobilized; 69 percent had been mobilized from the tracheobronchial lymph nodes. The biological half-life of indium was approximately 2.5 and 1.75 months for the lungs and tracheobronchial lymph nodes, respectively. A human study found that [111]InCl$_3$ or ^{111}In-diethylenetriamine pentaacetic acid (DTPA) accumulated in the major airways following a 5 to 10-min exposure by ultrasonic nebulization, with relatively poor (1.3 to 4.4 percent) alveolar deposition (Isitman *et al.*, 1974); the half-lives of ^{111}In-DTPA and [111]InCl$_3$ in the lung were 16 and 35 min, respectively.

Orally administered indium in the form of InP was poorly absorbed from the gastrointestinal tract and the total amount of indium retained by the tissues after 24 h was approximately 0.7 percent of the dose (Zheng *et al.*, 1994). After 96 h, indium was not detected in any of the tissues.

Another factor that strongly influences the distribution of indium is the pH of the solution used. Watanabe (1920) found that a majority of the dose remained in the blood when the pH of the indium solution was 3.5 and that at a pH of 10.5 approximately 80 percent of the dose was localized in the liver. More recent studies have shown that unless stabilized, InCl$_3$ injected at pH 4 or higher is hydrolyzed to In(OH)$_3$, behaves more like a colloidal preparation, and accumulates primarily in the liver, spleen, and other organs of the RES

(Smith *et al.*, 1978). At a pH of 4.0 or lower, intravenously administered ionic indium binds as a cation to the β-globulin transferrin of the blood with a half-life of 2 to 3.5 h (Hosain *et al.*, 1969; Cuaron *et al.*, 1974) with subsequent deposition in the kidney (Smith *et al.*, 1978; Fowler, 1986). While in the blood, the indium–transferrin complex is adsorbed to the surface membrane of both erythrocytes and reticulocytes (Beamish and Brown, 1974a).

Animal and human studies examining the placental transfer of indium have found that only a fraction of a percent is transferred to the fetus (Huddlestun *et al.*, 1969; van der Merwe *et al.*, 1970; Mahon *et al.*, 1973; Baltrukiewicz *et al.*, 1976). In baboons and humans, less than 0.1 to 0.5 percent crosses the placenta into the fetal circulation (Huddlestun *et al.*, 1969; van der Merwe *et al.*, 1970).

5.4.3 Metabolism

It is currently not known how indium enters non-phagocytic cells. Studies by Suzuki and Matsushita (1969) have shown that indium reacts with the phosphate or carboxyl groups and the protonated amino groups of the phospholipid monolayer of biomembranes and can also react with the phosphate groups of nucleic acids (Smith *et al.*, 1978). Suzuki and Matsushita (1969) also noted that indium was more reactive than mercury, cadmium, iron, manganese, calcium, and lithium. Beamish and Brown (1974b) determined that 35 percent of an intravenous dose of ^{111}In was localized to the lysosomal fraction of the liver and 14 to 24 percent to its nuclear fraction.

Ultrastructural and biochemical studies have shown that a single intraperitoneal injection of 10 to 40 mg In/kg (as $InCl_3$) disrupts the rough endoplasmic reticulum in hepatocytes (Fowler *et al.*, 1978, 1983) and renal proximal tubule cells (Woods and Fowler, 1982). In both cases this disruption included the loss of ribosomes. The subcellular changes following indium exposure were associated with a pronounced induction of microsomal heme oxygenase in liver, reduction of cytochrome P-450 levels and inhibition of several P-450-dependent microsomal enzyme activities (Woods *et al.*, 1979). In the kidney, δ-aminolevulinic acid dehydratase was markedly inhibited (Woods and Fowler, 1982). Other studies have shown that indium promotes the hydrolysis of glycyl-DL-leucine and inhibits ferroxidase (Huber and Frieden, 1970), erythrocyte δ-aminolevulinic acid dehydratase (Conner *et al.*, 1995) and isocitric dehydrogenase (Smith *et al.*, 1978). Many of indium's inhibitory effects can be antagonized with ferric dextran (Ferm, 1970). Studies by Beamish and Brown (1974b) have shown that indium does not follow the pathways of normal iron metabolism and does not bind to either hemoglobin or heme. However, both heme and non-heme-dependent enzymes do change in concert with structural changes in the endoplasmic reticulum and these biochemical changes occur prior to cellular necrosis or extensive mitochondrial damage (Fowler, 1988).

5.4.4 *Excretion*

Excretion is independent of the route of administration (intravenous, intratracheal, subcutaneous, intramuscular, or oral) with feces being the major route of elimination. Elimination is biphasic with an initial rapid phase followed by a long slow phase (Smith *et al.*, 1959; Castronovo and Wagner, 1971, 1973). The biological half-life of the rapid phase was estimated to be 1.9 days (intravenous $InCl_3$), 2 days (intravenous In_2O_3), 3 days (subcutaneous 114m-In), and 6.5 days (intramuscular 114m-In). For the slow phase the biological half-life was 6 to 9 days (intravenous $InCl_3$), 73.8 days (intravenous In_2O_3), 94 days (subcutaneous 114m-In) and 126 days (intramuscular 114m-In). Stern *et al.* (1967) determined the whole body half-life to be 14 to 15 days following intravenous administration of hydrated In_2O_3. Morrow *et al.* (1958) observed a biological half-life of 8 to 10 days in lungs of rats after inhalation exposure to indium sesquioxide particles.

Orally or subcutaneously administered indium salts that are transformed into colloidal forms are excreted primarily in the feces and only about 8 percent in the urine (Venugopal and Luckey, 1978). Indium salts that become hydrated and colloidal in the blood are quickly removed and deposited in the liver and spleen with 60 percent of the dose being excreted in the feces (Castronovo and Wagner, 1971, 1973). Following oral administration of an insoluble form of indium (InP), only 0.08 percent of the dose was excreted in the urine during a 240-h collection period (Zheng *et al.*, 1994).

A large amount of indium administered intratracheally as $In(OH)_3$ is excreted in the feces after ciliary removal from the lungs and bronchi and then swallowing. At 64 days postinstillation, 8 to 14 percent of the initial dose remained in the lungs with 50 percent of the dose having been excreted via feces and 8 percent in the urine (Smith *et al.*, 1978). The biological half-life for the slow and fast phases was 6 and 112 days, respectively. In a similar experiment, Smith *et al.* (1960) found that the excretion of $In(OH)_3$ following intratracheal intubation occurred mainly via the feces with only 1 percent of the absorbed dose present in the urine. In another experiment, 50 percent of the dose was excreted in the feces and 8 percent in the urine with the remainder retained in the tissues for 2 months following a single intratracheal administration of indium to rats (Venugopal and Luckey, 1978). Zheng *et al.* (1994) reported that following an intratracheal instillation of InP, 73 percent of the dose was excreted in the feces within the first 48h while less than 0.02 percent of the dose was excreted in the urine. It is hypothesized that InP, a particulate, is cleared from the lungs by the mucociliary pathway or by pulmonary macrophages.

5.5 General Toxicology

Although indium is considered a potential toxic element to humans, relatively little documented evidence concerning accidental or intentional exposure exists. Despite the increased use of indium by the semiconductor industry in the form of InP or InAs, its potential for toxicity as a result of occupational exposure has not been characterized. The only references in the literature which attributed adverse health effects to indium exposure were two Russian reports noting that workers exposed to indium compounds during its production complained of tooth decay, pains in joints and bones, nervous and gastrointestinal disorders, heart pains, and general debility (Podosinovskii, 1964, 1965). Except for general weakness, there does not seem to be any relation between these complaints and the toxic symptoms of indium poisoning observed in animal models. Because no other epidemiological studies have been conducted to substantiate these observations, it is likely that the adverse symptoms could have been caused by exposure to other toxic agents.

5.5.1 *Acute toxicity*

The following adverse effects have been observed in animals following acute administration of indium: weight loss, muscle tremors, localized convulsive motions, hind leg paralysis, nosebleeds, pulmonary edema, necrotizing pneumonia, renal and hepatic damage, and decreased lymphocyte count. Table 5.1 summarizes the acute toxicity of indium and its compounds following different routes of administration. For parenterally administered indium compounds, their toxicity was in the order of colloidal $In(OH)_3$ or colloidal hydrated In_2O_3 > water-soluble salts ($InCl_3$, In–citrate) > water-insoluble compounds (In_2O_3, $In(OH)_3$, or InSb).

As was observed with the distribution of indium compounds, the route of administration as well as the nature of the compound determines what the target tissue will be. For example, following intraperitoneal administration of $InCl_3$, the principle site of injury is the liver, while intravenous administration of the same compound will preferentially damage the kidneys (proximal convoluted tubules) (Smith *et al.*, 1978). This difference occurred because $InCl_3$ (pH 3) administered intravenously to mice became protein bound after injection, eventually accumulating in the kidney. The end result was that the treatment caused minor liver damage and severe kidney damage exhibited as the swelling of kidney tubule epithelium, presence of pyknotic nuclei, and eosinophilia, progressing to necrosis of proximal tubular lining (Castronovo and Wagner, 1971; Venugopal and Luckey, 1978).

The nature of the compound administered will also greatly influence the toxicity of indium. This is exemplified by studies showing that citrate-buffered $InCl_3$ was less toxic than its unbuffered form (Castronovo and Wagner, 1971). In the absence of citrate, $InCl_3$ hydrolyzes to hydrated $In(OH)_3$, a more toxic form of indium, in biological fluids and accumulates in the RES in a similar manner as In_2O_3.

In another study, colloidal In_2O_3 (LD_{50}: 0.32 mg In/kg) was approximately 40 times more toxic than ionic $InCl_3$ (LD_{50}: 12.8 mg In/kg) in mice (Castronovo

Table 5.1 Acute toxicity of indium and its compounds[a]

Compound	Animal	Dose/kg body weight Compound (mg)	Metal (mg)	Route[b]	Parameter
Indium	Rat	—	4200	oral	LD_{50}
In_2O_3	Mouse	479	396	IP	LD_{50}
	Mouse	5005	4136	IP	LD_{100}
	Rat	1156	955	IP	LD_{100}
In_2O_3 (hydrated)	Mouse	0.4	0.3	IV	LD_{50}
$In(OH)_3$	Mouse	0.9	0.6	IV	LD_{50}
	Mouse	1.6	1.1	IV	LD_{100}
$InCl_3$	Dog	1.0	0.5	IV	LD_{100}
	Mouse	24.3	12.6	IV	LD_{50}
	Mouse	5	2.6	IP	LD_{50}
	Rat	4200	2180	oral	LD_{50}
	Rat	7.9	4.1	IV	LD_{50}
	Rat	3.5	1.8	IP	LD_{50}
	Rat	6.4	3.3	IP	LD_{100}
	Rabbit	2138	1110	oral	LD_{50}
	Rabbit	0.6	0.3	IV	LD_{100}
	Rabbit	8.9	4.6	IP	LD_{100}
$In(NO_3)_3$	Mouse	3350	1279	oral	LD_{50}
	Mouse	7.5	2.9	IP	LD_{50}
	Mouse	100	38.2	IP	LD_{100}
	Rat	5.5	2.1	IP	LD_{50}
$In_2(SO_4)_3$	Rat	22.5	10	SC	LD_{50}
	Rat	28.3	12.5	SC	LD_{100}
	Rat	5.6	2.5	IV	LD_{50}
	Rat	28.5	12.6	IV	LD_{50}
	Rat	40.5	18	IP	LD_{100}
	Rabbit	2.5	1.1	SC	LD_{100}
	Rabbit	9.7	4.3	IP	LD_{100}
InP	Rat	200	158	IP	LD_{40}
$In(C_6H_5O_7)_3$	Mouse	600	101	SC	LD_{50}
InSb	Mouse	4770	1800	IP	LD_{50}
	Mouse	5974	2900	IP	LD_{100}

[a] Modified from Smith *et al.*, 1978; Venugopal and Luckey, 1978.
[b] IP, intraperitoneal; IV, intravenous; SC, subcutaneous.

and Wagner, 1971). An explanation for this difference was that the particulate form of indium, In_2O_3, was rapidly taken up by the cells of the RES, while the ionic form remained bound to plasma proteins. At a biological pH of 7.4, indium was insoluble and hydrated In_2O_3 was rapidly cleared from the blood by phagocytic RES cells. The phagocytosis of In_2O_3 concentrated indium in the liver and other RES organs (lymphoid tissue of the spleen, lymph nodes, and thymus), thereby enhancing its toxicity (Castronovo and Wagner, 1971). Further evidence of the role of phagocytosis in indium toxicity was presented in studies in which $Ga(OH)_3$ was used to inhibit RES phagocytosis of $In(OH)_3$ (Evdokimoff and Wagner, 1972a,b). In these studies, $Ga(OH)_3$ decreased the uptake of $In(OH)_3$ by the mouse liver which, in turn, decreased the hepatotoxic effects and increased the LD_{50} from 12.5 µg to 39 µg per mouse.

5.5.2   Irritation/sensitization

Indium has been shown to be non-irritating to humans and animals. In one study, disks coated with metallic indium strapped to the bare skin of rats and human volunteers for up to 30 days did not cause any irritation (McCord *et al.*, 1942). In another study, 5 percent solutions of $InCl_3$ and $In_2(SO_4)_3$ applied once a day for 13 days caused no observable effect in rabbits (McCord *et al.*, 1942).

Similar to the irritation studies, only two references in the literature have addressed the issue of sensitization. In the first study, indium-coated silver disks were implanted either subcutaneously, intramuscularly, or intraperitoneally in rats (Harrold *et al.*, 1943). Implantation of these disks induced only foreign-body reactions with no specific symptoms associated with indium toxicity. While radioisotopes of indium used for diagnostic X-ray scanning have relatively few toxic effects, a second study suggested that indium may have the potential to induce an anaphylactoid response (Raiciulescu *et al.*, 1972). This study found that 3 out of 770 people injected intravenously with [113]In for liver scanning developed anaphylactoid reactions within 20 min postinjection. This observation is questionable considering the extensive use of radiolabeled indium in the decades following this paper and the fact that there have been no other reports of anaphylactoid responses attributable to indium.

5.5.3   Subchronic toxicity

Using different routes of administration and several different indium-containing compounds, several studies have shown that the liver, kidney, and lung are target tissues. Furthermore, these studies also show that the toxic effects of indium are influenced by the route of administration, dose, compound, and animal model used.

A rabbit study found that animals fed 61 mg In/kg (as $In(NO_3)_3$) daily for 4.5 months had significant blood and kidney changes, as well as necrotic changes in the liver, moderate spermatogenic arrest, bone marrow hyperplasia, and congestion of the peritoneum (Smith *et al.*, 1978). In another study, rats fed a diet containing 4 percent $InCl_3$ (2 to 3 g In/kg per day) for 3 months had marked depression of growth (Downs *et al.*, 1959). In a second rat study, rats fed up to 2 to 3 g In/kg per day (as $InCl_3$) for 3 months showed anemia and increased levels of mononuclear cells and interalveolar granular exudate in the lungs (Smith *et al.*, 1978). Yoshikawa and Hasegawa (1971) reported that administration of 0.1 to 1 mg $In_2(SO_4)_3$/kg per day for 15 days to guinea pigs produced marked anemia and damage to the lungs, liver, and spleen. Feeding guinea pigs InSb at a 10 percent LD_{50} dose for 21 days caused blood changes, increased organ weights, and proliferative histiocyte reactions in the myocardium, liver, lungs, and kidneys (Smith *et al.*, 1978). Furthermore, rats fed 133 mg In/kg (as $In_2(SO_4)_3$) for 72 days showed growth depression and subacute bronchopneumonia (Smith *et al.*, 1978).

A study that ultimately determined the TLV of indium found that exposing rats for 4 h/day for 3 months to an average dose of 64 mg/m^3 In_2O_3 (53 mg In/m^3; mass median diameter = 0.5 μm) resulted in an atypical pulmonary inflammatory reaction, characterized by a paucity of cellular exudate, enlargement of the tracheobronchial lymph nodes, and alteration of the alveolar walls, which persisted for 12 weeks beyond the termination of exposure (Leach *et al.*, 1961). The lungs of the exposed rats were 3 to 5 times heavier than the controls. The lesion was further characterized by the absence of any change in severity and a lack of fibrosis.

There have been several studies which have failed to demonstrate any toxicity using water-soluble and -insoluble indium compounds. Venugopal and Luckey (1978) report that a study in which rats were fed 20 to 30 mg $In_2(SO_4)_3$/day for 30 days caused no toxic symptoms; however, after 80 days, the treated animals exhibited weight loss and skin roughening. Downs *et al.* (1959) observed no toxicity in rats fed a diet of 8 percent In_2O_3 for 3 months.

As part of the National Toxicology Program, the toxicity and carcinogenic potential of InP is being examined. To date, two subchronic (13 week) inhalation studies in which rats and mice were exposed to 1, 3, 10, 30, or 100 mg InP/m^3 for 6 h/day for 5 or 7 days/week have been completed (NTP, 1996a,b). Preliminary conclusions by the testing laboratory indicate that gross lesions were found on the lungs of all treated animals. The lungs were discolored and enlarged, with a three- to four-fold increase in lung weight, and lesions consisted of alveolar proteinosis and epithelium hyperplasia, chronic active inflammation, and interstitial fibrosis. In addition, moderate renal toxicity was noted in rats of the high dose group.

5.5.4 *Teratogenicity*

One teratology study has been reported in which the intraperitoneal administration of 0.5 to 1 mg/kg per day $In(NO_3)_3$ to hamsters on gestation day 8 resulted in an increase in skeletal defects (polydactaly, fusion) and, at doses of 2 mg/kg per day or more, an increased incidence of resorption with total litter loss (Ferm and Carpenter, 1970). It was hypothesized that indium may be a potent site-specific teratogen that alters enzyme-dependent limb bud differentiation.

5.5.5 *Chronic toxicity/carcinogenicity*

The following adverse effects have been observed following chronic administration of indium: weight loss; degenerative changes in the urogenital tract; inflammation and damage to the spleen, heart, and brain; necrotic changes in the renal convoluted tubules; and liver necrosis.

No studies designed to examine specifically the carcinogenic potential of indium were found in the literature. However, two studies examining the chronic toxic effects of indium in mice noted that there was no increase in tumor incidence compared with the controls (Schroeder and Mitchener, 1971; Smith *et al.*, 1978). In these studies, mice were administered drinking water containing 5 ppm indium as a chloride, acetate, or citrate throughout their lifetime, with an average dose of 0.33 mg In/kg per day. The authors noted that indium did affect the growth of the mice, but not their survival, mortality, or tumor incidence.

As part of the National Toxicology Program, a long-term inhalation study in rats and mice was initiated at the beginning of 1996. In these studies, the animals will be exposed to 0, 0.03, 0.1, or 0.3 mg/m^3 InP on a daily basis for 104 weeks.

5.6 Immunotoxicology

The effect of indium on the immune system or specific immune cells remains to be elucidated. Recent studies have observed that indium is capable of initiating a strong inflammatory response at very low doses. Furthermore, indium is capable of initiating a sustained inflammatory response resulting in permanent tissue damage even though it is cleared from the target tissues in a matter of days. It may be that while the majority of indium is removed in a very short period of time, a small portion of the administered dose binds to various intracellular and extracellular targets. As a result, the indium-containing complex has a longer biological half-life and is capable of chronically modulating the immune system.

Two recent investigations have examined the pulmonary effects of InCl$_3$ and the role the immune system plays in the subsequent inflammatory response (Blazka *et al.*, 1994a,b). In the first study, intratracheal administration of 3.25 µg In/kg to rats was capable of initiating an influx of inflammatory cells, and instillation of 1.3 mg In/kg resulted in a general pulmonary inflammatory response that was still evident after 56 days (Blazka *et al.*, 1994a). Furthermore, instillation of 1.3 mg In/kg resulted in a 32-fold increase in the total bronchoalveolar lavage (BAL) cell number with 67 percent of the cells being polymorphonuclear leukocytes (PMN) and 32 percent alveolar macrophages. In agreement with previous studies, clearance of indium from the lung was biphasic, with 87 percent of the dose eliminated from the lungs within 8 days of administration and less than 10 percent of the remaining indium being removed over the following 48 days.

In the second study, rats were exposed by nose-only inhalation to 0.2, 2, or $20\,mg\,InCl_3/m^3$ for 1 h (geometric means of the particles were 0.075, 0.133, and $0.75\,\mu m$, respectively) (Blazka *et al.*, 1994b). Exposure to $0.2\,mg\,InCl_3/m^3$ initiated an inflammatory response, and at $20\,mg\,InCl_3/m^3$ caused an eightfold increase in the recovered BAL fluid cell number. Commensurate with the level of lung injury 7 days after exposure, the animals in the high dose group developed a 'restrictive' pulmonary disease characterized by a decrease in lung volume and diffusion capacity along with increased airway reactivity (Blazka *et al.*, 1994b). In contrast to the first study, the pulmonary inflammatory effects of indium were reversible as evidenced by a compensatory increase in lung volume and carbon monoxide diffusing capacity, and decrease in lung inflammation suggested recovery from the lung injury. Furthermore, fibrosis did not develop after inhalation exposure to very low levels of indium, suggesting that a certain level of indium needs to be reached before irreversible tissue damage occurs. A possible explanation for the difference in inflammatory responses is that a significantly higher dose of indium was used in the first study. Pulmonary levels of indium 7 days after exposure were 5 to 10 times higher in intratracheally treated rats than in those exposed by inhalation to $20\,mg\,InCl_3/m^3$ for 1 h (Blazka *et al.*, 1994b).

Central to a fibrotic response is the release of alveolar macrophage-derived products such as tumor necrosis factor-α (TNF-α) and fibronectin that are in part responsible for fibroblast proliferation and enhanced collagen deposition. In addition, TNF-α recruits neutrophils and lymphocytes, increases the release of acute-phase protein, stimulates phagocytic cells to release reactive oxygen species, and stimulates fibroblast proliferation (Klempner *et al.*, 1978; Klebanoff *et al.*, 1986; Le and Vilcek, 1987). Fibronectin, a glycoprotein released by alveolar macrophages, acts as a chemotactic factor for fibroblasts, endothelial cells, and monocytes, stimulating the proliferation of fibroblasts, and has an important role in tissue repair by mediating the interactions between cells and extracellular matrix proteins (Bitterman *et al.*, 1983; Macarak and Howard, 1983; Driscoll *et al.*, 1992). It is generally accepted that prior to an increase in lung collagen, an influx of inflammatory cells occurs followed by a fibroproliferative response. This response is characterized by an increase in the number of polymorphonuclear leukocytes (PMN) and macrophages within the alveolar lumen, edema, and fibroblast hyperplasia.

Studies examining the role of inflammatory mediators have suggested that fibronectin and TNF-α have a central role in pulmonary inflammation and fibrosis (Rom *et al.*, 1987; Driscoll and Maurer, 1991). The inhalation of metals or inorganic dusts increases the release of fibronectin (Rom *et al.*, 1987; Driscoll and Maurer, 1991; Driscoll *et al.*, 1992). Blazka *et al.* (1994a) found that an intratracheal instillation of 1.3 mg In/kg resulted in a 1000- and 17-fold increase in fibronectin and TNF-α, respectively, within 48 h of treatment. Furthermore, administration of indium increased the levels of hydroxyproline, a biochemical marker of lung collagen, more than two-fold (Blazka *et al.*, 1994a). This study showed that indium can markedly change the structural integrity of the lung causing necrosis of the alveolar and bronchial/bronchiolar epithelial cells which in turn initiates inflammatory and repair processes that are responsible for the rapid development of fibrosis. In agreement with the results of the first investigation, Blazka *et al.* (1994b) found that exposure to considerably lower level of indium by inhalation was able to increase significantly BAL fluid fibronectin and TNF-α. Compared with other semiconductor compounds, such as gallium arsenide and silica, indium has a greater effect in increasing the lung weight and hydroxyproline levels (Chvapil *et al.*, 1979; Webb *et al.*, 1986; Blazka *et al.*, 1994a).

In contrast to the investigations with $InCl_3$, a water-soluble salt, studies with In_2O_3, an insoluble compound, found that higher doses were needed to evoke an inflammatory response and that no fibrosis resulted (Leach *et al.*, 1961; Stokinger, 1991). This may be due to the fact that, as a particulate, In_2O_3 is not as toxic or reactive with biomembranes. At higher doses, intratracheally administered In_2O_3 (41 mg In/rat) produced a granular dystrophy in liver and kidney cells, and alveolar membrane proliferation and fibrosis in the lungs (Smith *et al.*, 1978). An investigation examining the pulmonary toxicity of In_2O_3 found that rats exposed to 24 to 97 mg In_2O_3/m^3 for 4 h/day over a period of 3 months (for a total of 224 h) had widespread alveolar edema with no fibrosis (Leach *et al.*, 1961). The edema fluid contained only a few alveolar phagocytes, PMN, and nuclear debris.

5.7 Conclusions

Though indium has a relatively short biological half-life, indium-containing compounds are capable of damaging the liver, kidney, and lung. It is likely that, whether through direct or indirect mechanisms, indium activates the immune system. In the case of the lung, this may, in certain instances, result in the development of fibrosis. Because no specific biochemical markers reflecting indium toxicity have been identified and with the increasing potential of occupational exposure in industries that manufacture or utilize indium, further work to determine the actual risk to workers is needed.

Acknowledgments

The author wishes to express his appreciation to Drs Gabriela Adam-Rodwell, David Tonucci, and David Wilcox for their critical review of the manuscript and Dr Daniel Morgan for his assistance in the preparation of this manuscript.

References

ADAMSON, R.H., CANELLOS, G.P., and SIEBER, S.M. (1975). Studies on the antitumor activity of gallium nitrate (NSC-15200) and other group IIIa metal salts. *Cancer Chemother. Rep.* **59**, 599–610.

BALTRUKIEWICZ, Z., MARCINIAK, M., and URBANIAK, B. (1976). Distribution and retention of indium[113m] in the organism of pregnant rats. *Acta Phyiol. Pol.* **27**, 191–197.

BEAMISH, M.R. and BROWN, E.B. (1974a). A comparison of the behavior of ^{111}In and ^{59}Fe-labeled transferrin on incubation with human and rat reticulocytes. *Blood* **43**, 703–711.

(1974b). The metabolism of transferrin-bound ^{111}In and ^{59}Fe in the rat. *Blood* **43**, 693–701.

BITTERMAN, P.B., RENNARD, S.I., ADELBERG, S., and CRYSTAL, R. (1983). Role of fibronectin as a growth factor for fibroblasts. *J. Cell. Biol.* **97**, 1925–1932.

BLAZKA, M.E., DIXON, D., HASKINS, E., and ROSENTHAL, G.J. (1994a). Pulmonary toxicity to intratracheally-administered indium trichloride in Fischer 344 rats. *Fundam. Appl. Toxicol.* **22**, 231–239.

BLAZKA, M.E., TEPPER, J.S., DIXON, D., WINSETT, D.W., O'CONNOR, R.W., and LUSTER, M.I. (1994b). Pulmonary response of Fischer 344 rats to acute nose-only inhalation of indium trichloride. *Environ. Res.* **67**, 68–83.

CASTRONOVO, F.P. and WAGNER, H.N. (1971). Factors affecting the toxicity of the element indium. *Br. J. Exp. Path.* **52**, 543–559.

(1973). Comparative toxicity and pharmacodynamics of ionic indium chloride and hydrated indium oxide. *J. Nucl. Med.* **14**, 677–682.

CHVAPIL, M., ESKELSON, C.D., and OWEN, J.A. (1979). Early changes in the chemical composition of the rat lung after silica administration. *Arch. Environ. Health* **34**, 402–406.

CONNER, E.A., YAMAUCHI, H., and FOWLER, B.A. (1995). Alterations in the heme biosynthetic pathway from the III–V semiconductor metal, indium arsenide (InAs). *Chem.-Biol. Interact.* **96**, 273–285.

CUARON, A., GORDON, F., and RODRIGUEZ, C. (1974). Positive imaging of liver tumors by delayed hepatic scintiscanning with acidic ionic indium[113m] chloride. *Am. J. Roent. Radium Ther. Nucl. Med.* **122**, 318–326.

DOWNS, W.L., SCOTT, J.K., STEADMAN, L.T., and MAYNARD, E.A. (1959). *The Toxicity of Indium*. University of Rochester Atomic Energy Report No. UR-558. University of Rochester, New York.

DRISCOLL, K.E. and MAURER, J.K. (1991). Cytokine and growth factor release by alveolar macrophages: potential biomarkers of pulmonary toxicity. *Toxicol. Pathol.* **19**, 398–405.

DRISCOLL, K.E., MAURER, J.K., POYNTER, J., HIGGINS, J., ASQUITH, T., and MILLER, N.S. (1992). Stimulation of rat alveolar macrophage fibronectin release in a cadmium chloride model of lung injury and fibrosis. *Toxicol. Appl. Pharmacol.* **116**, 30–37.

Electronic Materials and Packaging (1994). *Emerging Technology Indium Phosphide: The Other III–V*. Publication No. 957. Research Information Ltd, Hempstead, Hertfordshire.

EPA (Environmental Protection Agency, Office of Toxic Substances) (1975). *Preliminary Investigation of Effects on the Environment of Boron, Indium, Nickel,*

Selenium, Tin, Vanadium, and Their Compounds. Vol. 2. Indium. Report No. EPA-560/2-75-005B. Washington, DC.

EVDOKIMOFF, V. and WAGNER, H.N. (1972a). Reduction of indium toxicity by blockade of the reticuloendothelial system. *J. Reticuloendothel. Soc.* **11**, 599–603.

———— (1972b). Hepatic phagocytosis as a mechanism for increasing heavy-metal toxicity. *J. Reticuloendothel. Soc.* **11**, 148–153.

FERM, V.H. (1970). Protective effect of ferric dextran on the embryopathic action of indium. *Experientia* **26**, 633–634.

FERM, V.H. and CARPENTER, S.J. (1970). Teratogenic and embryopathic effects of indium, gallium, and germanium. *Toxicol. Appl. Pharmacol.* **16**, 166–170.

FOWLER, B. (1986). Indium, in Friberg, L., Nordberg, G.F., and Vouk, V. (Eds) *Handbook on the Toxicology of Metals*, Vol, 2, pp. 267–275. New York: Elsevier Press.

FOWLER, B.A. (1988). Mechanism of indium, thallium, and arsine gas toxicity: relationships to biological indicators of cell injury, in Clarkson, T.W. (Ed.) *Biological Monitoring of Toxic Metals*, pp. 469–478. New York: Plenum Press.

FOWLER, B.A., KARDISH, R., and WOODS, J.S. (1978). Ultrastructural and biochemical studies of hepatic mitochondrial and microsomal structure following acute administration of indium. *Pharmacologist* **20**, 179.

———— (1983). Alteration of hepatic microsomal structure and function by indium chloride: ultrastructural, morphometric, and biochemical studies. *Lab. Invest.* **48**, 471–478.

HARROLD, G.C., MEEK, S.F., WHITMAN, N., and MCCORD, C.P. (1943). The physiologic properties of indium and its compounds. *J. Ind. Hyg. Toxicol.* **25**, 233–237.

HART, M.M. and ADAMSON, R.H. (1971). Antitumor activity and toxicity of salts of inorganic group 3A metals: aluminum, gallium, indium, and thallium. *Proc. Natl Acad. Sci. USA* **68**, 1623–1626.

HOSAIN, F., MCINTYRE, P.A., POULOSE, K, STERN, H.S., and WAGNER, H.N. (1969). Binding of trace amounts of ionic indium-113m to plasma transferrin. *Clin. Chim. Acta* **24**, 69–75.

HUBER, C.T. and FRIEDEN, E. (1970). The inhibition of ferroxidase by trivalent and other metal ions. *J. Biol. Chem.* **245**, 3979–3984.

HUDDLESTUN, J.E., MISHKIN, F.S., CARTER, J.E., DUBOIS, P.D., and REESE, J.C. (1969). Placental localization by scanning with indium-113m. *Radiology* **92**, 587–590.

ISITMAN, A.T., MANOLI, R., SCHMIDT, G.H., and HOLMES, R.A. (1974). An assessment of alveolar deposition and pulmonary clearance of radiopharmaceuticals after nebulization. *Am. J. Roent. Radium Ther. Nucl. Med.* **120**, 776–781.

KLEBANOFF, S.J., VADAS, M.A., HARLAN, J.M., SPARKS, L.H., GAMBLE, J.R., AGASTI, J.M., and WALTERSDORPH, A.M. (1986). Stimulation of neutrophils by tumor necrosis factor. *J. Immunol.* **136**, 4220–4225.

KLEMPNER, M.S., DINARELLO, C.A., and GALLIN, J.I. (1978). Human leukocytic pyrogen induces release of specific granule contents from human neutrophils. *J. Clin. Invest.* **61**, 1330–1336.

LE, J. and VILCEK, J. (1987). Tumor necrosis factor and interleukin-1: cytokines with multiple overlapping biological activities. *Lab. Invest.* **56**, 234–248.

LEACH, L.J., SCOTT, J.K., ARMSTRONG, R.D., STEADMAN, L.T., and MAYNARD, E.A. (1961). *The Inhalation Toxicity of Indium Sesquioxide in the Rat.* University of Rochester Atomic Energy Project Report No. UR-590. University of Rochester, New York.

LEWIS, J.L. (1992). Indium, in *Sax's Dangerous Properties of Industrial Materials*, Vol. 3, p. 1981. New York: Van Nostrand Reinhold.

MACARAK, E.J. and HOWARD, P.S. (1983). Adhesion of endothelial cells to extracellular matrix proteins. *J. Cell. Physiol.* **116**, 76–86.

MAHON, D.F., SUBRAMANIAM, G., and McAFEE, J.G. (1973). Experimental comparison of radioactive agents for studies of the placenta. *J. Nucl. Med.* **14**, 651–659.

McCORD, C.P., MEEK, S.F., HARROLD, G.C., and HEUSSNER, C.E. (1942). The physiologic properties of indium and its compounds. *J. Ind. Hyg.* **24**, 243–254.

Mineral Commodity Summaries (1996). *Indium.* US Bureau of Mines, Washington DC.

MORROW, P.E., GIBB, F.R., CLOUTIER, R., CASARETT, L.J., and SCOTT, J.K. (1958). *Fate of Indium Sesquioxide and of Indium[114] Trichloride Hydrolysate Following Inhalation in Rats.* University of Rochester Atomic Energy Project Report No. UR-508. University of Rochester, New York.

NTP (National Toxicology Program) (1996a). *13-Week Subchronic Inhalation Toxicity Study of Indium Phosphide – Mice.* (Abstract.) National Toxicology Program, Research Triangle Park, NC.

(1996b). *13-Week Subchronic Inhalation Toxicity Study of Indium Phosphide – Rats.* (Abstract.) National Toxicology Program, Research Triangle Park, NC.

PEREZ-ALBUERNE, E.A. and TYAN, Y.S. (1980). Photovoltaic materials. *Science* **208**, 902–907.

PODOSINOVSKII, V.V. (1964). Work hygiene problems in the production and use of indium and gallium and their toxicological characteristics. *Mosk. Med. Inst.*, 110–112.

(1965). The toxicity of indium compounds of industrial importance. *Gig. Sanit.* **30**, 28–34.

RAICIULESCU, N., NICULESCU-ZINCA, D., and STOICHITA-PAPILLIAN, M. (1972). Anaphylactoid reactions induced by In-113m and Au-198 radiopharmaceuticals used for liver scanning. *Rev. Roum. Med. Intern.* **9**, 55–60.

ROM, W.N., BITTERMAN, P.B., RENNARD, S.I., CANTIN, A., and CRYSTAL, R.G. (1987). Characterization of the lower respiratory tract inflammation of nonsmoking individuals with interstitial lung disease associated with chronic inhalation of inorganic dusts. *Am. Rev. Respir. Dis.* **136**, 1429–1434.

SCHROEDER, H.A. and MITCHENER, M. (1971). Scandium, chromium, gallium, yttrium, rhodium, palladium and indium in mice: effects on growth and life span. *J. Nutr.* **101**, 1431–1438.

SMITH, G.A. and SCOTT, J.L. (1957). *Metabolism of Indium by Subcutaneous Injection.* The University of Rochester Atomic Energy Report No. UR-507. University of Rochester, New York.

SMITH, G.A., THOMAS, R.G., and SCOTT, J.K. (1959). *Distribution and Excretion of Indium-114m After Multiple Intramuscular Injection to the Rat.* University of Rochester Atomic Energy Report No. UR-554. University of Rochester, New York.

(1960). The metabolism of indium after administration of a single dose to the rat by intratracheal, subcutaneous, intramuscular and oral injection. *Health Phys.* **4**, 101–108.

SMITH, G.A., THOMAS, R.G., BLACK, B., and SCOTT, J.K. (1957). *The Metabolism of Indium[114m] Administered to the Rat by Intratracheal Intubation.* University of Rochester Atomic Energy Report No. UR-500. University of Rochester, New York.

SMITH, I.C., CARSON, B.L., and HOFFMEISTER, F. (1978). *Trace Elements in the Environment, Vol. 5. Indium.* Ann Arbor, MI: Ann Arbor Publishers.

STERN, H.S., GOODWIN, D.A., WAGNER, H.N., and KRAMER, H.H. (1966). In^{113m} – a short lived isotope for liver scanning. *Nucleonics* **24**, 57–59.

STERN, H.S., GOODWIN, D.A., SCHEFFEL, U., and KRAMER, H.H. (1967). In^{113m} – for blood pool and brain scanning. *Nucleonics* **25**, 62–65.

STOKINGER, H.E. (1991). Indium, in Clayton, G.D. and Clayton, F.E. (Eds) *Patty's Industrial Hygiene and Toxicology*, Vol. 2A, pp. 1654–1661. New York: John Wiley and Sons.

SUZUKI, Y. and MATSUSHITA, H. (1969). Interaction of metal ions with phospholipid monolayer and their acute toxicity. *Ind. Health* **7**, 143–154.

THAKUR, M.L., GUNSAEKERA, S., and MERRICK, M.V. (1973). Indium radionucleotide helps detect cancers. *Chem. Eng. News* **14**, 12.

VAN DER MERWE, E.J., LOTTER, M.G., VAN HEERDEN, P.D., SLABBER, C.F., and BESTER, J. (1970). Absorbed dose calculations for ^{113m}In placental scanning. *J. Nucl. Med.* **11**, 31–35.

VENUGOPAL, B. and LUCKEY, T.D. (Eds) (1978). Toxicity of group III metals, in *Metal Toxicity in Mammals*, Vol. 2, pp. 116–121. New York: Plenum Press.

VIGNOLI, L., POURSINES, Y., OLIVER, H., and MERLAND, R. (1946). A study of experimental intoxication by indium. *Arch. Mal. Prof.* **7**, 356–360.

WATANABE, K. (1920). The effects of pH on the In-113m distribution in order to simplify the preparation of In-113m for blood pool and liver scanning. *Jpn. J. Nucl. Med.* **7**, 8–15.

WEBB, D.B., WILSON, S.E., and CARTER, D.E. (1986). Comparative pulmonary toxicity of gallium arsenide, gallium (III) oxide, or arsenic (III) oxide intratracheally-instilled into rats. *Toxicol. Appl. Pharmacol.* **82**, 405–416.

WOODS, J.S. and FOWLER, B.A. (1982). Selective inhibition or renal delta-amino-levulinic acid dehydratase by indium: biochemical and ultrastructural studies. *Exp. Molec. Pathol.* **36**, 306–315.

WOODS, J.S., CARVER, G.T., and FOWLER, B.A. (1979). Altered regulation of hepatic heme metabolism by indium chloride. *Toxicol. Appl. Pharmacol.* **49**, 455–461.

YOSHIKAWA, H. and HASEGAWA, T. (1971). Experimental indium poisoning. *Igaku To Seibutsugaku* **83**, 45–48.

ZHENG, W., WINTER, S.M., KATTNIG, M.J., CARTER, D.E., and SIPES, I.G. (1994). Tissue distribution and elimination of indium in male Fischer 344 rats following oral and intratracheal administration of indium phosphide. *J. Toxicol. Environ. Health* **43**, 483–494.

6

Lead

MICHAEL J. McCABE, JR

Institute of Chemical Toxicology, Wayne State University, 2727 Second Avenue, Detroit, MI 48201-2654, USA

> I have made you a tester of metals
>   and my people the ore,
> that you may observe
>   and test their ways.
> They are all hardened rebels,
>   going about to slander.
> They are bronze and iron;
>   they all act corruptly.
> The bellows blow fiercely
>   to burn away the **lead** with fire,
> but the refining goes on in vain;
>   the wicked are not purged out.
> They are called rejected silver,
>   because the Lord has rejected them.
>   (*Jeremiah 6:27–30*)

6.1   Introduction

Man has mined and used lead since antiquity. Despite its widespread utility and its commercial importance, lead has long endured a low status among metals as the above verses cited from the Old Testament, where lead is metaphorically compared negatively to sin, attest. This negativity has persisted to the present day, when the recognized toxicity of lead is the premier environmental health issue for children in the United States and in many other countries. There is a common misconception that adverse health effects due to environmental lead exposure are a problem of the past. However, based on new scientific data it is believed that human populations are potentially affected at lower levels than previously thought to be toxic, and organ systems, such as the immune system, are also potentially affected. It is estimated that 3 to 4 million American children exhibit excessive blood lead levels as defined by the Centers for Disease Control (i.e. less than $10\,\mu g\,Pb/dl$ of whole blood, CDC, 1991), and the World Health Organization has concluded, based on prospective logitudinal studies in human pediatric populations and experimental animal studies in rodents and non-human primates, that exposures producing blood lead levels as low as $10\,\mu g\,Pb/dl$ adversely impact cognitive functions (WHO, 1995). The positions of the CDC and the WHO are based on many epidemiological findings which collectively assert that intoxication with lead at subclinical levels, previously thought to be harmless, significantly impairs intelligence and neurobehavioral development, particularly in young children. Although most concern and emphasis has been placed on the dangers of lead exposure among preschool-aged children, there has also been a gradual lowering of the occupational toxicological standards in many countries. No comparable epidemiological database either in a pediatric lead-exposed population or an occupationally exposed cohort exists which characterizes the human immune system as a target for lead toxicity. However, based on animal studies and the available information indicating that lead affects human immune parameters, the immune system as a target for subclinical lead toxicity has been proposed and investigated by several laboratories. Detailed epidemiological data on the potential immunotoxicity of lead are lacking and are obviously needed.

6.2   History

6.2.1   *Sources and uses of lead – lead was a necessary evil*

An extensive and surprisingly well informed historical literature on the usage and toxicity of lead dating back to Greco–Roman times exists. Moreover, small, decorative, metallic lead beads have been found in excavations dating back from 7000 to 6500 BC. Since lead is never found naturally in its metallic state, this is evidence that the earliest civilizations had learned to smelt lead from its ore, galena (PbS), to produce the metal. In addition to its decorative uses, lead compounds were used for a variety of cosmetic and internal and external medicinal purposes in numerous ancient civilizations. Although anecdotal with respect to the larger problem of low-dose lead toxicity in modern Western civilization, lead usage has persisted in the present day medicinal folklore of many cultures, particularly in some developing nations. The Egyptians, Phoenicians, Greeks, and Hebrews all used lead in small quantities, but extensive mining and usage of lead began with the Romans. Historical environmental records in glaciers, polar ice caps, freshwater and marine sediments, and tree rings detailing lead production over the past 5000 years substantiate the extensive mining and usage of lead by the Romans and later during the Industrial Revolution (Davidson and Rabinowitz, 1992). Interestingly, with respect to the problem of modern day lead toxicity, the toxic effects of lead have been recognized for almost as long as lead has been used. Several anecdotes pertaining to the importance of lead in shaping Western civilization have been described in historical accounts from the Roman era. For example, the Roman invasion of Britain was partly due to the need for lead-containing ores, and the fall of the Roman empire has been partly attributed to the intoxication of the aristocracy with lead from drinking wine laced with lead to impede souring (Gilfillan, 1965; Nriagu, 1983). Food and beverage containers fashioned from ceramics or pewter remain a fairly common means of accidental exposure to lead. Despite the recognized toxicity of lead in both the ancient and the modern worlds, its usage and the problems accompanying its usage have persisted. How then could such a problem persist?

Over the past 25 years the use of lead in commercial products such as gasoline, paint, ceramic glazes, solders, and plumbing supplies has decreased steadily due to legislation such as the clean air act and the lead in paint act. The commercial importance of lead is based on its relative abundance, ease of casting, high density, low melting point (327°C), low strength, ease of fabrication, acid resistance, electrochemical reaction with sulfuric acid, and chemical stability. So, the reason that the issue of lead toxicity has persisted is that lead has some rather unique chemical properties that have important industrial applications, which have justified its usage.

In most cases, namely petrol fuels and paint pigments, alternative compounds have been found that serve similar functions as lead once did. At present, the main industrial use of lead is in the production of storage batteries, which can lead to occupational exposures of battery workers. Interestingly, in this exposure population, several reports have demonstrated that neutrophil function as measured by chemotactic response, chemiluminescence activity, or myeloperoxidase-dependent killing is significantly suppressed in workers with elevated blood lead levels (Hager-Malecka *et al.*, 1986; Governa *et al.*, 1987; Bergeret *et al.*, 1990; Valentino *et al.*, 1991; Queiroz *et al.*, 1993, 1994).

6.2.2   *The legacy of lead usage*

Once lead is mined and introduced into the environment, it stays there. Today, the three primary sources of lead exposure in the United States are: (i) lead-based paint found in older homes, (ii) lead in urban soil and dust, and (iii) lead in drinking water. Emission of lead compounds from automobiles during this century was the single largest anthropogenic source of lead in history. Gasoline containing tetraethyl lead to boost octane ratings was placed on the market in the United States beginning in 1923 despite opposition by leading scientists and public health officials (Lin, 1992). By 1975, consumption of lead in gasoline in the United States approached 170 000 tons. Thus, for nearly 60 years prior to the lead-in-gasoline phasedown that took place during the late 1970s and early 1980s, large quantities of lead were redistributed into the environment. The legacy of this lead redistribution is that dust and soil, particularly in urban environments, have elevated lead levels resulting in contamination of the food chain and drinking water. The importance of leaded gasoline as an environmetal source of lead exposure is highlighted by the National Health and Nutrition Examination Survey (NHANES II) data, which demonstrated that as lead consumption in gasoline declined from 1976 to 1980, the mean blood lead level of the US population dropped markedly (Pirkle *et al.*, 1994).

Although exposure to leaded gasoline was reduced dramatically within the past 15 years, exposure to lead-based paint on urban housing stocks has remained a leading cause of lead exposure among children in the United States. Lead-based paint (containing up to 50 percent lead) was used extensively until the 1940s. In the early 1950s the lead content of paint available to the general public was reduced without being controlled by statute; however, lead compounds continued to be used in some pigments and as drying agents for oil-based paints. Lead pigments performed exceptionally well in many cases. In fact, advertisements from publications of the World War II era touted that white lead primer 'lasted and lasted'. While this truth in advertising might be laudable, the legacy of this lead usage remains an important public heath problem. Federal regulations on the lead content of paints began with the enactment of the Lead-Based Paint Poisoning Prevention Act in 1971 and the Consumer Product Safety Commission establishing maximum allowable levels of lead in paint to 0.5 and 0.06 percent in 1973 and 1978, respectively. Based on a 1987 American Housing Survey, an estimated 74 percent of all occupied housing units built before 1980 have lead-based paint somewhere in the dwelling. It is estimated that approximately 13 million American children under 7 years of age were exposed to lead via this source through ingestion of paint chips, or via the swallowing of lead in house dust and soil during normal hand-to-mouth activities. Lead exposure caused by dust and fumes generated during the renovation of older homes containing lead-based paint is a problem for children from all backgrounds. Lead-based paint on imported toys also remains a problem and an important, although infrequent, source of exposure. While young, poor children from urban environments are obviously at higher risk from lead exposure and its adverse health effects, lead exposure affects children of all socio-economic backgrounds. The misconception that childhood lead exposure is confined to the poor arises from three basic misunderstandings: (i) ignorance about the extent of environmental contamination, (ii) the erroneous concept of a 'normal' blood level, and (iii) the viewing of lead poisoning as a clinical disease. While this section has touched on the first issue, pertaining to the extent and sources of lead in the environment, the latter two issues are discussed in the following sections.

6.3 General Toxicology

6.3.1 *Lead poisoniong is an insidious disease*

Lead has no physiological role in the human body or, for that matter, in any biological process; there is no such thing as a 'normal' blood lead level. Until the late 1960s, the upper limit of 'normal' blood lead in children and in adults was thought to be 60μg Pb/dl and 80μg Pb/dl, respectively. In 1970, after much debate amongst public health officals, the upper limit for children was dropped to 40μg Pb/dl. In 1975, based on new clinical and epidemiological data the Centers for Disease Control lowered the limit to 30μg Pb/dl; in 1985, they lowered it again to 25μg Pb/dl; and in 1991, further lowered the limit to the current 10μg Pb/dl level. To some, it is not surprising to find adverse health effects at lower and lower blood lead levels, since the typical body burdens of lead in modern Americans (i.e. less than 10μg Pb/dl) are at least 300 times greater than those found in native North American populations prior to European settlement (Patterson *et al.*, 1991). Again, this higher background body burden of lead in modern man is due to the persistence of lead in the enviroment, affording the opportunity for exposure, due to the extensive use of lead in the recent past. There is a perception that lead intoxication is a disease for the 'history books' (Needleman, 1991). While this is mostly true for overt clinical disease, several reasons support the view that subclinical lead intoxication will have important emphasis in tomorrow's health policy and that the adverse effects of lead are relevant public health issues today. First, as stated above, there is no such thing as a normal blood lead level, and the influence of 'normal' body burdens of lead on physiological systems is, at least, scientifically relevant. Second, despite the continued decrease in the usage and production of lead and the decline in blood lead levels, the continued lowering of the CDC blood lead intervention level coupled with the realization that lead causes toxic injury at these low levels indicates that lead intoxication remains clinically relevant. Third, the principle target organs of subclinical lead toxicity have been widened to include organs, such as the immune system, that historically have not been considered as part of the clinical features of lead intoxication. Finally, individuals whose immune systems, for example, are affected by low-level lead toxicity are likely to be asymptomatic, especially given the lack of proven and consistent biomarkers of lead intoxication.

6.3.2 *Organ systems overtly affected by lead –*
clinical manifestations of lead toxicity

The hematopoietic, renal, and central nervous systems in humans are all well recognized and well characterized as being affected by lead intoxication at levels requiring clinical intervention and management. These organ systems as targets for high-dose lead toxicity are interesting from the perspective of the potential immunotoxicity of lead, in that red blood cells, the kidney, and the CNS are all well-established targets of autoimmune disease; however, the involvement of lead in the exacerbation of any human autoimmune disease is clearly speculative at this time. Nevertheless, there is clearly a precedent for the involvement of metals (e.g. mercury, nickel) in several autoimmune disease models. In erythrocytes, anemia is a clinical manifestation of lead poisoning. The anemia is produced by two mechanisms – impairment of heme biosynthesis and accelerated destruction of erythrocytes. Lead inhibits heme biosynthesis at two enzymatic steps. The first being the cytoplasmic enzyme δ-aminolevulinic acid dehydratase (i.e. ALA-D), which catalyzes the rate-limiting step in heme biosynthesis, and the second being the mitochondrial enzyme ferrochelatase, which catalyzes the transfer of iron from ferritin into protoporphyrin to form heme. Consequently, elevated levels of ALA in serum or of coproporphyrin in urine are additional markers of lead toxicity.

Extensive research has been conducted on examining the central nervous system as a site of lead toxicity, and excellent reviews on the subject of lead neurotoxicology are available (Bressler and Goldstein, 1991). Animal studies and human epidemiological data support the notion that lead exposure at extremely low and environmentally relevant levels results in impaired cognitive functions, especially with the developmental exposures of children. Importantly, many of the biochemical/molecular and cellular mechanisms proposed to account for lead neurotoxicology are applicable to the immune system. These include effects of lead on cell surface integrins, intracellular Ca^{2+}, and protein kinase C, to mention a few (Cookman *et al.*, 1987; Markovac and Goldstein, 1988; Laterra *et al.*, 1992; Long *et al.*, 1994). It is unknown whether lead modulates these processes in lymphoid tissues.

There are reports suggesting a potential linkage between occupational lead nephropathy and an autoimmune kidney disease in humans occupationally exposed to lead, suggesting that lead may exacerbate autoimmune kidney disease (Wedeen *et al.*, 1979; Wedeen, 1982). The renal involvement in lead toxicity is manifested as chronic nephropathy, which may progress to generalized kidney failure; proximal tubule cells appear to be the kidney cells most sensitive to lead. Intoxication evidenced by relatively low blood lead levels (i.e. approximately 25 μg Pb/dl) may be associated with reduced serum 1α,25-dihydroxyvitamin-D_3 levels, since a significant negative correlation between blood lead levels and circulating 1α,25-dihydroxyvitamin-D_3 levels in lead-intoxicated children has been reported (Rosen *et al.*, 1980; Mahaffey *et al.*, 1982). Given the overall serum chemistry of these lead-intoxicated children (i.e. elevated serum levels of parathyroid hormone and 25-hydroxyvitamin-D_3), it has been hypothesized that the renal 1α-hydroxylase enzyme is inhibited by lead. No direct evidence for this hypothesis has been found, but it remains viable and testable. The 1α-hydroxylase has also been found in interferon-γ-stimulated macrophages, meaning that this potential target enzyme may be affected by lead in immunocompetent cells as well, and may result in impaired interaction of macrophages and T_{H1} lymphocytes. Additionally, since it has become increasingly recognized that 1α,25-dihydroxyvitamin-D_3 is a paracrine regulator of cellular functions in a much broader range of cells than initially thought, investigations of the effects of lead on 1α,25-dihydroxyvitamin-D_3-modulated immune functions, particularly T-lymphocyte reactivities, seems warranted especially since these functions are inhibited by lead in other cell types. For example, 1α,25-dihydroxyvitamin-D_3-mediated production of the bone matrix protein osteocalcin by osteoblasts, and osteoblast–osteoclast coupling, are markedly impaired by low concentrations of lead *in vitro* (Long *et al.*, 1990). Skeletal lead is important for three general reasons. First, bone serves as the major storage depot (i.e. 90 percent or more of the total body burden) for lead, which can be released under various physiological states or pathological conditions. Second, skeletal lead burdens are potentially a more useful indicator of lead exposure than are blood lead level measurements. No immunotoxicological studies have made use of this correlation, which may be important for future assessments of the relative importance of lead immunotoxicology. Third, the skeleton itself serves as an important site of low-dose lead toxicity (Pounds *et al.*, 1991).

Finally, a growing body of experimental evidence indicates that high doses of lead impair reproductive functions in laboratory animals and in humans. These observations have been the root causes of concern and consideration of the potential adverse effects of lead exposure on the developing fetus. Again, the findings that lead causes neurological damage to the fetus at low blood lead levels has led the field with regard to addressing the issue of developmental lead toxicity. No studies have been reported that directly or adequately address the effects of lead intoxication on the developing immune system.

6.4   Immunotoxicology

6.4.1   *The immune system as a target for lead toxicity*

In recent years, several independent laboratories have accumulated data suggesting that the immune system is a target for low-dose lead toxicity, and many reviews on the subject have been published (McCabe, Jr, 1994; McCabe, Jr and Lawrence, 1994; Lawrence and McCabe, Jr, 1995). The collective approach of these investigators in the field of lead immunotoxicology research has been varied and all-inclusive, but they have presented provocative data that support the general premise that immune reactivities are affected by environmentally relevant levels of lead. The literature contains accounts of the effects of lead on the immune systems of a variety of animal models, including humans, mice, rats, rabbits, cows, and fish, to name a few. Furthermore, *in vivo* and *in vitro* approaches have been adopted to test the effects of lead on immune parameters, and influences of lead on humoral and cell-mediated immunity have been documented. Finally, effects of lead on all four major immunocompetent cells, T-lymphocytes, B-lymphocytes, macrophage/monocytes, and neutrophils have been described.

Attempts to understand the mechanism(s) involved in lead immuno-modulation have produced many conflicting results. The reasons for these conflicting results are likely due to differences in several aspects of the experimental systems employed (e.g. different species, strains, form of metal, route of exposure, immune parameter assayed), as well as the complexity of the immune system itself. Analogous problems exist for the interpretation of data related to many chemicals and drugs thought to modulate immune function. A report by Luster *et al.* (1988) details the immunological assays that can be used to predict whether a xenobiotic is an immunotoxicant. Importantly, two specific immune function assays, the antibody-forming cell response to the T-lymphocyte-dependent antigen SRBC (sheep red blood cell) and host resistance to the obligate intracellular pathogen *Listeria monocytogenes*, which are believed to be the most accurate predictors to test the potential immunotoxicity of a compound, are significantly affected by exposure to lead. A decreased host resistance to environmental or experimental pathogens is consistent with the hypothesis that lead alters the activities of one or more components of the immune response resulting in impaired defense against the invading pathogen and disease exacerbation. Despite the testing of this general hypothesis by several laboratories, a consensus delineating the mechanisms and immune components most sensitive to lead has not been advanced. Furthermore, while *in vivo* whole-animal studies and *in vitro* cell culture experiments have demonstrated that the immune system is affected by exposure to lead and provided some insight into the potential mechanisms involved, it has not been established whether human health is adversely affected by any lead-induced modulation of immune reactivities. As such, the mechanisms involved in lead immunomodulation are poorly characterized, and the consequences from lead modulation of immune reactivities upon the health of individuals at risk for lead exposure/intoxication are not known. Nevertheless, given the accumulation of a substantial body of evidence indicating that murine immune reactivities are altered by lead both *in vivo* and *in vitro*, further consideration of the human immune system as a target for low-dose lead toxicity is warranted. This chapter aims to review the animal data that support the contention that the immune system may be a relevant target organ for low-dose lead toxicity, and to highlight the idea that T-lymphocyte function, in particular, is a critical target of immunomodulation by lead.

6.4.2 In vivo *studies*

Effects of lead exposure on host resistance

Numerous animal studies have considered whether lead intoxication alters host resistance to infectious agents (Hemphill *et al.*, 1971; Koller, 1973; Gainer,

1974, 1977; Cook *et al.*, 1975; Lawrence, 1981a; Kowolenko *et al.*, 1991). An increase in lethal susceptibility to a variety of infectious disease agents has been consistently demonstrated in lead-intoxicated animals (reviewed by McCabe, Jr, 1994), suggesting that lead compromises certain aspects of protective immunity. Animal models to assess the effects of lead on host resistance have utilized various host species and strains, chemical forms of lead, routes of exposure to lead, doses of lead, infectious agents, and routes of exposure to the infectious agent. Despite the variations in experimental designs and approaches, the overwhelming majority of data support the fact that lead impairs host resistance. Furthermore, there is some evidence that lead exposure may alter host resistance in humans (Ewers *et al.*, 1982; Jaremin, 1990; Horiguchi *et al.*, 1992).

Despite the abundance of data pointing to reduced host resistance to infection in lead-burdened animals, the direct immune indices that are modulated by lead for most of these infectious disease models remain uncharacterized. The most extensive assessments of the influence of lead on host resistance have been conducted by Lawrence's laboratories, which have shown that mortality and the recovery of viable bacteria from the spleen and liver are markedly increased in *Listeria*-infected CBA/J mice exposed to lead via the drinking water (Lawrence, 1981a; Kowolenko *et al.*, 1991). These studies utilized a rational experimental design, in that the route of lead administration, as well as the levels of lead intoxication achieved, approximated that from 'natural' exposures. Resistance to *Listeria monocytogenes*, a Gram-positive obligate intracellular bacteria, has been widely used as an endpoint for assessing changes in host resistance. The pathogenesis and role of the immune system in limiting infection have been well established for *Listeria*. *In vivo* assessment of the influence of a 4-week exposure to various concentrations of lead in the drinking water (i.e. 0, 0.08, 0.4, 2, and 10 mM) revealed that cell-mediated immunity, measured as an increase in the recovery of viable *Listeria* from the spleen, was suppressed at the highest lead exposure levels. In addition, the incidence and rate of mortality were increased in *Listeria*-infected mice exposed to 0.4 to 10 mM lead solutions. Similar results were obtained with mice exposed to lead for 2 weeks, indicating that the exposure time to lead in the drinking water was not critical. It is well established from pharmacokinetic studies in rodents that, at constant lead exposure levels, blood lead concentrations remain fairly constant throughout the exposure period. Recovery from *Listeria* infection depends on the phagocytosis of the organism by macrophages and the subsequent activation of the macrophage to a bactericidal state following interaction with sensitized T-lymphocytes, particularly T_H1 lymphocytes (North, 1970; Buchmeir and Schrieber, 1985; Dalton *et al.*, 1993). In tests of the *in vivo* and *in vitro* effects of lead exposure on *Listeria* phagocytosis, it was determined that phagocytosis of *Listeria* by either peritoneal or splenic macrophages was not significantly affected by lead intoxication (Kowolenko *et al.*, 1991). Other reports support the concept that lead intoxication does not influence afferent functions of macrophages, including phagocytosis. For example, *in vitro* addition of lead to non-induced peritoneal macrophage cultures does not affect macrophage binding and phagocytosis of SRBC or of opsonized SRBC (Lawrence, 1981b,c). On the other hand, the effects of inhalation exposure on the phagocytic uptake of opsonized polystyrene latex particles by rabbit pulmonary macrophages are that phagocytic competence decreases with increasing exposure time to lead oxide (Zelikoff *et al.*, 1993). This a good example where a variation in the experimental design, particularly with regard to the route of exposure to lead, produces strikingly different results. Therefore, it is impossible to generalize that macrophage phagocytosis is an immune parameter that is sensitive or insensitive to lead intoxication. What may be more meaningful for both risk assessment and collective understanding of the mechanisms of lead immunomodulation is to consider the routes of exposure to lead by selected human populations and to consider which data sets are most relevant to that particular exposed population. For example, exposure to respirable lead particles in the air more likely occurs in occupational settings where the target population is composed chiefly of adults.

An important mode of resistance to *Listeria* in infected mice is the early mobilization, recruitment, and activation of bone marrow-derived and splenic macrophages (North, 1970; Bennett and Baker, 1977). Lead intoxication clearly inhibits macrophage development and recruitment during the early phase of responsiveness to *Listeria* by impairing the ability of bone marrow-derived macrophage precursors to proliferate and to form colonies in response to colony stimulating factor-1 (CSF-1; Kowolenko *et al.*, 1988, 1989). Exposure to lead via injection has also been reported to alter bone marrow cellularity (Schlick and Friedberg, 1982). Interestingly, serum levels of the monocytic-differentiating cytokine CSF-1 are increased in lead-intoxicated mice, yet the monocytic lineage cells remain unresponsive to these heightened levels (Kowolenko *et al.*, 1989). Similarly, bone marrow cells remain unresponsive to CSF-1 *in vitro*, as demonstrated by no changes in the levels of colony formation (Kowolenko *et al.*, 1988). The lack of bone marrow-derived macrophage development in lead-intoxicated, *Listeria*-challenged mice may contribute to the increased mortality observed in this model. However, since it is also well established that a successful immune response to *Listeria* involves a dynamic interaction between macrophages and T_H1 lymphocytes, and since interferon-γ (IFN-γ), a major T_H1-derived effector cytokine, has been unequivocally implicated as the predominant cytokine controlling macrophage anti-listericidal activity, the relevant primary cell target to lead in the *Listeria* model may be a helper T-lymphocyte subset. The effects of lead intoxication on the later stages of listeriosis, during which time T_H1-derived cytokines such as IFN-γ and macrophages control the immune response to and the clearance of *Listeria*, has gone unexplored. Lead has been shown to inhibit the activities of T_H1 lymphocytes both *in vivo* and *in vitro* (Heo *et al.*, 1996; McCabe, Jr and Lawrence, 1991).

In addition to macrophage development/recruitment and T_H1 activation, resistance to *Listeria* and other obligate intracellular pathogens may be compromised via a direct effect of lead on macrophage bactericidal functions. Lead has been reported to inhibit the intracellular killing of *Leishmania enrietti* by infected macrophages activated with lipopolysaccharide (LPS) and IFN-γ (Mauël *et al.*, 1989). The decreased killing of leishmania parasites correlated with decreased respiratory burst activity; lead has been reported to depress the oxidative response of macrophages treated with phorbol myristic acetate, as well as with zymosan or latex particles (Castranova *et al.*, 1980; Hilbertz *et al.*, 1986). However, a modified respiratory burst response of macrophages is not a particularly sensitive indicator of lead immunotoxicity, as it only occurs at relatively high lead levels (i.e. 30 μM or more). Inducible production of nitric oxide by T-lymphocyte activated murine macrophages appears to be sensitive to lead at low concentrations *in vitro* (Tian and Lawrence, 1995). Nitric oxide plays a role in host resistance to *Listeria* (Beckerman *et al.*, 1993).

Effects of lead on delayed-type hypersensitivity

In addition to tests on the effects of lead exposure *in vivo* on humoral immunity, the effect of chronic lead exposure on the T-lymphocyte-mediated and monocyte-dependent delayed-type hypersensitivity (DTH) reaction has been investigated. One investigation demonstrated a strong, dose-dependent, negative correlation between blood lead levels and DTH reactivity to SRBC challenge in BALB/c mice (Müller *et al.*, 1977). The lead exposure regimen for this study consisted of injecting mice intraperitoneally once daily for 30 days with various amounts of lead acetate (0.025 to 0.25 mg). The doses of lead and the route of administration of lead given in this exposure protocol were the same as those in the study by Hemphill *et al.* (1971), where lead was shown to suppress mouse resistance to *Salmonella typhimurium*. On the 30th day of consecutive daily lead injections, the mice were sensitized intravenously with SRBC. To assess the DTH response, the mice were then challenged 4 days post-sensitization by injection of SRBC into the right hind footpad; footpad swelling was then measured 0, 24, and 48 h later. The results of this study demonstrated that DTH reactivity to SRBC was significantly suppressed in lead-exposed mice relative to control mice, and that the reduced footpad swelling was inversely correlated with elevated blood lead levels. Since DTH responsiveness is a measure of cell-mediated immunity involving the participation of T_H1 cells that secrete interleukin-2 (IL-2), IFN-γ, and tumor necrosis factor-α (TNF-α), the observed effects of lead on the DTH reactivity of mice to SRBC is consistent with the idea that T-lymphocyte regulatory functions were targeted by lead. However, since DTH is also characterized as a local inflammatory reaction due to a large influx of non-specific inflammatory cells in response to T-lymphocyte-derived cytokines, it is also possible that lead suppressed the local inflammatory response and monocytic infiltration characteristics of the afferent phase of the DTH response. A major drawback of the experimental design adopted by Müller *et al.* (1977) is that administration of lead via daily, bolus, intraperitoneal injections constitutes an irrelevant route of exposure. Studies ongoing in the author's laboratory, wherein BALB/c mice are exposed to lead acetate via the drinking water prior to sensitization and challenge with SRBC, have confirmed the observation that reduced DTH reactivity in lead-exposed mice is inversely correlated with blood lead concentrations (McCabe, unpublished observation). In a study performed by Faith *et al.* (1979), the effects of chronic low-level pre- *and* post-natal lead exposure on cell-mediated immunity, including DTH responsiveness to purified protein derivative, were assessed. The lead administration regimen in this report consisted of exposing female weanling rats to lead acetate (0, 25, or 50 ppm for 7 weeks) in their drinking water, followed by mating with untreated males and continued exposure to lead throughout gestation and lactation. The offspring of the pregnant dams were continuously exposed to lead for 35 to 45 days after weaning and then the analysis of DTH reactivity was initiated. In this case, DTH reactivity was assessed by intraperitoneal injection with Freund's complete adjuvant followed, 7 days later, by intraperitoneal injection with ^{3}H-thymidine and challenge 24 h later in the ear lobe with purified protein derivative (PPD). Consistent with the interpretation of the study performed by Müller *et al.* (1977) described above, the DTH reactivity to PPD was markedly reduced in lead-intoxicated rats and the reduced association of ^{3}H-thymidine with tissue plugs obtained from the challenged ear was inversely correlated with blood lead levels. Two additional reports have tested the effects of lead on DTH reactivity as an indicator of cell-mediated immunity. Laschi-Locquerie *et al.* (1984) demonstrated that subcutaneous injection of Swiss–Webster mice with various doses of lead suppressed contact hypersensitivity to picryl chloride, and Descotes *et al.* (1984), using an experimental design comparable with that of Müller *et al.* (1977), suggested that the effect of lead on DTH reactivity to SRBC depends on which lead salt (i.e. acetate, carbonate, chloride, nitrate, or oxide) is injected. Hence, by a variety of exposure routes and doses, lead suppresses DTH reactivity to several test antigens. However, a mechanism accounting for the lead-suppressed DTH reactivity has not been advanced. A hypothesis under investigation in the author's laboratory is that the reduced DTH reactivity in lead-intoxicated animals is due to an inhibition of T_H1 activation, the production of T_H1-derived cytokines, and the effector functions of T_H1 lymphocytes. This hypothesis is based on the observations that lead inhibits antigen presentation to T_H1 lymphocytes and their proliferation (McCabe, Jr and Lawrence, 1991), and that DTH reactivity is largely controlled by T_H1-regulatory activities. Furthermore T-cell DTH activity is important in host defense against obligate intracellular pathogens, which as described above is also impaired by lead intoxication.

122

6.4.3 In vitro *studies*

Effects of lead on humoral immunity

Lead displays immunoenhancing effects on certain lymphocyte activities *in vitro*. For example, numerous studies have investigated the effects of lead on B-lymphocyte proliferation and terminal differentiation into antibody-secreting plasma cells *in vitro*. However, since the differentiation of B-lymphocytes into antibody-producing cells is often a T-lymphocyte-dependent process, it is often difficult to discern if the effects of lead on antibody production are a direct effect on B-lineage cells or an indirect effect on regulatory T-lymphocytes. Given the mutual dependency of T-lymphocytes and B-lymphocytes on each other during T-lymphocyte-dependent B-lymphocyte activation, it is not surprising that evidence of lead modulation of both cell types abounds in the published literature. Nevertheless, lead produces a two- to three-fold increase in the number of antigen-specific IgM-secreting plaque-forming cells (PFC) from Mischell–Dutton cultures containing normal spleen cells from a variety of mouse strains (Blakley and Archer, 1981; Lawrence, 1981a–c; Warner and Lawrence, 1986a,b; McCabe, Jr and Lawrence, 1991). In the Mischell–Dutton assay system, enhancement of the PFC response to sheep erythrocytes by lead could be due to direct stimulation of B-lymphocytes, or indirect effects of lead on T-helper or -suppressor lymphocytes, or upon macrophage activities. Humoral immunity of human lymphocytes exposed *in vitro* to lead is also enhanced (Borella and Giardino, 1991).

A decade of research aimed at determining the relevant, primary target cell or mechanism whereby lead stimulates the PFC response has revealed that no single mechanism can account for the enhancement in the response in the presence of lead; it has a multitude of effects on the various subsets of immunocompetent cells. Lead does have direct effects on B-lymphocytes. Most noteworthy, using highly enriched, resting, splenic B-lymphocytes obtained from several different mice strains, it has been shown that exposure to low concentrations of lead *in vitro* markedly increases the cell surface expression of the major histocompatibility complex class II molecules (i.e. I-A and I-E), as determined by flow cytometric analysis, and the total cellular content of the major histocompatibility complex (MHC) class II molecules, as determined by cell fractionation and Western blot analysis (McCabe, Jr and Lawrence, 1990; McCabe, Jr et al., 1991). The increased cell surface density of MHC class II molecules on murine B-lymphocytes caused by lead is selective, in that there is no corresponding increase in total protein synthesis and MHC class I molecule expression. Furthermore, the only other metal found to increase membrane MHC class II expression was mercury. The extent of the increase in membrane MHC class II is comparable with that caused by other conventional B-lymphocyte activators, such as anti-IgM, interleukin-4, and LPS. The lead-induced increase in MHC class II is not due to IL-4 activity from residual T-lymphocytes, since neutralizing antibodies to IL-4 do not prevent the lead-induced effect. The significance of the increased Ia on B-lymphocytes, which does not occur on macrophages, is unknown. It remains to be determined if the increased Ia alters: (i) the ability of B-lymphocytes to form antigen-specific conjugates with T-helper lymphocytes; (ii) the ability of B-lymphocytes to present antigen to T-helper lymphocytes; or (iii) intracellular B-lymphocyte signaling pathways evoked in response against ligation of cell surface Ia molecules. In all likelihood, the increased B-lymphocyte surface Ia facilitates T-lymphocyte to B-lymphocyte communication via cognate interactions; however, lead also enhances factor-mediated T-helper lymphocyte help for B-lymphocyte differentiation (McCabe, Jr and Lawrence, 1991). Additional experiments hinted that a potential mechanism by which lead increases cell surface MHC class II expression was that the intracellular processing and transport of the α and β chains of the Ia complex were altered by lead. Consequently, the levels of the invariant chain (Ii), which serves as an intracellular chaperon for Ia, are also found to be increased in lead-treated B-lymphocytes (McCabe, Jr *et al.*, 1991). Interestingly, a number of different post-translational modifications to the Ia/Ii complex occur during intracellular passage including sialylation (Holt *et al.*, 1985), and desialylation of membrane Ia chains reportedly increases their recognition by T-lymphocytes (Cowing and Chapdelaine, 1983; Frohman and Cowing, 1985; Taiara *et al.*, 1986; Taiara and Nariuchi, 1988). It is noteworthy that low-level lead exposure impairs normal desialylation of neural cell adhesion molecules (Cookman *et al.*, 1987), which is developmentally regulated and coincident with postnatal synaptogenesis. Therefore, it seems likely that the molecular relevance of lead modulation of B-lymphocyte membrane MHC class II serves to alter subsequent T-/B-lymphocyte interactions.

Effects of lead on T-lymphocyte function

Studies attempting to understand the effects of lead on T-lymphocyte function have produced some conflicting results that may be due to differences in the experimental models employed. Experimental models to assess T-lymphocyte function as a target of lead immunotoxicity have utilized different species or strains, forms of lead, doses or concentrations of lead, routes of exposure, and T-lymphocyte parameters. Another, perhaps more important, reason that studies aimed at determining the mechanism whereby lead modulates T-lymphocyte function have produced conflicting results is that the effects of lead on T-lymphocyte functions are varied.

Lead stimulates the proliferation of mouse CD4+ T-lymphocytes (Warner and Lawrence, 1986a,b). The lead-induced T-lymphocyte proliferation was shown to require the presence of Ia+ cells in culture and was blocked by the inclusion in the cultures of neutralizing antibodies to either CD4 or Ia. The Ia denotes the I-region associated loci of the mouse MHC and encodes the two class II molecules, I-A and I-E. As in an autologous mixed-lymphocyte response, peak proliferation induced by lead occurred late in culture (day 5 or later), was cell density dependent, and required the presence of both T-lymphocytes and Ia+-antigen presenting cells. The lead-induced proliferating cell type was determined to be a T-lymphocyte by examination of the ability of lead to induce cell cycle entry. Cell cycle status was assessed by acridine orange staining and flow cytometric analysis. Furthermore, simultaneous examinations of cell surface phenotype and cell cycle progression indicated that the predominant cell type proliferating in response to lead was a T-lymphocyte, and cell surface phenotype analysis showed enhanced recovery of T-lymphocytes in lead-stimulated cultures. Although lead stimulates a slight enhancement of the synthesis/secretion of the T-lymphocyte growth factor, IL-2 (Warner and Lawrence, 1988), it does not account for the marked stimulation of T-lymphocyte proliferation in the presence of lead.

Interleukin-2 is the predominant autocrine growth factor for a subset of CD4+ helper T-lymphocytes known as T_H1 cells. Lead intoxication *in vitro* at moderate concentrations (i.e. $10\mu M$) results in a striking and significant inhibition of the proliferation of several established T_H1 clones stimulated with cognate antigen in association with antigen-presenting cells. Several studies have supported the fact that lead intoxication *in vitro* inhibits antigen presentation (Kowolenko *et al.*, 1988; Smith and Lawrence, 1988); however, the proliferation of established T_H2 clones stimulated by their cognate antigen in association with antigen-presenting cells is either significantly or slightly enhanced in the presence of lead depending upon the individual T_H2 clone being assayed (McCabe, Jr and Lawrence, 1991). The helper activity for humoral immunity provided by T_H2 cells is enhanced by lead *in vitro*, suggesting that a true dichotomy in the effects of lead on T_H1 and T_H2 functions may exist. In support of this notion, a recent report indicates that lead exposure either *in vitro* or *in vivo* modifies the production of T-lymphocyte-derived cytokines in a manner predicted by the known immunomodulatory effects of lead on T_H1 and T_H2 proliferation (Heo *et al.*, 1996). Specifically, injection of mice with $PbCl_2$ ($50\mu g$, three times, subcutaneously), followed by *ex vivo* analysis of cytokine production by activated T-lymphocytes revealed that the lead-intoxicated mice had increased production of stimulated IL-4, a T_H2-derived cytokine, and reduced production of IFN-γ, a T_H1-derived cytokine and a major T_H1 immune effector cytokine (Heo *et al.*, 1996). It remains to be determined if intoxication with lead by a more natural route of exposure, such as intoxication via the drinking water, modulates T-helper lymphocyte cytokine production and activities in a similar manner. As described above, the effects of *in vivo* exposure to lead on host resistance to obligate intracellular pathogens, such as *Listeria*, and on DTH reactivity are consistent with the notion that T_H1-mediated immune effector functions are targeted by lead. Similarly, the enhancement of B-lymphocyte differentiation in the presence of lead is consistent with the notion of lead stimulation of T_H2 activity.

Evaluation of the potential direct or indirect effects of lead on suppressor T-lymphocyte function has received little attention. In one intriguing report, concanavalin A-induced suppressor cell activity from humans with occupational lead exposure was found to be significantly ($p \le 0.02$) increased (Cohen *et al.*, 1989). No other experimental indications of lead influencing T-suppressor lymphocyte function have been reported. However, it should be pointed out that the cross-regulation of T_H-lymphocyte effector functions and cytokine release by T_H1 and T_H2 cells is, in a sense, a functional suppression.

6.5   Summary

The toxicity of lead on the immune system is like that produced in other target tissues and organ systems in that it does not appear to cause a unique disease or pathological state, but rather it modulates immunoregulatory functions by many cellular, biochemical, and molecular processes. That the immune system is a relevant and potentially important target for lead toxicity has been fueled largely by *in vivo* studies wherein intoxication with lead decreases host resistance to infectious pathogens. The most realistic of these studies have demonstrated that exposure to relatively low and environmentally relevant concentrations of lead via the drinking water markedly suppresses host resistance and averts disease resolution to sublethal challenges of pathogenic microbes. Extensive data have been accumulated which address the cellular and molecular mechanisms whereby lead influences immune reactivities. It is the opinion of the author that the direct and indirect effects of lead on T-lymphocyte function are likely to be the most meaningful to human health.

References

BECKERMAN, K.P., ROGERS, H.W., CORBETT, J.A., SCHREIBER, R.D., MCDANIEL, M.L., and UNANUE, E.R. (1993). Release of nitric oxide during the T-cell independent pathway of macrophage activation. *J. Immunol.* **150**, 888–895.

BENNETT, M. and BAKER, E.E. (1977). Marrow-dependent cell function in early stages of infection with *Listeria monocytogenes*. *Cell. Immunol.* **33**, 203–210.

BERGERET, A., POUGET, E., TEDONE, R., MEYGRET, T., CADOT, R., and DESCOTES, J. (1990). Neutrophil functions in lead-exposed workers. *Human Exp. Toxicol.* **9**, 231–233.

BLAKLEY, B.R. and ARCHER, D.L. (1981). The effect of lead acetate on the immune response in mice. *Toxicol. Appl. Pharmacol.* **61**, 18–26.

BORELLA, P. and GIARDINO, A. (1991). Lead and cadmium at very low doses affect *in vitro* immune responses of human lymphocytes. *Environ. Res.* **55**, 165–177.

BRESSLER, J.P. and GOLDSTEIN, G.W. (1991). Mechanisms of lead neurotoxicity. *Biochem. Pharmacol.* **41**, 479–484.

BUCHMEIR, N.A. and SCHRIEBER, N.D. (1985). Requirement of endogenous interferon-gamma production for the resolution of *Listeria monocytogenes* infection. *Proc. Natl Acad. Sci. USA* **82**, 7404–7408.

CASTRANOVA, V., BOWMAN, L., REASOR, M.J., and MILES, P.R. (1980). Effects of heavy metal ions on selected oxidative metabolic processes in rat alveolar macrophages. *Toxicol. Appl. Pharmacol.* **53**, 14–23.

CDC (Center for Disease Control) (1991). *Preventing Pb Poisoning in Young Children: A Statement by CDC*. Department of Human Health Series Publication, Atlanta.

COHEN, N., MODAI, D., GOLIK, A., WEISSGARTEN, J., PELLER, S., KATZ, A., AVERBUKH, Z., and SHAKED, U. (1989). Increased concanavalin A-induced suppressor cell activity in humans with occupational lead exposure. *Environ. Res.* **48**, 1–6.

COOK, J.A., HOFFMANN, E.O., and DiLUZIO, N.R. (1975). Influence of lead and cadmium on the susceptibility of rats to bacterial challenge. *Proc. Soc. Exp. Biol. Med.* **150**, 741–747.

COOKMAN, G.R., KING, W., and REGAN, M. (1987). Chronic low-level lead exposure impairs embryonic to adult conversion of neural cell adhesion molecule. *J. Neurochem.* **49**, 399–403.

COWING, C. and CHAPDELAINE, J. (1983). T-cells discriminate between Ia antigens expressed on allogeneic accessory cells and B-cells: a potential function for carbohydrate side chains on Ia molecules. *Proc. Natl Acad. Sci. USA* **80**, 6000–6004.

DALTON, D.K., PITTS-MEEK, S., KESHAV, S., FIGARI, I.S., BRADLEY, A., and STEWART, T.A. (1993). Multiple defects of immune cell function in mice with disrupted interferon-gamma genes. *Science* **259**, 1739–1742.

DAVIDSON, C.I. and RABINOWITZ, M. (1992). Lead in the environment: from sources to human receptors, in Needleman, H.L. (Ed.) *Human Lead Exposure*, pp. 65–86, Boca Raton, FL: CRC Press.

DESCOTES, J., EVREUX, J.C., LASCHI-LOCQUERIE, A., and TACHON, P. (1984). Comparative effects of various lead salts on delayed hypersensitivity in mice. *J. Appl. Toxicol.* **4**, 265–266.

EWERS, U., STILLER-WINKLER, R., and IDEL, H. (1982). Serum immunoglobulin, complement C3, and salivary IgA levels in lead workers. *Environ. Res.* **29**, 351–357.

FAITH, R.E., LUSTER M.I., and KIMMEL, C.A. (1979). Effect of chronic developmental lead exposure on cell-mediated immune functions. *Clin. Exp. Immunol.* **35**, 413–420.

FROHMAN, M. and COWING, C. (1985). Presentation of antigen by B cells: functional dependence of radiation dose, interleukins, cellular activation, and differential glycosylation. *J. Immunol.* **134**, 2269–2275.

GAINER, J.H. (1974). Lead aggravates viral disease and represses the antiviral activity of interferon inducers. *Environ. Health Perspect.* **7**, 113–119.

(1977). Effects of heavy metals and of deficiency of zinc on mortality rates in mice infected with encephalomyocarditis virus. *Am. J. Vet. Res.* **38**, 869–872.

GILFILLAN, S.C. (1965). Lead poisoning and the fall of Rome. *J. Occup. Med.* **7**, 53–60.

GOVERNA, M., VALENTINO, M., and VISONA, I. (1987). *In vitro* impairment of human granulocyte functions by lead. *Arch. Toxicol.* **59**, 421–425.

HAGER-MALECKA, B., LUKAS, A., SZCZEPANSKI, Z., FRYDRYCH, J., SLIWA, F., LUKAS, W., and ROMANSKA, K. (1986). Granulocyte viability test in children from an environment with heavy metal pollution. *Acta Paediat. Hungary* **27**, 227–231.

HEMPHILL, F.E., KAEBERLE, M.L., and BUCK, W.B. (1971). Lead suppression of mouse resistance to *Salmonella typhimurium*. *Science* **172**, 1031–1032.

127

HEO, Y., PARSONS, P.J., and LAWRENCE, D.A. (1996). Lead differentially modifies cytokine production *in vitro* and *in vivo*. *Toxicol. Appl. Pharmacol.* **138**, 149–157.

HILBERTZ, U., KRAMER, U., DERUITER, N., and BAGINSKI, B. (1986). Effects of cadmium and lead on oxidative metabolism and phagocytosis by mouse peritoneal macrophages. *Toxicology* **39**, 47–57.

HOLT, G.D., SWIEDLER, S.J., FREED, J.H., and HART, G.W. (1985). Murine Ia-associated invariant chain's processing to complex oligosaccharide forms and its dissociation from the I-A complex. *J. Immunol.* **135**, 399–407.

HORIGUCHI, S., ENDO, G., KIYOTA, I., TERAMOTO, K., SHINAGAWA, K., WAKITANI, F., TANAKA, H., KONISHI, Y., KIYOTA, A., OTA, A., and FUKUI, M. (1992). Frequency of cold infections in workers at a lead refinery. *Osaka City Med. J.* **38**, 79–81.

JAREMIN, B. (1990). Immunological humoral responsiveness in men occupationally-exposed to lead. *Bull. Inst. Marine Trop. Med. Gdynia* **41**, 1–4.

KOLLER, L.D. (1973). Immunosuppression produced by lead, cadmium, and mercury. *Am. J. Vet. Res.* **34**, 1457–1458.

KOWOLENKO, M., TRACY, L., MUDZINSKI, S., and LAWRENCE, D.A. (1988). Effect of lead on macrophage function. *J. Leukocyte Biol.* **43**, 357–364.

KOWOLENKO, M., TRACY, L., and LAWRENCE, D.A. (1989). Lead-induced alterations of *in vitro* bone marrow cell responses to colony stimulating factor-1. *J. Leukocyte Biol.* **45**, 198–206.

(1991). Early effects of lead on bone marrow cell responsiveness in mice challenged with *Listeria monocytogenes*. *Fundam. Appl. Toxicol.* **17**, 75–82.

LASCHI-LOCQUERIE, A., DESCOTES, J., TACHON, P., and EVREUX, J.C. (1984). Influence of lead acetate on hypersensitivity experimental study. *J. Immunopharmacol.* **6**, 87–93.

LATERRA, J., BRESSLER, J.P., INDURTI, R.R., BELLONI-OLIVA, L., and GOLDSTEIN, G.W. (1992). Inhibition of astroglia-induced endothelial differentiation by inorganic lead: a possible role for protein kinase C. *Proc. Natl Acad. Sci. USA* **89**, 10748–10752.

LAWRENCE, D.A. (1981a). *In vivo* and *in vitro* effects of lead on humoral and cell-mediated immunity. *Infect. Immun.* **31**, 136–143.

(1981b). Heavy metal modulation of lymphocyte activities. II. Lead, an *in vitro* mediator of B-cell activation. *Int. J. Immunopharmacol.* **3**, 153–161.

(1981c). Heavy metal modulation of lymphocyte activities. I. *In vitro* effects of heavy metals on primary humoral immune responses. *Toxicol. Appl. Pharmacol.* **57**, 439–451.

LAWRENCE, D.A. and MCCABE, M.J., JR (1995). Immune modulation by toxic metals, in Goyer, R.A., Klaassen, C.D., and Waalkes, M.P. (Eds), *Metal Toxicology*, pp. 305–337, New York: Academic Press.

LIN, F. (1992). Modern history of lead poisoning: a century of discovery and rediscovery, in Needleman, H.L. (Ed.), *Human Lead Exposure*, pp. 23–44, Boca Raton, FL: CRC Press.

LONG, G.J., ROSEN, J.F., and POUNDS, J.G. (1990). Lead impairs the production of osteocalcin by rat osteosarcoma (ROS 17/2.8) cells. *Toxicol. Appl. Pharmacol.* **106**, 270–277.

LONG, G.J., ROSEN, J.F., and SCHANNE, F.A. (1994). Lead activation of protein kinase C from rat brain. Determination of free calcium, lead, and zinc by ^{19}F-NMR. *J. Biol. Chem.* **269**, 834–837.

LUSTER, M.I., MUNSON, A.E., and THOMAS, P. (1988). Development of a testing battery to assess chemical-induced immunotoxicity: National Toxicology Program's guidelines for immunotoxicity evaluation in mice. *Fundam. Appl. Toxicol.* **10**, 2–19.

MAHAFFEY, K.R., ROSEN, J.F., CHESNEY, R.W., PEELER, J.T., SMITH, C.M., and DELUCA, H.F. (1982). Association between age, blood lead concentration, and serum 1,25-dihydroxyvitamin D_3 in children. *Am. J. Clin. Nutr.* **35**, 1327–1331.

MARKOVAC, J. and GOLDSTEIN, G.W. (1988). Picomolar concentrations of lead stimulate protein kinase C. *Nature* **334**, 71–73.

MAUËL, J., RANSIJN, A., and BUCHMILLER-ROUILLER, Y. (1989). Lead inhibits intracellular killing of *Leishmania* parasites and extracellular cytolysis of target cells by macrophages exposed to macrophage activating factor. *J. Leukocyte Biol.* **45**, 401–409.

MCCABE, M.J., JR (1994). Mechanisms and consequences of immunomodulation by lead, in Dean, J.H., Luster, M.I., Munson, A.E., and Kimber, I. (Eds), *Immunotoxicology and Immunopharmacology*, 2nd Edn, pp. 143–162, New York: Raven Press.

MCCABE, M.J., JR and LAWRENCE, D.A. (1990). The heavy metal lead exhibits B-cell-stimulatory factor activity by enhancing B-cell Ia expression and differentiation. *J. Immunol.* **145**, 671–677.

(1991). Lead, a major environmental pollutant, is immunomodulatory by its differential effects on CD4+ T-cell subsets. *Toxicol. Appl. Pharmacol.* **111**, 13–23.

(1994). Effects of metals on lymphocyte development and function, in Schook L.B. and Laskin, D.L. (Eds), *Xenobiotics and Inflammation*, pp. 193–216, New York: Academic Press.

MCCABE, M.J., JR, DIAS, J.A., and LAWRENCE, D.A. (1991). Lead influences translational or posttranslational regulation of Ia expression and increases invariant chain expression in mouse B cells. *J. Biochem. Toxicol.* **6**, 269–276.

MÜLLER, S., GILLERT, K.-E., KRAUSE, C., GROSS, U., L'AGE-STEHR, J., and DIAMANSTEIN, T. (1977). Suppression of delayed type hypersensitivity of mice by lead. *Experientia* **33**, 667–668.

NEEDLEMAN, H.L. (1991). Childhood lead poisoning: a disease for the history texts. *Am. J. Publ. Health* **81**, 685–687.

NORTH, R.J. (1970). The relative importance of blood monocytes and fixed macrophages to the expression of cell-mediated immunity to infection. *J. Exp. Med.* **132**, 521–534.

NRIAGU, J.O. (1983). Saturnine gout among roman aristocrats. Did lead poisoning contribute to the fall of the empire? *N. Engl. J. Med.* **308**, 660–663.

PATTERSON, M. and SHIRAHATA, H. (1991). Natural skeletal levels of lead in *Homo sapiens* uncontaminated by technological lead. *Sci. Total Environ.* **107**, 205–236.

PIRKLE, J.L., BRODY, D.J., GUNTER, E.W., KRAMER, R.A., PASCHAL, D.C., FLEGAL, K.M., and MATTE, T.D. (1994). The decline in blood lead levels in the United States. The National Health and Nutrition Examination Surveys (NHANES). *J. Am. Med. Assoc.* **272**, 284–291.

POUNDS, J.G., LONG, G.J., and ROSEN, J.F. (1991). Cellular and molecular toxicity of lead in bone. *Environ. Health Perspect.* **91**, 17–32.

QUEIROZ, M.L.S., ALMEIDA, M., GALLAO, M.I., and H'EHR, N.F. (1993). Defective neutrophil function in workers occupationally exposed to lead. *Pharmacol. Toxicol.* **72**, 73–77.

QUEIROZ, M.L.S., COSTA, F.F., BINCOLETTO, C., PERLINGEIRO, R.C.R., DANTAS, D.C.M., CARDOSA, M.P., and ALMEIDA, M. (1994). Engulfment and killing capabilities of neutrophils and phagocytic splenic function in persons occupationally exposed to lead. *Int. J. Immunopharmacol.* **16**, 239–244.

ROSEN, J.F., CHESNEY, R.W., HAMSTRA, A., DeLUCA, H.F., and MAHAFFEY, K.R. (1980). Reduction in 1,25-dihydroxyvitamin D₃ in children with increased lead absorption. *N. Engl. J. Med.* **302**, 1128–1131.

SCHLICK, E. and FRIEDBERG, K.D. (1982). Bone marrow cells of mice under the influence of low lead doses. *Arch. Toxicol.* **49**, 227–236.

SMITH, K.L. and LAWRENCE, D.A. (1988). Immunomodulation of *in vitro* antigen presentation by cations. *Toxicol. Appl. Pharmacol.* **96**, 476–484.

TAIARA, S. and NARIUCHI, H. (1988). Possible role of neuraminidase in activated T-cells in the recognition of allogeneic Ia. *J. Immunol.* **141**, 440–446.

TAIARA, S., KAKIUCHI, T., MINAMI, M., and NARIUCHI, H. (1986). The regulatory role of sialic acids in the response of class II reactive T-cell hybridomas to allogeneic B-cells. *J. Immunol.* **137**, 2448–2454.

TIAN, L. and LAWRENCE, D.A. (1995). Lead inhibits nitric oxide production *in vitro* by splenic macrophages. *Toxicol. Appl. Pharmacol.* **132**, 156–163.

VALENTINO, M., GOVERNA, M., MARCHISEPPE, I., and VISONA, I. (1991). Effects of lead on polymorphonuclear leukocyte (PMN) functions in occupationally-exposed workers. *Arch. Toxicol.* **65**, 685–688.

WARNER, G.L. and LAWRENCE, D.A. (1986a). Stimulation of murine lymphocyte responses by cations. *Cell. Immunol.* **101**, 425–439.

(1986b). Cell surface and cell cycle analysis of metal-induced murine T-cell proliferation. *Eur. J. Immunol.* **16**, 1337–1347.

(1988). The effect of metals on IL-2-related lymphocyte proliferation. *Int. J. Immunopharmacol.* **10**, 629–637.

WEDEEN, R.P. (1982). The role of lead in renal failure. *Clin. Exp. Dial. Apher.* **6**, 113–146.

WEDEEN, R.P., MALLIK, D.K., and BETUMAN, V. (1979). Detection and treatment of occupational lead nephropathy. *Arch. Int. Med.* **139**, 53–57.

WHO (World Health Organization) (1995). *Environmental Health Criteria 165. Inorganic Lead.* World Health Organization, Geneva.

ZELIKOFF, J.T., PARSONS, E., and SCHLESINGER, R.B. (1993). *Environmental Research* **62**, 207–222.

7

Mercury

PIERLUIGI E. BIGAZZI

Department of Pathology, University of Connecticut Health Center, Farmington, CT 06030, USA

7.1 History

In this review we will discuss some aspects of the toxicology of the element 'mercury (Planet *Mercury*) . . . known to ancient Chinese and Hindus; found in Egyptian tombs of 1550 BC' (Hammond, 1970). Mercuric sulfide was indeed one of the chemicals identified in fluid preserving the body of a Han lady who died about 180 BC (Majno, 1975). Similarly, mercuric sulfide (cinnabar) was used by Celsus as an antiseptic (Majno, 1975). Names of planets were first used in the Middle Ages to define certain metals and the name 'Mercurius' was introduced in medieval Latin by alchemists who were probably impressed with mercury's unusual property of quick and easy flow (*hydrargyrum* – quick silver) and thought it had magic properties. Through the name of the planet, this metal acquired some connection with ancient gods involved in science and other less lofty pursuits. The Egyptian Thoth, scribe of the gods and himself god of wisdom and learning, was reputed to have founded alchemy and other occult sciences (Fowden, 1993). His Greek name was Hermes Trismegistus and somehow was identified by the Greeks with their Hermes, the messenger of the gods who was also the god of science, eloquence, cunning, protector of commerce, and guide of the dead to Hades. The Etruscan Turms was derived from the Greek Hermes (Pallottino, 1984) and so was the Roman Mercurius (Mercury), who also was the messenger of the gods and the god of commerce, manual skill, eloquence, cleverness, travel, and thievery. Considering these celestial connections and the current popularity of the so-called 'T_H1/T_H2 paradigm', it is not surprising that a recent editorial has asked the question: 'Mercury: god of T_H2 cells?' (Mathieson *et al.*, 1991). We will use that query as a starting point for this review of mercury's immunotoxicity, keeping in mind

131

the possibility that this compound may resemble those gods of antiquity and have a variety of functions.

7.2 Occurrence

Mercury has three chemical forms, elemental mercury, inorganic mercury, and organic mercury (reviewed in WHO, 1989, 1990, 1991; Gerhardsson and Skerfving, 1996; Hamada and Osame, 1996; Zelikoff and Cohen, 1996). It is widely present in the environment, with a major natural source of environmental mercury pollution provided by degassing from the Earth's crust. Its levels are also increasing as a consequence of discharges from various industries, medical and scientific waste, and the processing of raw ores. In addition, mercury vapor released from dental amalgam fillings may be a major source of inorganic mercury in humans.

7.3 General Toxicology

The toxic effects of mercury on a variety of organs and tissues have recently been discussed in depth by various investigators, therefore we will refer to those reviews for a more complete analysis (Goyer and Cherian, 1995; Miura *et al.*, 1995; Verity, 1995; Chang, 1996; Cory-Slechta, 1996; Fowler, 1996; Hamada and Osame, 1996; Massaro, 1996; Nieminen and Lemasters, 1996; Weiner and Nylander, 1996). Methyl mercury and mercury vapor are clinically the most important chemical forms of mercury. Minamata disease has focused the world's attention on the epidemic intoxication of large numbers of subjects by methylmercury pollution and its effects on the central nervous system and fetal development (Hamada and Osame, 1996). Major clinical manifestations of methylmercury poisoning are sensory impairment, ataxia, visual symptoms (e.g. peripheral visual loss), and hearing impairment. Pathologic lesions are detected in both central nervous system and peripheral nerves. Fetal poisoning results in retardation of physical and mental development. Chronic mercury vapor poisoning may occur after occupational exposure in a variety of workers and causes mental changes (e.g. irritability, difficulty of concentration, insomnia), tremor, and nephrotoxicity (Hamada and Osame, 1996). On the other hand, inorganic mercury salt poisoning, which was once a common cause of renal failure, is now less common (Solez, 1983). The compound was ingested accidentally or for suicidal purposes; in addition, exposure could result from its use as an antiseptic or to induce abortion. Death might occur within the first day, but survivors would become oliguric and develop what was defined as 'acute tubular necrosis'. In the days when mercurial diuretics (a type that is no longer used) were part of the treatment in congestive heart failure, several cases of nephrotic syndrome were also recorded (Heptinstall, 1983). This pathology could also occur in workers of the mercury industry and as the result of the use of inorganic mercury in teething powders or ointments, an effect now quite rare. The potential health risks of mercury from dental amalgam are still a source of controversy (Fung and Molvar, 1992; Jokstad *et al.*, 1992; Molin, 1992; Olsson and Bergman, 1992; Eley and Cox, 1993; Langworth *et al.*, 1993; Weiner and Nylander, 1996).

7.4 Immunotoxicology

The adverse effects of mercury on various components of the immune system have been reviewed recently (Bigazzi, 1994, 1996; Lawrence and McCabe, 1995; Pelletier and Druet, 1995; Exon *et al.*, 1996; Kimber and Basketter, 1996; Sharma and Dugyala, 1996; Zelikoff and Smialowicz, 1996). However, in the present chapter we will devote particular attention to the immunotoxicology of mercury, which is extremely interesting because of its polymorphic and at times confusing characteristics. Exposure to low concentrations of mercury (in its various chemical forms) can depress or stimulate the immune system and even induce autoimmune disease in various animal species, through mechanisms that are still poorly understood.

Early studies of immunotoxicity focused on changes in resistance to micro-organisms in experimental animals exposed to mercury (reviewed by Sharma and Dugyala, 1996). Treatment with mercuric chloride ($HgCl_2$) resulted in a decreased resistance to pseudorabies virus in New Zealand rabbits and to encephalomyocarditis virus in CD-1 mice. Recent studies of BALB/c mice have shown that $HgCl_2$ at a single dose of 20 µg aggravated a generalized infection with herpes simplex virus type 2 (Ellermann-Eriksen *et al.*, 1994; Christensen *et al.*, 1996). Examination of the course of the infection indicated that the early, non-specific defense mechanisms were affected and that mercury might interfere with the production of cytokines by macrophages. Overall, changes in resistance to infection have suggested that components of the immune system may be affected by the metal and have focused the attention of investigators on immune cells. Studies of mercury effects on lymphocytes, macrophages, and other cells of the immune system are relevant to human health not only because they may elucidate the cellular and molecular basis of mercury-induced immunotoxicity, but also may help in understanding similar effects by other xenobiotics.

7.4.1 *Immunotoxicity of mercury on T-lymphocytes*

In vitro *studies*

Initial investigations of *in vitro* mercury effects were performed on unseparated populations of human lymphocytes and showed that this metal was capable of inducing transformation and mitoses in peripheral blood lymphocytes from patients with allergy to mercury as well as normal controls (Schopf *et al.*, 1967, 1969; Caron *et al.*, 1970). Mercury enhanced human lymphocyte stimulation after 6 days of culture and also stimulated the production of β$_2$-microglobulin by lymphocytes (Ohsawa and Kimura, 1979). Leukocyte aggregation and stimulation were demonstrated using concentrations of mercury of 10 µg/ml (Hutchinson *et al.*, 1976). Human peripheral blood lymphocytes were found to proliferate after exposure to 30 µM HgCl$_2$ for 5 to 6 days (Warner and Lawrence, 1986). Experiments using murine lymphocytes also showed that treatment with mercury stimulates cell proliferation. However, other investigators have reported inhibitory effects; *in vitro* exposure of human and murine lymphocytes to mercury compounds decreased their response to phytohemagglutinin (PHA) and concanavalin A (ConA) stimulation (reviewed by Sharma and Dugyala, 1996).

In conclusion, most of the older studies on human and animal lymphocytes have demonstrated either a stimulatory or a inhibitory effect of mercury on T-lymphocyte function. With a few exceptions, those studies did not use well-characterized populations of T-lymphocytes. On the other hand, experiments with human T-lymphocytes purified by rosette formation and treated *in vitro* with HgCl$_2$ or methylmercuric chloride (MeHgCl) showed a monocyte-dependent inhibition of PHA- and phorbol myristate acetate (PMA)-induced cell proliferation and a decreased production of interleukin-2 (IL-2) (Shenker *et al.*, 1992). More recently, Jiang and Möller (1995) have investigated the ability of HgCl$_2$ to activate murine lymphocytes *in vitro* and found that at a very low concentration (10 µM) it could induce an increased rate of DNA synthesis, with a maximum at 4 to 6 days of incubation. The most significant increase in thymidine uptake was induced in spleen cells from A.SW and BALB/c mice, that is, animals from strains that can respond *in vivo* to mercury. Mercury induced activation of both CD4$^+$ and CD8$^+$ T-lymphocytes in responder mice, whereas only CD8$^+$ T-lymphocytes were activated in non-responder (DBA/2) mice. The presence of accessory cells was necessary for this effect to occur, as demonstrated by the lack of response when adherent cells were removed. Similarly, CD4$^+$ T-lymphocytes were crucial for HgCl$_2$-induced activation, since anti-CD4 monoclonal antibody (mAb) completely inhibited cell proliferation after exposure to mercury. Another recent study has examined the *in vitro* effects of HgCl$_2$ on various T-lymphocyte lines and in particular the murine 2B4.11 T-lymphocyte hybridoma (Aten *et al.*, 1995a). Depending on the cell line, treatment with mercury induced time- and dose-dependent cell death, with typical features of both apoptosis and necrosis. Expression of human *Bcl*-2 inhibited HgCl$_2$-induced apoptosis. Recombinant IL-2 or a combined treatment with IL-4 and monoclonal antibody to IL-4 significantly diminished cell death of 2B4.11 T-lymphocytes caused by mercury treatment (Prigent *et al.*, 1995).

In vivo *studies*

Similarly to the *in vitro* findings, the *in vivo* effects of mercury treatment have varied from stimulation to inhibition, depending on animal strain, route of administration, and chemical form of mercury (reviewed by Sharma and Dugyala, 1996). For example, BALB/c mice exposed to methylmercury in their diet for 12 weeks showed an increased T-lymphocyte response to ConA stimulation. In contrast, a decrease in PHA- and ConA-induced stimulation of spleen cells was observed in Swiss and $B_6C_3F_1$ hybrid mice treated with $HgCl_2$. More recently, the injection of a low dose of mercury into Brown Norway rats caused a marked decrease in the ConA-induced generation of interferon-γ (IFN-γ)-producing cells from splenocyte cultures prepared 1 h after mercury administration (van der Meide *et al.*, 1993). Injection of Lewis rats with $HgCl_2$ caused no inhibitory effects on splenic IFN-γ production. The presence of the reduced form of glutathione in the culture medium was essential, because in its absence the number of IFN-γ-producing cells was severely reduced in all cultures. The same group has observed that nitric oxide suppresses IFN-γ production in the spleen of $HgCl_2$-exposed Brown Norway rats (van der Meide *et al.*, 1995).

7.4.2 *Immunotoxicity of mercury on B-lymphocytes*

In vitro *studies*

As previously described for T-lymphocytes, earlier investigations used unfractionated lymphocytes treated with mercury and B-lymphocyte mitogens (reviewed by Exon *et al.*, 1996). Similarly to the results obtained with T-lymphocytes, some investigators reported inhibition, others stimulation of B-lymphocytes. Such results can be explained by the use of different strains of animals, different chemical forms of mercury, and the lack of purification of B-lymphocytes. A more recent study has been performed on human B-lymphocytes, isolated from peripheral blood by first removing monocytes using counterflow centrifugal elutriation and T-lymphocytes by E-rosette formation (Shenker *et al.*, 1993). Both $HgCl_2$ and MeHgCl caused a dose-dependent reduction in B-lymphocyte proliferation in the presence or absence of monocytes. Mercury also inhibited the ability of these cells to synthesize IgM and IgG. In contrast to most other reports, another group of investigators has used resting B-lymphocytes to determine the effects of mercury on quiescent lymphocytes (Daum *et al.*, 1993). Mercury inhibited both RNA and DNA synthesis by B-lymphocytes from DBA/2J mice, with an IC_{50} of 50 ± 12 nM for RNA synthesis and 120 ± 20 nM for DNA synthesis. At 200 nM, mercury arrested cell cycle progression, reducing entry into $G1_a$, $G1_b$, and S/G2 to 35, 10, and 12 percent of the control lipopolysaccharide (LPS)-induced response. The inhibition of RNA synthesis was rapid and irreversible. In addition, mercury had significant inhibitory effects on Ig isotype expression. The concentrations of mercury capable of reducing Ig isotype production to 50 percent of controls were 10 nM for IgG_3, 0.10 µM for IgG_{2b}, 0.28 for IgG_1, 0.39 for IgG_{2a}, and 0.45 for IgM. Thus, the IgG_3 isotype was the most sensitive to the effects of mercury suggesting the possibility that this metal acts on a discrete subset of B-lymphocytes. These results are of great interest because they indicate that $HgCl_2$ has direct effects on B-lymphocytes, independently of its stimulation or inhibition of cytokine production by T-lymphocytes.

In vivo *studies*

Early reports showed that administration of methylmercury to rabbits and mice suppressed antibody responses to influenza virus and sheep red blood cells (SRBC) (reviewed by Exon *et al.*, 1996). In contrast, exposure of BALB/c mice to methylmercury resulted in increased activity of LPS-stimulated B-lymphocytes. Similarly, polyclonal activation of B-lymphocytes was observed in mercury-treated SJL/N and C57Bl/6J mice. Significantly increased IgE and IgG levels were found in A.SW and increased IgE levels in C57Bl/6J mice. Other investigations demonstrated that $HgCl_2$ induces a striking increase of total serum IgE in Brown Norway rats, with values that reached a maximum of 6300 µg/ml serum (Prouvost-Danon *et al.*, 1981). This large amount of IgE is likely the result of polyclonal stimulation, since IgE antibodies specific for renal GBM (glomerular basement membrane) or other antigens have not been detected to date. Lewis rats do not have such an increase in IgE. The serum levels of IgM, IgG_1, IgG_{2b}, IgG_{2c}, and IgA were also increased in Brown Norway rats treated with mercury, whereas IgG_{2a} was not increased compared with controls (Pelletier *et al.*, 1988a,b). In contrast to IgE, the high levels of IgG and some IgG isotypes probably reflected the production of autoantibodies to laminin and other autoantigens.

7.4.3 *Immunotoxicity of mercury on other immune cells*

The literature on the immunotoxic effects of mercury on immune cells other than T- and B-lymphocytes is quite sparse (reviewed by Zelikoff and Smialowicz, 1996). This deficiency can be mostly attributed to the mindset of many immunologists, who especially in earlier years focused their attention on lymphocytes. Recent progress showing the importance of macrophages and other cells in both induction and effector phases of the immune response will hopefully lead to rapid progress in this area.

Macrophages

The immunotoxic action of mercury on macrophages is obviously of great interest, considering the various roles these cells play in antigen presentation and inflammation. To date, only a few studies of mercury effects on macrophages have been published. In general, inorganic mercury was reported to have inhibitory effects on the production of superoxide and free radicals and on phagocytosis (reviewed by Zelikoff and Smialowicz, 1996). In our laboratory we have investigated the *in vitro* effects of $HgCl_2$ on peritoneal macrophages from inbred Brown Norway and Lewis rats and found that, depending on its concentration, mercury causes *in vitro* stimulation of H_2O_2 release from Lewis but not Brown Norway rat macrophages and inhibits erythrophagocytosis of Lewis and Brown Norway 'resident' peritoneal macrophages (Contrino *et al.*, 1992). More recently, it has been observed that treatment with $HgCl_2$ induces *in vivo* and *in vitro* secretion of IL-1 by murine macrophages (Zdolsek *et al.*, 1994). Higher gene expression of IL-12 has been reported in spleen cells of Lewis rats, again suggesting the involvement of macrophages (Mathieson and Gillespie, 1996). Finally, in depth studies of mercury immunotoxicity on Langerhans cells, dendritic cells, and antigen-presenting cells in general are still lacking.

Polymorphonuclear leukocytes (PMN)

These cells, so important in acute inflammation, have not traditionally received much consideration by cellular immunologists. However, studies of the immunotoxic effects of mercury on PMN are of great interest, since these cells not only release free radicals and a variety of enzymes, but are also cytokine producers. As previously mentioned for macrophages, such investigations have been rather sparse. In our laboratory, we have evaluated the *in vitro* effects of $HgCl_2$ on human PMN functions (Contrino *et al.*, 1988). $HgCl_2$ consistently suppressed adherence, polarization, chemotaxis, and erythrophagocytosis; low $HgCl_2$ concentrations significantly enhanced chemiluminescence and stimulated H_2O_2 production. Other laboratories have demonstrated inhibitory effects on microbicidal activities and respiratory burst of human PMN (Malamud *et al.*, 1985; Baginski, 1988). These effects were later confirmed by the finding that methylmercury, mercuric chloride, and silver lactate decrease superoxide anion formation and chemotaxis of human PMN (Obel *et al.*, 1993). Investigations of PMN from experimental animals have been even scarcer than for human PMN. Our laboratory has examined the *in vitro* effects of mercury on peritoneal PMN from inbred Brown Norway and Lewis rats and found that $HgCl_2$ stimulates *in vitro* H_2O_2 release from Lewis, but not Brown Norway, PMN (Contrino *et al.*, 1992).

Mast cells

Until recently, effects of mercury on mast cells had not received much attention. Fortunately, there is increasing interest in this type of immune cells, which (in addition to their well-known involvement in allergic reactions) provide a source of IL-4 and therefore may be involved in immunoregulation.

It has recently been reported that $HgCl_2$ increases the sensitivity of peritoneal mast cells from Brown Norway rats to degranulation by a monoclonal anti-IgE (Oliveira *et al.*, 1994). Conversely, mast cells from Lewis rats exhibit very little degranulation by anti-IgE and no enhancement by $HgCl_2$. Additional experiments using a human leukemic mast cell line (HMC-1) have shown that $HgCl_2$ inhibits cell proliferation as well as expression of mRNA for IL-8, tumor necrosis factor-α (TNF-α), and IL-4 (Warbrick *et al.*, 1995).

Natural killer (NK) cells

These cells have an early and important role in host resistance; therefore, their possible impairment by mercury and other xenobiotics may result in increased susceptibility to disease and more severe pathologic effects (reviewed by Zelikoff and Smialowicz, 1996). Unfortunately, as with macrophages, PMN, and mast cells, there are few studies of the immunotoxic effects of mercury on NK cells. Treatment of BALB/c mice with methylmercury resulted in a 44 percent reduction of NK cell activity (Ilback *et al.*, 1991). Similarly, after exposure to methylmercury *in utero* and during lactation, splenic NK cell activity was suppressed in 15-day-old rats (Ilback *et al.*, 1991).

7.4.4 Mercury-induced allergy

To date, the best known immunologic alteration induced in humans by inorganic mercury exposure is allergy, which can occur as an anaphylactic reaction or as a delayed-type hypersensitivity response (contact dermatitis) (reviewed by Kimber and Basketter, 1996). In particular, thimerosal (used as an antiseptic and preservative) is a well-recognized cause of this disorder. However, lichenoid contact dermatitis in red tattoo (mercuric sulfide) areas or adjacent to amalgam fillings has also been reported (Schrallhammer-Benkler *et al.*, 1992). Its appearance seems highly dependent on still undefined, individual susceptibility traits.

7.4.5 Mercury-induced autoimmunity

As previously mentioned, reports of autoimmune disease induced in humans by exposure to mercury are quite rare (Bigazzi, 1988, 1994, 1996; Kosuda and Bigazzi, 1996). A nephrotic syndrome (possibly immune complex mediated) was frequently observed in the past when patients were treated with mercurial diuretics for congestive heart failure, a therapy that has since been discontinued. The use of skin ointments or creams containing mercury has provided a few examples of nephrotic syndrome with immune deposits in the renal glomerular basement membrane. In addition, there have been occasional reports of mercury-induced immune complex-mediated glomerulonephritis and systemic autoimmune disease. In conclusion, mercury (absorbed through the skin, respiratory, or gastrointestinal systems) may cause autoimmune responses and disease in humans. However, studies of this environmental hazard are usually quite difficult because low doses of mercury may be absorbed over a long period of time and escape notice. Once the autoimmune disease develops, a retrospective study may not reveal the etiologic agent or establish a cause–effect relationship. On the other hand, investigations using experimental animals have provided excellent evidence that administration of mercury causes autoimmunity in rabbits, rats, and mice (Goldman *et al.*, 1991; Bigazzi, 1992).

Rabbits

Mercury treatment of outbred rabbits results in the formation of renal immune deposits characterized by two-stage kinetics. The first stage shows linear deposits of IgG at the level of both glomerular basement membrane (GBM) and tubular basement membrane (TBM). Eluates from these deposits contain autoantibodies to laminin (Fukatsu *et al.*, 1987). The second stage shows granular deposits of IgG both in GBM and TBM. These deposits also contain autoantibodies that react most strongly with laminin. Thus, the major target organ is the kidney, which exhibits membranous glomerulonephropathy, whereas other organs and tissues show minor effects or seem to be spared. Circulating autoantibodies against GBM/laminin, present in outbred NZW rabbits, may have a pathogenic role as demonstrated by their correlation with proteinuria. Activation of the complement cascade (C3 has been noted in renal immune deposits) may cause the proteinuria in rabbits with autoantibodies to laminin.

Rats

Repeated exposure to relatively low doses of $HgCl_2$ results in autoimmune responses to renal antigens in three inbred strains of rats (Brown Norway, MAXX, and DZB), whereas 17 other strains are resistant (Druet *et al.*, 1977; Henry *et al.*, 1988; Aten *et al.*, 1992a,b). Rats from the susceptible strains experience membranous glomerulonephropathy, characterized by proteinuria and the production of autoantibodies to epitopes of the renal glomerular basement membrane, including laminin. This disorder has a self-limiting course somewhat similar to that of 'monophasic' experimental allergic encephalomyelitis in rats: autoimmune responses reach their peak approximately 2 weeks after the beginning of mercury treatment, then regress spontaneously in spite of continuous administration of mercury. Brown Norway rats that have recovered from this glomerulonephritis are subsequently resistant to additional treatment with mercury. In spite of extensive investigations, the mechanisms of immunoregulation in this animal model are still uncertain.

Mercury-treated Brown Norway, MAXX, and DZB rats produce high levels of autoantibodies to laminin. Autoantibodies to other autoantigens (e.g. type IV collagen, heparan sulfate proteoglycan, entactin, thyroglobulin) have also been detected in Brown Norway rats after exposure to mercury, but they are present in lower concentrations and for shorter periods of time. In addition, they do not appear to correlate with disease. Rats of the PVG strain produce autoantibodies to nuclear antigens after mercury treatment, but no antibodies to laminin or other autoantibodies have been reported (Weening *et al.*, 1978, 1980, 1981). A deficiency in IL-2 production was observed in Brown Norway rats after mercury treatment (Baran *et al.*, 1988). Other effects of mercury include splenomegaly, lymph node hyperplasia, and thymic atrophy in Brown Norway rats. Changes in peripheral lymphocyte subpopulations of mercury-treated rats have also been reported. In our laboratory, we have found that mercury-treated Brown Norway rats had a decrease of RT6[+] cells (Kosuda *et al.*, 1991, 1993). This decrease correlated inversely with autoimmune responses to GBM/laminin.

Renal immune deposits observed after mercury treatment in Brown Norway and MAXX rats are similar to those observed in outbred rabbits and are characterized by two-stage kinetics. The first stage shows linear deposits of IgG at the level of both GBM and TBM. Eluates from these deposits contain autoantibodies to laminin (Fukatsu *et al.*, 1987; Bigazzi *et al.*, 1989; Aten *et al.*, 1992a,b; 1995b). The second stage shows granular deposits of IgG in both GBM and TBM. These deposits also contain autoantibodies that react most strongly with laminin (Fukatsu *et al.*, 1987). A granular pattern (*not* preceded by the linear staining) is observed in GBM of mercury-treated PVG rats (Weening *et al.*, 1981). Histopathologically, all rat models of mercury autoimmunity show membranous glomerulonephropathy, but lesions in other tissues were not usually reported (Aten *et al.*, 1992a,b). However, recent publications from some laboratories have described inflammatory processes in various organs and tissues of mercury-treated Brown Norway rats (see below, under Mercury and graft-versus-host disease). The immunogenetics of mercury-induced autoimmunity have been carefully investigated in rats (Goldman *et al.*, 1991). Rats of the RT-1[n] haplotype are susceptible to anti-GBM/laminin autoimmunity, whereas those of the RT-1[l] haplotype are resistant, and rats with other haplotypes (RT-1[c,a,k,f,b]) show intermediate susceptibility. Susceptibility to the first phase (anti-GBM antibodies) depends on several genes, one of which is RT-1-linked, whereas the second phase (immune complex-type glomerulonephritis) depends on one major RT-1-linked gene or cluster of genes, with a role for other non-RT-1-linked genes controlling the magnitude of the response.

The pathogenesis of mercury-induced autoimmunity in rats is not completely clear. The major target organ in all models is the kidney, which exhibits membranous glomerulonephropathy, whereas other organs and tissues show minor effects or seem to be spared. Circulating autoantibodies against GBM/laminin, present in Brown Norway (as well as MAXX and DZB) rats, may have a pathogenic role, as demonstrated by their correlation with proteinuria. Activation of the complement cascade or direct autoantibody effects (when there is scarce or no binding of complement components) may cause the proteinuria in animals with autoantibodies to laminin. Rats with circulating autoantibodies to nuclear and nucleolar antigens probably experience proteinuria through the renal deposition of immune complexes and complement activation. Lymphocytes and macrophages, often present in the renal interstitium, may be an additional damaging factor, but direct proof of their pathogenic involvement is lacking.

Results from *in vivo* depletion of (possibly regulatory) T-lymphocyte subpopulations have been inconclusive. Depletion of CD8$^+$ T-lymphocytes was obtained in Brown Norway rats by a combination of thymectomy and treatment with a monoclonal antibody (mAb) against rat CD8 (Mathieson *et al.*, 1991). The lack of CD8$^+$ T-lymphocytes did not affect the initial induction and spontaneous regression of renal autoimmune response following exposure to mercury. More recently, the role of CD45RB/RC^{high} T-lymphocytes has been investigated in Brown Norway rats *in vivo* depleted of these cells by the administration of the mAb OX22 (Mathieson *et al.*, 1993). The elimination of CD45RB/RC^{high} T-lymphocytes did inhibit a graft-versus-host-like syndrome observed in mercury-treated Brown Norway rats (see below under Mercury and graft-versus-host disease) (Mathieson *et al.*, 1992, 1993). However, there was no significant difference in titer of anti-GBM antibodies between OX22-treated animals and controls, both exposed to mercury, suggesting that renal autoimmunity and graft-versus-host-like syndrome may be controlled by different mechanisms. Similarly negative were the results obtained in Lewis rats, that were 'resistant' to the autoimmune effects of mercury (i.e. they did not produce autoantibodies to laminin or experience glomerulonephritis). It was suggested that CD8$^+$ T-lymphocytes might have been responsible for this lack of response to mercury (Pelletier *et al.*, 1990). However, Lewis rats depleted of their CD8$^+$ T-lymphocytes did not develop the autoimmune abnormalities observed after mercury treatment in Brown Norway rats.

Extremely low doses of mercury have been found effective in Brown Norway rats. However, the mechanisms by which this metal stimulates the cells of the immune system to cause autoimmunity are still uncertain. T-lymphocytes play a central role, as shown by the absence of mercury-induced effects in T-lymphocyte-deprived animals. The $\alpha_4\beta_1$-integrin that has essential functions in leukocyte migration is also involved in autoimmunity caused by HgCl$_2$. Brown Norway rats treated with monoclonal antibodies to this integrin had decreased levels of autoantibodies to the renal GBM, did not show immunoglobulins bound to the kidney *in vivo*, and had a drastic reduction in proteinuria (Molina *et al.*, 1995). Interestingly, this treatment did not inhibit the production of antibodies to single-stranded DNA. Various subpopulations of T-lymphocytes are affected by mercury. The RT6$^+$ subset of T-lymphocytes (that has a regulatory role in other rat models of autoimmune disease) decreases in mercury-treated Brown Norway rats. B-lymphocytes may be

141

activated by mercury directly or indirectly through IL-4 production by T-lymphocytes. Another effect of mercury is an increased expression of MHC class II molecules on B-lymphocytes, detected as early as 3 days after the first injection of the metal. The presence of hyper-IgE has suggested that IL-4 could be an important mediator of T-lymphocyte dependent B-lymphocyte activation. T_H2 hyperactivity may be associated with decreased T_H1 functions (e.g. impaired ability to produce IL-2 *in vitro*).

Mice

Mercuric chloride administered to mice of the H-2^s phenotype (A.SW, B10.S, SJL/J, etc.) induces the production of antinucleolar and antinuclear autoantibodies (Eneström and Hultman, 1984; Goter Robinson *et al.*, 1984; Hultman and Eneström, 1986, 1987, 1988, 1989, 1992; Hultman *et al.*, 1989, 1992; Reuter *et al.*, 1989; Saegusa *et al.*, 1990; Hultman and Johansson, 1991). The former react with a 34 kDa nucleolar protein (U3 RNP protein, also defined as 'fibrillarin') and occasionally with other nucleolar proteins of 60 to 70 and 10 to 15 kDa. The latter (antichromatin and/or antihistone antibodies) may occur in a small percentage of H-2^s and H2^k mice after mercury treatment. However, autoantibodies against nucleolar antigens are usually present in higher concentrations. No autoantibodies to kidney, thyroid, or skin have been observed in mercury-treated mice (Hultman *et al.*, 1992). The antifibrillarin autoantibodies seem to recognize the same epitopes as the autoantibodies from certain patients with scleroderma (Reimer, 1990). Thus, exposure to mercury results in a murine model that allows the study of environmental factors in scleroderma and other autoimmune diseases. Antinucleolar autoimmune responses of H-2^s mice last up to 10 to 12 weeks or longer (Hultman *et al.*, 1989). Other effects of mercury include lymph node hyperplasia in H-2^s mice. Very low doses of mercury have been found effective in SJL/J mice, but treatment needed to be more prolonged than with rats.

Recent experiments have shown that athymic SJL/J mice treated with mercury did not produce antinucleolar antibodies and did not experience systemic immune complex deposition (Hultman *et al.*, 1995). A similar lack of mercury-induced autoimmune responses was observed in euthymic SJL/J mice depleted of CD4$^+$ T-lymphocytes. However, when SJL/J mice were treated with mercury and then injected with mAb against CD4$^+$ T-lymphocytes, there was no decline in the titer of antinucleolar antibodies in spite of a severe reduction of CD4$^+$ lymphocytes (Hultman *et al.*, 1995). In conclusion, the induction of autoimmunity by mercury was strictly dependent on the presence of T-lymphocytes, but once autoimmunity was long-standing, it was no longer susceptible to therapy by CD4$^+$ T-lymphocyte depletion. Another recent study has shown that mercury strongly activated CD4$^+$ T-lymphocytes from responder, but not from non-responder, mice (Jiang and Möller, 1996). On the other hand, CD8$^+$ T-lymphocytes were activated by mercury in all strains, irrespective of their responder/non-responder status (Jiang and Möller, 1996).

These observations indirectly confirm previous evidence that had suggested IL-4 as an important mediator of T-lymphocyte-dependent B-lymphocyte activation. Increased levels of IL-4 mRNA have been detected within CD4$^+$ T-lymphocytes of mercury-treated H-2^s mice and treatment with antimouse IL-4 monoclonal antibody reportedly caused a switch in the immunoglobulin subclass of antifibrillarin antibodies (Ochel *et al.*, 1991). Thus, at least in H-2^s mice, T$_H$2 cells secreting IL-4 may be responsible for B-lymphocyte stimulation. T$_H$2 hyperactivity may be associated with decreased T$_H$1 functions (e.g. impaired ability to produce IL-2 *in vitro*). Highly susceptible strains such as A.SW, SJL, and B10.S (H-2^s haplotype) do not develop contact dermatitis in an ear swelling test (a form of delayed-type hypersensitivity reaction); on the other hand, H-2^d mice (resistant or low responders) develop contact dermatitis. Therefore, it was suggested that mercury induces an imbalance between T$_H$1 and T$_H$2 cells (Goldman *et al.*, 1991).

7.4.6   Mercury and graft-versus-host disease (GVHD)

A well-known major complication associated with allogeneic bone marrow transplantation is graft-versus-host disease (Deeg and Cottler-Fox, 1990; Ferrara and Burakoff, 1990; Hess, 1990; Rolink *et al.*, 1990; Shulman, 1990; Snover, 1990; Vossen and Heidt, 1990). Acute GVHD rapidly produces dermatitis (rash and desquamation), hepatitis, gastroenteritis, pancytopenia, aplastic anemia, and hypogammaglobulinemia. The skin presents necrosis of epithelial cells, with vacuolar degeneration of the basal layer, often with a modest lymphoid infiltrate. There is necrosis of gastrointestinal epithelial cells (i.e. 'exploding crypt cell'), the regenerating compartment of the small and large intestine, and the neck region of the stomach. Hepatocellular involvement with hepatitis-like, portal inflammation and infiltration of lymphocytes into biliary epithelium is also observed; involvement of other organs (kidney included) is uncommon.

Chronic GVHD is characterized by persistent lymphoid hyperplasia, hypergammaglobulinemia, production of autoantibodies (usually antinuclear), and pathologic lesions reminiscent of systemic lupus erythematosus (SLE) and other types of collagen diseases. The skin and its appendages are the structures most commonly affected by chronic GVHD, with lymphocytic infiltration along the dermal–epidermal junction, around hair follicles and other dermal appendages. Late changes include epidermal atrophy, decreased elasticity, dermal fibrosis, and loss of hair follicles, and sweat glands. Gastrointestinal involvement is mostly limited to the squamous epithelium of the esophagus and a cholestatic hepatitis. A sicca syndrome and an obstructive airway disease are both characterized by inflammatory infiltrates. Other autoimmune-like manifestations include polymyositis, arthritis, or serositis.

However, involvement of the kidneys and their vessels is controversial. Apart from renal damage caused by irradiation or cyclosporin, direct GVHD effects on the kidney may be quite rare.

Both types of GVHD can be experimentally induced in mice and rats (Hess, 1990; Florquin and Goldman, 1994). In selected parent and F_1 combinations, GVHD induced by injection of parental lymphocytes into adult non-irradiated F_1 hybrid mice results in a chronic autoimmune syndrome resembling human SLE (Rolink *et al.*, 1990). For example, (C57Bl/10 × DBA/2)F_1 mice injected with DBA/2 T-lymphocytes develop high levels of serum immunoglobulins (especially IgG and IgE) and IgG antibodies to nuclear antigens (double-stranded DNA and various nuclear and nucleolar proteins, including histones and fibrillarin), red cells, mitochondria, skin basement membranes, and the brush border of proximal renal tubules. An immune complex-mediated glomerulonephritis occurs in these animals. A similar phenomenon occurs in host-versus-graft disease (HVGD).

Based on clinical similarities between allogeneic diseases and the adverse effects of the anti-convulsant diphenylhydantoin, it was first proposed that a GVHD-like mechanism might account for some adverse reactions induced by xenobiotics (Gleichmann *et al.*, 1984). This concept has later been extended to autoimmunity induced by mercury, gold salts, and penicillamine (Goldman *et al.*, 1991). Interestingly, adoptive transfer of lymphocytes from AO rats into (AO × Brown Norway)F_1 hybrids, pre-injected with complete Freund's adjuvant, results in a GVHD with glomerulopathy that shares several immunopathologic features with $HgCl_2$-induced autoimmunity in rats (Aten *et al.*, 1992a,b). IgM, IgG_1, and IgG_{2a} are bound to the kidney glomerular basement membrane in a linear fashion and, when eluted, they react with renal basement membranes and especially laminin. Cellular infiltrates were found in the liver, tongue, pancreas, and salivary glands of recipients. Similarly, Brown Norway rats injected neonatally with (Brown Norway × Lewis)F_1 hybrid spleen cells exhibit a polyclonal B-lymphocyte activation, with hypergammaglobulinemia (mostly IgE and IgG_1), production of anti-laminin and anti-DNA antibodies, and renal immunoglobulin deposits (Dubey *et al.*, 1992).

Mathieson *et al.* (1992) have reported that $HgCl_2$-treated Brown Norway rats develop inflammation and ulceration of the skin with subepidermal mononuclear cell infiltrate. The livers of these rats are enlarged and show periportal mononuclear cell infiltrates; there are also gross hemorrhagic lesions of the gut, with intense submucosa inflammation and a leukocytoclastic vasculitis. In addition, these rats develop inflammatory polyarthritis that may be mediated by CD8[+] T-lymphocytes (Kiely *et al.*, 1995a,b, 1996). They also produce antibodies to myeloperoxidase (Esnault *et al.*, 1992). Passive transfer of serum from diseased animals did not lead to tissue injury in recipient rats, and intravenous immunoglobulins or depletion of CD4[+] T-lymphocytes did not have significant effects on the systemic vasculitis (Qasim *et al.*, 1993). Another group of investigators has later shown that $HgCl_2$-treated Brown

Norway rats develop a focal adenitis in several different glands (parotid, submandibular, lachrymal, and thyroid) (Peszkowski *et al.*, 1993). The inflammatory process detected in a variety of organs has been interpreted as mimicking chronic GVHD (Mathieson *et al.*, 1992) and confirming the suggestion that a common mechanism may be responsible for allogeneic reactions and the effects of certain xenobiotics (Goldman *et al.*, 1991). Recent studies in mice have shown that $HgCl_2$ and diphenylhydantoin stimulate IgG production to TNP-Ficoll and TNP-OVA, providing additional support to the GVHD hypothesis for chemical-induction of autoimmunity (Albers, 1996).

The evidence briefly summarized above is intriguing and worthy of additional consideration. Older publications from other laboratories never mentioned the occurrence of a GVHD in $HgCl_2$-treated Brown Norway rats, either because the various lesions did not occur or were not detected. Recently, we have carefully examined a variety of tissues from Brown Norway rats injected with mercury and found no inflammatory changes in the thyroid, pancreas, intestine, skin, liver, parotid, or lung (Bigazzi *et al.*, unpublished data). The discrepancy between our observations and the reports by some groups is still unexplained.

7.4.7 *Mercury-induced immunosuppressive effects*

Administration of mercury may result in immunosuppression of autoimmune disease, a phenomenon that is usually reviewed in a cursory and uncritical fashion. Yet it has the potential of providing very useful insights into the immunotoxic mechanisms of this metal. The best example described to date is provided by Lewis rats, which when treated with $HgCl_2$ do not develop experimental autoimmune disease after immunization with organ-specific antigens (Pelletier *et al.*, 1987a,b, 1988a,b). These findings are of great interest because they might indicate new therapeutic modalities for autoimmune disease. Indeed, Druet's group has suggested that 'to seek chemicals that mimic mercury effects in Lewis rats could be an interesting approach for therapeutics of multiple sclerosis in humans. . . . Such treatments could be applied to different autoimmune diseases' (Castedo *et al.*, 1993). More recently, the 'use of T_H2-inducing compounds such as subtoxic doses of mercuric chloride' has been listed as one of possible intervention strategies to reduce development of T_H1 lymphocytes in organ-specific autoimmune diseases (Liblau *et al.*, 1995).

The first demonstration of protective effects was provided in Lewis rats treated with mercury and then immunized with renal brush border antigen gp330 in complete Freund's adjuvant (Pelletier *et al.*, 1987a,b). These animals did not experience Heymann's nephritis, had low antibody responses to gp330, and also showed significantly lower responses to sheep red blood cells. Similarly, it was shown that mercury attenuated or even prevented the clinical manifestations of experimental allergic encephalomyelitis in Lewis rats and inhibited both myelin basic protein (MBP)-induced proliferative responses of

T-lymphocytes and antibody responses to MBP. This activity of $HgCl_2$ was explained as due to a non-antigen-specific immunosuppression mediated by an increase in the number of $CD8^+$ T-lymphocytes, at that time defined as 'suppressor/cytotoxic cells' (Pelletier *et al.*, 1987a,b). $CD8^+$ T-lymphocyte depletion reportedly abrogated this suppression and eliminated the protective effects of mercury on experimental allergic encephalomyelitis (Pelletier *et al.*, 1990).

However, later flow cytometry (FCM) studies of lymphocyte subpopulations of $HgCl_2$-treated Lewis rats have not confirmed a significant increase in percentage and/or number of $CD8^+$ T-lymphocytes (Kosuda *et al.*, 1993). Alternative explanations of the protective effects have been proposed, such as some interference with the immunization process, suggested by the observation that Brown Norway rats immunized with MBP and treated with mercury were protected against experimental allergic encephalomyelitis (Levine and Saltzman, 1988). Thus, pretreatment of rats with $HgCl_2$ before immunization with auto-antigens in adjuvants might actually cause changes in the draining lymph nodes or somehow affect the 'afferent' phase (possibly autoantigen presentation) of the experimentally induced autoimmune response.

As a second possibility, the prolonged mercury treatment used in the protocols of Druet *et al.* (1977) may have resulted in toxic effects capable of generating a non-specific suppression. In fact, the number of $HgCl_2$ injections administered to Lewis rats in the original protection studies was much higher than the routine treatment used to induce autoimmunity in Brown Norway rats. Inhibition of experimental allergic encephalomyelitis was obtained in animals that had received a total of 18 to 30 injections in a period of 6 to 8 weeks, for a cumulative dose of 1.8 to $3\,mg\,HgCl_2/100\,g$ body weight of the rat (compared with the usual $600\,\mu g\,HgCl_2/100\,g$ body weight in Brown Norway rats) (Pelletier *et al.*, 1988a,b). $HgCl_2$-induced down-modulation of Heymann's nephritis was also obtained in Lewis rats injected with $HgCl_2$ for 8 weeks (Pelletier *et al.*, 1987a,b). Such protracted exposure to the metal might have toxic effects, which could be interpreted as non-specific immunosuppression. However, the length of mercury exposure does not seem to be necessarily associated with the immunosuppressive effects of this metal. In later studies, Druet *et al.* (1977) have demonstrated that after 14 days of $HgCl_2$ treatment, T-lymphocytes from Lewis rats did not proliferate *in vitro* in response to ConA, whereas proliferation was obtained after elimination of $CD8^+$ T-lymphocytes (Pelletier *et al.*, 1990). In addition, $CD8^+$ T-lymphocytes capable of reacting with $CD4^+$ lymphocytes have also been detected in Lewis rats injected with $HgCl_2$ for 7 days (Castedo *et al.*, 1993).

Suppressor and helper T-lymphocytes were titrated in mercury-treated Lewis rats (Pelletier *et al.*, 1991; Rossert *et al.*, 1991). A high frequency of $CD8^+$ T-suppressor lymphocytes and a 10-fold lower frequency of MBP-specific, T-helper lymphocytes were detected in these rats. In addition, a third cell type allowing the proliferative response of T-helper cells in spite of the presence of

T-suppressor cells and interpreted as 'contrasuppressor cells' was detected. Additional investigations by the same groups showed that HgCl$_2$ induced autoreactive anti-class II CD4$^+$ T-lymphocytes in both Brown Norway and Lewis rats. T-lymphocyte lines derived from Lewis CD4$^+$ T-lymphocytes of this type proliferated in the presence of normal class II-bearing cells, secreted IL-2, and did not induce B-lymphocytes to produce immunoglobulins. Transfer of one of these lines into normal Lewis rats led to the appearance of CD8$^+$ T-lymphocytes responsible for a non-antigen-specific immunosuppression that induced complete protection from experimental allergic encephalomyelitis. In this case protection was eliminated after treatment with an mAb to CD8$^+$ T-lymphocytes. These cells were defined as 'anti-ergotypic' following Irun Cohen's terminology (Lohse *et al.*, 1989). The interpretation of these findings was in agreement with the prevailing paradigms of that period; however, more recent investigations have questioned the regulatory role of CD8$^+$ T-lymphocytes. In other rat models of autoimmunity, such as insulin-dependent diabetes mellitus of DP BB rats (Mordes *et al.*, 1987; Crisá *et al.*, 1992; Whalen *et al.*, 1994) and irradiated, thymectomized PVG rats (Fowell and Mason, 1993), it is a CD4$^+$, not a CD8$^+$, T-lymphocyte that directly prevents autoimmunity.

Therefore, recent findings obtained in rats with experimental autoimmune uveoretinitis are of great interest. This ocular autoimmune disease can be experimentally induced by immunization with the soluble retinal S-antigen. Rats immunized with the S-antigen also develop an autoimmune pinealitis. F$_1$ hybrids between Brown Norway and Lewis rats, when treated with mercury, were protected against the development of experimental autoimmune uveoretinitis or pinealitis (Saoudi *et al.*, 1991). HgCl$_2$ injections, started 1 week before experimental immunization, prevented clinical and histological manifestations of autoimmune uveoretinitis and moreover protected against pinealitis. Circulating antibodies to S-antigen were detected in both HgCl$_2$-treated and water-treated controls, that is, in this case of mercury-induced protection, there was no apparent suppression of antibody production. The inhibition of autoimmune uveoretinitis was not mediated by CD8$^+$ T-lymphocytes, as indicated by depletion studies.

Lymph node cells from protected rats had a considerably reduced ability to adoptively transfer autoimmune uveoretinitis to naive rats (Saoudi *et al.*, 1993). Those cells did not produce significant levels of IL-2 and IFN-γ, but had high levels of IL-4 mRNA. T-lymphocytes from animals with autoimmune uveoretinitis and capable of adoptively transferring the disease did produce high levels of IL-2 and IFN-γ and had lower levels of IL-4 mRNA (Saoudi *et al.*, 1993). The IgG isotype profile of antibodies to S-antigen showed that rats immunized with S-antigen and mercury treated had higher levels of IgG$_1$ (an isotype probably driven by T$_H$2 lymphocytes), whereas similarly immunized water-treated rats produced IgG$_{2a}$ and IgG$_{2b}$ antibodies. The IgG$_{2b}$ isotype is probably T$_H$1 dependent in the rat (Saoudi *et al.*, 1993). In conclusion, the protection effect was interpreted as demonstrating an activation of CD4$^+$ T$_H$2 lymphocytes, capable of protecting from a T_H1-mediated autoimmune disease. The authors concluded that 'the development of drugs that selectively activate a given T_H subset might therefore be of value in influencing the other T_H subset'. Manipulations of the T_H1/T_H2 lymphocyte balance have indeed been suggested by various investigators as an approach to treat human autoimmune disease (Liblau *et al.*, 1995; Adorini *et al.*, 1996; Nicholson and Kuchroo, 1996; Röcken *et al.*, 1996).

However, the protective effects of mercury have been observed only in *experimentally* induced autoimmune diseases such as Heymann's nephritis, encephalomyelitis and uveoretinitis. In addition, they were noted only in one strain of rats (Lewis or the F_1 hybrids between Lewis and Brown Norway rats). Until recently, there were no reports of protection in other rat strains against *naturally occurring* (so-called 'spontaneous') autoimmune disease. For this reason, we have investigated the possible immunosuppressive effects of $HgCl_2$ in insulin-dependent diabetes mellitus and thyroiditis of BB rats (Kosuda *et al.*, 1997). The spontaneous expression of autoimmunity in BB rats appears to be determined largely by the relative balance of autoreactive effector T-lymphocytes and regulatory T-lymphocytes (Rossini *et al.*, 1993). This regulatory/effector T-lymphocyte balance may be characterized by a predominance of T_H1-type lymphocytes, since recent studies of cytokine message expression in islets and thyroids of DP and RT6-depleted DR rats have detected the presence of IFN-γ and IL-12 mRNA, suggesting a T_H1-type inflammatory response (Zipris *et al.*, 1996).

Our results show that the administration of mercury had no inhibitory effects on the development and progression of insulitis and clinical insulin-dependent diabetes mellitus in DP BB rats. Similarly, the incidence of thyroiditis (another autoimmune disease spontaneously occurring in these rats) was not affected. On the other hand, long-term exposure of BB rats to mercury stimulated autoantibody responses to thyroglobulin and laminin. These effects confirm previous findings in $(NZB \times NZW)F_1$ hybrid mice that after treatment with mercury developed a polyclonal B-lymphocyte cell activation, with high levels of IgG_1, IgG_3, and IgE immunoglobulins, autoantibodies of different specificities, and deposition of immune complexes in the kidneys (Al-Balaghi *et al.*, 1996). Autoantibodies to laminin produced by BB rats were most commonly of the IgG_1 and IgG_{2a} subclasses. However, a considerable number of sera also contained antibodies of the IgG_{2b} and IgG_{2c} isotypes. To date, there is little information on the regulation of IgG isotype expression in rats. However, it has been suggested that rat IgG_{2a} (similarly to the stimulation of mouse IgG_1) may be induced by IL-4 (Benbernou *et al.*, 1993), whereas IgG_{2b} may be driven by IFN-γ and/or IL-2 (Saoudi *et al.*, 1993). It has also been hypothesized that mercury stimulates the production of T_H2-type cytokines in Brown Norway rats and T_H1-type cytokines in Lewis rats (Dubey *et al.*, 1991; Goldman *et al.*, 1991; Pelletier *et al.*, 1992). Our findings suggest that autoantibody production in BB rats may be activated by both T_H1- and T_H2-type lymphocytes (or an alternative type). As suggested by the IgG isotype profile of antibodies to laminin, exposure to mercury may stimulate cytokine production but does not seem to change this pattern of response in BB rats. Thus, the effects of mercury in BB rats may not conform to the schematic division between type 1 and type 2 cytokine activity. Other recent investigations have raised questions about overly simplistic interpretations of the 'T_H1/T_H2 paradigm' (Nickerson *et al.*, 1994; Kelso, 1995; Aebischer and Stadler, 1996; Carter and Dutton, 1996; Strom *et al.*, 1996).

In summary, the immunosuppressive effects of $HgCl_2$ may be a unique strain-dependent characteristic of the Lewis rat strain. This is not surprising, since we already know that the autoimmune effects of mercury are genetically controlled and vary with the inbred strain of rats and mice (Bigazzi, 1992). The mercury-induced $CD8^+$ T-lymphocyte-mediated immunosuppressive effects reported in Lewis rats and the $CD4^+$ T_H2-mediated effects of (Lewis × Brown Norway)F_1 hybrids are obviously dependent on distinct subsets of immunoregulatory cells, able to modulate in different ways the activity of the autoreactive cell population. In contrast, mercury can induce, but not inhibit, autoimmunity in Brown Norway and BB rats that are deficient in certain immunoregulatory cells and are susceptible to either experimental or spontaneous autoimmune disease. Thus, the utilization of chemicals that mimic mercury effects as therapeutic immunosuppressive agents of animal models of autoimmune disease is probably dependent on the immunomodulatory mechanisms of the various rat or mouse strains and may actually help in dissecting the variety of possible immunoregulatory networks. On the other hand, the extension of such treatment to human autoimmune disease is clearly more difficult and problematic.

7.5 The 'Mercury Paradigm'

The toxic effects of mercury compounds on central and peripheral nervous system, kidneys, and other organs have been well known for a long time. Less appreciated has been the activity of mercury on the various components of the immune system. As for other xenobiotics, immunotoxicity of mercury often occurs at extremely low doses (much lower than for other organs and tissues) and, therefore, may be a more subtle and insidious phenomenon. As briefly summarized in this review, mercury acts, directly or indirectly, on T- and B-lymphocytes, macrophages, PMN, and mast cells. These effects have been well demonstrated both *in vitro* and *in vivo*, but are still poorly understood.

There are many ways in which mercury can interact with T-lymphocytes (reviewed by Griem and Gleichmann, 1995). The first is independent of antigen and might be similar to the activity of conventional mitogens. Early studies showed that oxidation or reduction of thiol or disulfide groups modulates murine T-lymphocyte proliferation. Mercury may dimerize the thiol group-bearing proteins on the cell membrane and induce tyrosine phosphorylation and cell death (Rahman *et al.*, 1993). *In vitro* treatment with $HgCl_2$ of spleen cells or thymocytes from C57Bl/6 mice resulted in aggregation of trans-membrane CD4, CD3, CD45, and Thy-1 (Nakashima *et al.*, 1994). The non-receptor protein tyrosine kinase p56^{lck} was also aggregated and activated, with extensive protein tyrosine phosphorylation. These phenomena were observed at a high concentration of HgCl$_2$ (0.5 mM) that caused progressive cell death within 5 to 20 min; lower concentrations did not appear to produce extensive aggregation of surface antigens or p56^{lck}. On the other hand, treatment with 0.01 and 0.001 mM HgCl$_2$ resulted in prolonged DNA synthesis and increased IL-2 release. These findings were interpreted as suggesting a novel mechanism of disregulation of the ligand-dependent receptor function (Nakashima *et al.*, 1994), but should be carefully re-examined at concentrations that are not rapidly cytotoxic.

Methylmercury has also been shown to cause a rapid increase in intracellular free Ca^{2+} levels in rat splenocytes, whereas HgCl$_2$ increases Ca^{2+} influx from extracellular sources (Tan *et al.*, 1993). In addition to its mitogen-like activity, mercury may alter self-antigens or form metal–protein complexes leading to immune responses to new or cryptic epitopes. For example, Kubicka-Muranyi *et al.* (1995) have shown that some CD4$^+$ T-lymphocyte hybridomas obtained from HgCl$_2$-treated mice specifically responded to fibrillarin bound to mercury, whereas others reacted against untreated fibrillarin. Finally, mercury might behave as a hapten and sensitize T-lymphocytes when bound to proteins.

Mechanisms of mercury interaction with T-lymphocytes are not mutually exclusive. In addition, they do not restrict other activities, such as inhibition of certain immunoregulatory cells or stimulation of certain cytokines. Thus, mercury effects on B-lymphocytes, macrophages, and mast cells, for example, may have as much relevance as its activity on T-lymphocytes. Increased production of IL-4 has been noted in both T-lymphocytes and mast cells from Brown Norway rats. Other cytokines, such, IL-12 and IL-1, are usually produced by macrophages, keratinocytes, B-lymphocytes, and other non-T-lymphocytes. Higher gene expression of IL-12 has been reported in spleen cells of Lewis rats. *In vitro* treatment with HgCl$_2$ enhances secretion of IL-1 by murine macrophages. We are just beginning to appreciate the possibility that the induction of mercury immunotoxicity may be dependent on a variety of antigen-presenting cells and their products, leading to different types of polarization of T-lymphocytes.

Thus, we started this review with an evocation of the multi-talented ancient Gods associated with mercury and conclude it with an increased recognition of the diverse immunotoxic activities of this metal. The initial question was: 'Is mercury the God of T$_H$2 cells?' The answer seems to be a definite 'yes', but then we must add 'and also the god of macrophages, mast cells, B-lymphocytes, NK cells, and possibly T$_H$1 and CD8'. Assuming the distinction between T$_H$1 and T$_H$2 is still valid, it would be too limiting to attribute to mercury only an effect on T$_H$2 lymphocytes. Depending on the genetic makeup of the animal exposed to mercury, T$_H$2 and/or T$_H$1 stimulation or, better, induction of type 1 and type 2 cytokines (Carter and Dutton, 1996), may occur. A prevalence of one or the other is possible, again based on immunogenetic and pharmacogenetic differences.

What about mercury's effects on the human immune system? A distinguished immunotoxicologist once told me that too much effort had been spent studying the effects of mercury on the immune system, considering the sparse evidence of mercury-induced immunopathology in humans. I do not completely agree with that statement. It is true that to date we do not see many cases of human autoimmune disease induced by mercury. In addition, several recent studies of workers previously exposed to mercury have not detected persistent immunological consequences (Ellingsen *et al.*, 1994). However, one cannot discount the occasional patient with autoimmune responses and/or disease (Lindqvist *et al.*, 1974; Röger *et al.*, 1992; Schrallhammer-Benkler *et al.*, 1992) and the enormous amount of evidence obtained from experimental animals, showing the diverse and polymorphic pathologic consequences of mercury exposure. It is clear that mercury does act on the immune system of rabbits, mice, and rats. *In vitro* evidence and occasional *in vivo* findings indicate that the human immune system is also susceptible to mercury's activities. The current paucity of mercury-induced human pathology may just be related to lack of sufficient exposure or the bias of individual susceptibility traits.

In conclusion, because of its rather unique range of immunotoxic properties, mercury should be considered as a 'paradigm' (Kuhn, 1970) of xenobiotic immunotoxicity. *In vitro* and *in vivo* studies of the pleiotropic immunotoxic effects of this metal can provide extremely valuable information that can be applied to the possible mechanisms of action of other xenobiotics, whether they be suppressive or stimulatory of the various components of the immune system.

Acknowledgments

Our research on mercury-induced autoimmunity is supported by USPHS Grant ES03230.

References

ADORINI, L., GUÉRY, J.C., and TREMBLEAU, S. (1996). Manipulation of the T_H1/T_H2 cell balance: an approach to treat human autoimmune diseases? *Autoimmunity* **23**, 53–68.

AEBISCHER, I. and STADLER, B.M. (1996). T_H1-T_H2 cells in allergic responses: at the limits of a concept. *Adv. Immunol.* **61**, 341–403.

AL-BALAGHI, S., MÖLLER, E., MÖLLER, G., and ABEDI-VALUGERDI, M. (1996). Mercury induces polyclonal B-cell activation, autoantibody production and renal immune complex deposits in young (NZB × NZW)F$_1$ hybrids. *Eur. J. Immunol.* **26**, 1519–1526.

ALBERS, R. (1996). *Chemical-induced autoimmunity. Immune disregulation assessed with reporter antigens.* Unpublished PhD thesis, Utrecht University, The Netherlands.

ATEN, J., STET, R.J.M., WAGENAAR-HILBERS, J.P., WEENING, J.J., FLEUREN, G.J., and NIEUWENHUIS, P. (1992a). Glomerulopathy induced by graft-versus-host reaction in the rat. Requirement of donor CD4[+] T-lymphocytes and MHC class II incompatibility at the lymphoid compartment. *Scand. J. Immunol.* **35**, 93–105.

ATEN, J., VENINGA, A., BRUIJN, J.A., PRINS, F.A., DE HEER, E., and WEENING, J.J. (1992b). Antigenic specificities of glomerular-bound autoantibodies in membranous glomerulopathy induced by mercuric chloride. *Clin. Immunol. Immunopathol.* **63**, 89–102.

ATEN, J., PRIGENT, P., PONCET, P., BLANPIED, C., CLAESSEN, N., DRUET, P., and HIRSCH, F. (1995a). Mercuric chloride-induced programmed cell death of a murine hybridoma. I. Effect of the proto-oncogene Bcl-2. *Cell. Immunol.* **161**, 98–106.

ATEN, J., VENINGA, A., COERS, W., SONNENBERG, A., TIMPL, R., CLAESSEN, N., VAN EENDENBURG, J.D.H., DE HEER, E., and WEENING, J.J. (1995b). Autoantibodies to the laminin P1 fragment in $HgCl_2$-induced membranous glomerulopathy. *Am. J. Pathol.* **146**, 1467–1480.

BAGINSKI, B. (1988). Effect of mercuric chloride on microbicidal activities of human polymorphonuclear leukocytes. *Toxicology* **50**, 247–256.

BARAN, D., LANTZ, O., DOSQUET, P., SFAKSI, A., and DRUET, P. (1988). Interleukin-2 production in Brown-Norway rats with $HgCl_2$-induced autoimmune disease: paradoxical *in vivo* versus *in vitro* findings. *Clin. Exp. Immunol.* **73**, 401–405.

BENBERNOU, N., MATSIOTA-BERNARD, P., and GUENOUNOU, M. (1993). Antisense oligonucleotides to interleukin-4 regulate IgE and IgG_{2a} production by spleen cells from *Nippostrongylus brasiliensis*-infected rats. *Eur. J. Immunol.* **23**, 659–663.

BIGAZZI, P.E. (1988). Autoimmunity induced by chemicals. *J. Toxicol. Clin. Toxicol.* **26**, 125–156.

(1992). Lessons from animal models: the scope of mercury-induced autoimmunity. *Clin. Immunol. Immunopathol.* **65**, 81–84.

(1994). Autoimmunity and heavy metals. *Lupus* **3**, 449–453.

(1996). Autoimmunity induced by metals, in Chang, W. (Ed.) *Toxicology of Metals*, pp. 835–852, Boca Raton, FL: CRC Lewis Publishers.

BIGAZZI, P.E., MICHAELSON, J.H., and POTTER, N.T. (1989). Epibodies in autoimmunity: antisera against autoantibodies to the renal glomerular basement membrane react with idiotypes as well as with autoantigens. *Autoimmunity* **5**, 3–16.

CARON, G.A., POUTALA, S., and PROVOST, T.T. (1970). Lymphocyte transformation induced by inorganic and organic mercury. *Int. Arch. Allergy* **37**, 76–87.

CARTER, L. and DUTTON, R.W. (1996). Type 1 and type 2: a fundamental dichotomy for all T-cell subsets. *Curr. Opin. Immunol.* **8**, 336–342.

CASTEDO, M., PELLETIER, L., ROSSERT, J., PASQUIER, R., VILLAROYA, H., and DRUET, P. (1993). Mercury-induced autoreactive anti-Class II T-cell line protects from experimental autoimmune encephalomyelitis by the bias of CD8[+] antiergotypic cells in Lewis rats. *J. Exp. Med.* **177**, 881–889.

CHANG, L.W. (1996). Toxico-neurology and neuropathology induced by metals, in Chang, L.W. (Ed.) *Toxicology of Metals*, pp. 511–535, Boca Raton, FL: CRC Lewis Publishers.

CHRISTENSEN, M.M., ELLERMANN-ERIKSEN, S., RUNGBY, J., and MOGENSEN, S.C. (1996). Influence of mercuric chloride on resistance to generalized infection with herpes simplex virus type 2 in mice. *Toxicology* **114**, 57–66.

CONTRINO, J., KOSUDA, L.L., MARUCHA, P., KREUTZER, D.L., and BIGAZZI, P.E. (1992). The *in vitro* effects of mercury on peritoneal leukocytes (PMN and macrophages) from inbred Brown Norway and Lewis rats. *Int. J. Immunopharmacol.* **14**, 1051–1059.

CONTRINO, J., MARUCHA, P., RIBAUDO, R., FERENCE, R., BIGAZZI, P.E., and KREUTZER, D.L. (1988). Effects of mercury on human polymorphonuclear leukocyte function *in vitro*. *Am. J. Pathol.* **132**, 110–118.

CORY-SLECHTA, D.A. (1996). Comparative neurobehavioral toxicology of heavy metals, in Chang, L.W. (Ed.) *Toxicology of Metals*, pp. 537–560, Boca Raton, FL: CRC Lewis Publishers.

CRISÁ, L., MORDES, J.P., and ROSSINI, A.A. (1992). Autoimmune diabetes mellitus in the BB rat. *Diabetes/Metabol. Rev.* **8**, 9–37.

DAUM, J.R., SHEPHERD, D.M., and NOELLE, R.J. (1993). Immunotoxicology of cadmium and mercury on B-lymphocytes. I. Effects on lymphocyte function. *Int. J. Immunopharmacol.* **15**, 383–394.

DEEG, H.J. and COTTLER-FOX, M. (1990). Clinical spectrum and pathophysiology of acute graft-vs-host disease, in Burakoff, S.J., Deeg, H.J., Ferrara, J., and Atkinson, K. (Eds) *Graft-vs-Host Disease*, pp. 311–335, New York: Marcel Dekker.

DRUET, E., SAPIN, C., GÜNTHER, E., FEINGOLD, N., and DRUET, P. (1977). Mercuric chloride-induced anti-glomerular basement membrane antibodies in the rat. Genetic control. *Eur. J. Immunol.* **7**, 348–351.

DUBEY, C., BELLON, B., and DRUET, P. (1991). T_H1- and T_H2-dependent cytokines in experimental autoimmunity and immune reactions induced by chemicals. *Eur. Cytokine Network* **2**, 147–152.

DUBEY, C., KUHN, J., WISSING, M., NISOL, F., CHAVEZ, M., BAZIN, H., GOLDMAN, M., DRUET, P., and BELLON, B. (1992). Susceptibility and resistance to autoimmunity following neonatal injection of semi-allogeneic spleen cells in rats. *J. Autoimmun.* **5**, 629–640.

ELEY, B.M. and COX, S.W. (1993). The release, absorption and possible health effects of mercury from dental amalgam: a review of recent findings. *Br. Dent. J.* **175**, 355–362.

ELLERMANN-ERIKSEN, S., CHRISTENSEN, M.M., and MOGENSEN, S.C. (1994). Effect of mercuric chloride on macrophage-mediated resistance mechanisms against infection with herpes simplex virus type 2. *Toxicology* **93**, 269–287.

ELLINGSEN, D.G., GAARDER, P.I., and KJUUS, H. (1994). An immunological study of chloralkali workers previously exposed to mercury vapor. *Acta Pathol. Microbiol. Immunol. Scand.* **102**, 170–176.

ENESTRÖM, S. and HULTMAN, P. (1984). Immune-mediated glomerulonephritis induced by mercuric chloride in mice. *Experientia* **40**, 1234–1240.

ESNAULT, V.L., MATHIESON, P.W., THIRU, S., OLIVEIRA, D.B., and LOCKWOOD, C.M. (1992). Autoantibodies to myeloperoxidase in Brown Norway rats treated with mercuric chloride. *Lab. Invest.* **67**, 114–120.

EXON, J.H., SOUTH, E.H., and HENDRIX, K. (1996). Effects of metals on the humoral immune response, in Chang, L.W. (Ed.) *Toxicology of Metals*, pp. 797–810, Boca Raton, FL: CRC Lewis Publishers.

FERRARA, J. and BURAKOFF, S.J. (1990). The pathophysiology of acute graft-vs-host disease in a murine bone marrow transplant model, in Burakoff, S.J., Deeg, H.J., Ferrara, J., and Atkinson, K. (Eds) *Graft-vs-Host Disease*, pp. 9–29, New York: Marcel Dekker.

FLORQUIN, S. and GOLDMAN, M. (1994). Allogeneic diseases, in Cohen, I.R. and Miller, A. (Eds) *Autoimmune Disease Models. A Guidebook*, pp. 291–301, San Diego: Academic Press.

FOWDEN, G. (1993). *The Egyptian Hermes. A Historical Approach to the Late Pagan Mind*, Princeton, NJ: Princeton University Press.

FOWELL, D. and MASON, D. (1993). Evidence that the T-cell repertoire of normal rats contains cells with the potential to cause diabetes. Characterization of the $CD4^+$ T-cell subset that inhibits this autoimmune potential. *J. Exp. Med.* **177**, 627–636.

FOWLER, B.A. (1996). The nephropathology of metals, in Chang, L.W. (Ed.) *Toxicology of Metals*, pp. 721–729, Boca Raton, FL: CRC Lewis Publishers.

FUKATSU, A., BRENTJENS, J.R., KILLEN, P.D., KLEINMAN, H.K., MARTIN, G.R., and ANDRES, G.A. (1987). Studies on the formation of glomerular immune deposits in Brown Norway rats injected with mercuric chloride. *Clin. Immunol. Immunopathol.* **45**, 35–47.

FUNG, Y.K. and MOLVAR, M.P. (1992). Toxicity of mercury from dental environment and from amalgam restorations. *J. Toxicol. Clin. Toxicol.* **30**, 49–61.

GERHARDSSON, L. and SKERFVING, S. (1996). Concepts on biological markers and biomonitoring for metal toxicity, in Chang, L.W. (Ed.) *Toxicology of Metals*, pp. 81–107, Boca Raton, FL: CRC Lewis Publishers.

GLEICHMANN, E., PALS, S.T., ROLINK, A.G., RADASZKIEWICZ, T., and GLEICHMANN, H. (1984). Graft-versus-host reactions: clues to the etiopathology of a spectrum of immunological diseases. *Immunol. Today* **5**, 324–332.

GOLDMAN, M., DRUET, P., and GLEICHMANN, E. (1991). T_H2 cells in systemic autoimmunity: insights from allogeneic diseases and chemically-induced autoimmunity. *Immunol. Today* **12**, 223–227.

GOTER ROBINSON, C.J., ABRAHAM, A.A., and BALASZ, T. (1984). Induction of anti-nuclear antibodies by mercuric chloride in mice. *Clin. Exp. Immunol.* **58**, 300–306.

GOYER, R.A. and CHERIAN, M.G. (1995). Renal effects of metals, in Goyer, R.A., Klaassen, C.D., and Waalkes, M.P. (Eds) *Metal Toxicology*, pp. 389–412, San Diego: Academic Press.

GRIEM, P. and GLEICHMANN, E. (1995). Metal ion induced autoimmunity. *Curr. Opinion Immunol.* **7**, 831–838.

HAMADA, R. and OSAME, M. (1996). Minamata disease and other mercury syndromes, in Chang, L.W. (Ed.) *Toxicology of Metals*, pp. 337–351, Boca Raton, FL: CRC Lewis Publishers.

HAMMOND, C.R. (1970). The elements, in Weast, R.C. (Ed.) *Handbook of Chemistry and Physics*, pp. B4–B38, Cleveland, OH: The Chemical Rubber Co.

HENRY, G.A., JARNOT, B.M., STEINHOFF, M.M., and BIGAZZI, P.E. (1988). Mercury-induced renal autoimmunity in the MAXX rat. *Clin. Immunol. Immunopathol.* **49**, 187–203.

HEPTINSTALL, R.H. (1983). The nephrotic syndrome, in Hepinstall, R.H. (Ed.) *Pathology of the Kidney*, pp. 637–740, Boston: Little, Brown and Company.

HESS, A.D. (1990). Syngeneic graft-vs-host disease, in Burakoff, S.J., Deeg, H.J., Ferrara, J., and Atkinson, K. (Eds) *Graft-vs-Host Disease*, pp. 95–107, New York: Marcel Dekker.

HULTMAN, P. and ENESTRÖM, S. (1986). Localization of mercury in the kidney during experimental acute tubular necrosis studied by the cytochemical silver amplification method. *Br. J. Exp. Pathol.* **67**, 493–503.

(1987). The induction of immune complex deposits in mice by peroral and parenteral administration of mercuric chloride: strain dependent susceptibility. *Clin. Exp. Immunol.* **67**, 283–292.

(1988). Mercury induced antinuclear antibodies in mice: characterization and correlation with renal immune complex deposits. *Clin. Exp. Immunol.* **71**, 269–274.

(1989). Mercury induced B-cell activation and antinuclear antibodies in mice. *J. Clin. Lab. Immunol.* **28**, 143–150.

(1992). Dose–response studies in murine mercury-induced autoimmunity and immune-complex disease. *Toxicol. Appl. Pharmacol.* **113**, 199–208.

HULTMAN, P. and JOHANSSON, U. (1991). Strain differences in the effect of mercury on murine cell-mediated immune reactions. *Food Chem. Toxicol.* **29**, 633–638.

HULTMAN, P., ENESTRÖM, S., POLLARD, K.M., and TAN, E.M. (1989). Anti-fibrillarin autoantibodies in mercury-treated mice. *Clin. Exp. Immunol.* **78**, 470–477.

HULTMAN, P., BELL, L.J., ENESTRÖM, S., and POLLARD, K.M. (1992). Murine susceptibility to mercury. I. Autoantibody profiles and systemic immune deposits in inbred, congenic and intra-H-2 recombinant strains. *Clin. Immunol. Immunopathol.* **65**, 98–109.

HULTMAN, P., JOHANSSON, U., and DAGNAES-HANSEN, F. (1995). Murine mercury-induced autoimmunity: the role of T-helper cells. *J. Autoimmun.* **8**, 809–823.

HUTCHINSON, F., MACLEOD, T.M., and RAFFLE, E.J. (1976). Leucocyte aggregation and lymphocyte transformation induced by mercuric chloride. *Clin. Exp. Immunol.* **26**, 531–533.

ILBACK, N.G., SUNDBERG, J., and OSKARSSON, A. (1991). Methyl mercury exposure via placenta and milk impairs natural killer (NK) cell function in newborn rats. *Toxicol Lett.* **58**, 149–158.

JIANG, Y. and MÖLLER, G. (1995). *In vitro* effects of $HgCl_2$ on murine lymphocytes. I. Preferable activation of $CD4^+$ T-cells in a responder strain. *J. Immunol.* **154**, 3138–3146.

(1996). Unresponsiveness of $CD4^+$ T-cells from a non-responder strain to $HgCl_2$ is not due to $CD8^+$-mediated immunosuppression: an analysis of the very early activation antigen CD69. *Scand. J. Immunol.* **44**, 565–570.

JOKSTAD, A., THOMASSEN, Y., BYE, E., CLENCHAAS, J., and AASETH, J. (1992). Dental amalgam and mercury. *Pharmacol. Toxicol.* **70**, 308–313.

KELSO, A. (1995). T_{H1} and T_{H2} subsets: paradigms lost? *Immunol. Today* **16**, 374–379.

KIELY, P.D., GILLESPIE, K.M., and OLIVEIRA, D.B. (1995a). Oxpentifylline inhibits tumor necrosis factor-α mRNA transcription and protects against arthritis in mercuric chloride-treated Brown Norway rats. *Eur. J. Immunol.* **25**, 2899–2906.

KIELY, P.D., THIRU, S., and OLIVEIRA, D.B.G. (1995b). Inflammatory polyarthritis induced by mercuric chloride in the Brown Norway rat. *Lab. Invest.* **73**, 284–293.

KIELY, P.D., O'BRIEN, D., and OLIVEIRA, D.B. (1996). Anti-CD8 treatment reduces the severity of inflammatory arthritis, but not vasculitis, in mercuric chloride-induced autoimmunity. *Clin. Exp. Immunol.* **106**, 280–285.

KIMBER, I. and BASKETTER, D.A. (1996). Contact hypersensitivity to metals, in Chang, L.W. (Ed.) *Toxicology of Metals*, pp. 827–833, Boca Raton, FL: CRC Lewis Publishers.

KOSUDA, L.L. and BIGAZZI, P.E. (1996). Chemical-induced autoimmunity, in Smialowicz, R.J., and Holsapple, M.P. (Eds) *Experimental Immunotoxicology*, pp. 419–465, Boca Raton, FL: CRC Press.

KOSUDA, L.L., WAYNE, A., NAHOUNOU, M., GREINER, D.L., and BIGAZZI, P.E. (1991). Reduction of the RT6.2$^+$ subset of T-lymphocytes in Brown Norway rats with mercury-induced renal autoimmunity. *Cell. Immunol.* **135**, 154–167.

KOSUDA, L.L., GREINER, D.L., and BIGAZZI, P.E. (1993). Mercury-induced renal autoimmunity. Changes in RT6$^+$ T-lymphocytes of 'susceptible' and 'resistant' rats. *Environ. Health Perspect.* **101**, 178–185.

(1997). Effects of HgCl$_2$ on the expression of autoimmune responses and disease in diabetes-prone (DP) BB rats. *Autoimmunity* (in press).

KUBICKA-MURANYI, M., GRIEM, P., LÖBBEN, B., ROTTMAN, N., LÜHRMANN, R., BEYER, K., and GLEICHMANN, E. (1995). Mercuric chloride-induced auto-immunity in mice involves an upregulated presentation of altered and unaltered nucleolar self antigen. *Int. Arch. Allergy Immunol.* **108**, 1–10.

KUHN, T.S. (1970). *The Structure of Scientific Revolutions.* Chicago: University of Chicago Press.

LANGWORTH, S., ELINDER, C.G., and SUNDQUIST, K.G. (1993). Minor effects of low exposure to inorganic mercury on the human immune system. *Scand. J. Work Environ. Health* **19**, 405–413.

LAWRENCE, D.A. and McCABE, M.J., JR (1995). Immune modulation by toxic metals, in Goyer, R.A., Klaassen, C.D., and Waalkes, M.P. (Eds) *Metal Toxicology*, pp. 305–337, San Diego: Academic Press.

LEVINE, S. and SALTZMAN, A. (1988). Suppression of experimental allergic encephalomyelitis in rats by mercuric chloride. *Chem.-Biol. Interact.* **69**, 17–21.

LIBLAU, R.S., SINGER, S.M., and McDEVITT, H.O. (1995). T$_{H1}$ and T$_{H2}$ CD4$^+$ T-cells in the pathogenesis of organ-specific autoimmune diseases. *Immunol. Today* **16**, 34–38.

LINDQVIST, K.J., MAKENE, W.J., SHABA, J.K., and NANTULYA, V. (1974). Immunofluorescence and electron microscopic studies of kidney biopsies from patients with nephrotic syndrome, possibly induced by skin lightening creams containing mercury. *E. Afr. Med. J.* **51**, 168–169.

LOHSE, A.W., MOR, F., KARIN, N., and COHEN, I.R. (1989). Control of experimental autoimmune encephalomyelitis by T-cells responding to activated T-cells. *Science* **244**, 820–822.

MAJNO, G. (1975). *The Healing Hand. Man and Wound in the Ancient World.* Cambridge, MA: Harvard University Press.

MALAMUD, D., DIETRICH, S.A., and SHAPIRO, I.M. (1985). Low levels of mercury inhibit the respiratory burst in human polymorphonuclear leukocytes. *Biochem. Biophys. Res. Commun.* **128**, 1145–1151.

MASSARO, E.J. (1996). The developmental cytotoxicity of mercurials, in Chang, L.W. (Ed.) *Toxicology of Metals*, pp. 1047–1081, Boca Raton, FL: CRC Lewis Publishers.

MATHIESON, P.W. and GILLESPIE, K.M. (1996). Cloning of a partial cDNA for rat interleukin-12 (IL-12) and analysis of IL-12 expression *in vivo. Scand. J. Immunol.* **44**, 11–14.

MATHIESON, P.W., STAPLETON, K.J., OLIVEIRA, D.B., and LOCKWOOD, C.M. (1991). Immunoregulation of mercuric chloride-induced autoimmunity in Brown Norway rats: a role for CD8$^+$ T-cells revealed by *in vivo* depletion studies. *Eur. J. Immunol.* **21**, 2105–2109.

MATHIESON, P.W., THIRU, S., and OLIVEIRA, D.B.G. (1992). Mercuric chloride-treated Brown Norway rats develop widespread tissue injury including necrotizing vasculitis. *Lab. Invest.* **67**, 121–129.

(1993). Regulatory role of OX22^{high} T-cells in mercury-induced autoimmunity in the Brown Norway rat. *J. Exp. Med.* **177**, 1309–1316.

MIURA, K., NAGANUMA, A., HIMENO, S., and IMURA, N. (1995). Mercury toxicity, in Goyer, R.A. and Cherian, M.G. (Eds) *Toxicology of Metals*, pp. 163–187, Berlin: Springer-Verlag.

MOLIN, C. (1992). Amalgam – fact and fiction. *Scand. J. Dent. Res.* **100**, 66–73.

MOLINA, A., SÁNCHEZ-MADRID, F., BRICIO, T., MARTÍN, A., BARAT, A., ALVAREZ, V., and MAMPASO, F. (1995). Prevention of mercuric chloride-induced nephritis in the Brown Norway rat by treatment with antibodies against the α4 integrin. *J. Immunol.* **153**, 2313–2320.

MORDES, J.P., GALLINA, D.L., HANDLER, E.S., GREINER, D.L., NAKAMURA, N., PELLETIER, A., and ROSSINI, A.A. (1987). Transfusions enriched for W3/25$^+$ helper/inducer-T-lymphocytes prevent spontaneous diabetes in the BB/W rat. *Diabetologia* **30**, 22–26.

NAKASHIMA, I., PU, M., NISHIZAKI, A., ROSILA, I., MA, L., KATANO, Y., OHKUSU, K., RAHMAN, S.M., ISOBE, K., HAMAGUCHI, M., and SAGA, K. (1994). Redox mechanism as alternative to ligand binding for receptor activation delivering disregulated cellular signals. *J. Immunol.* **152**, 1064–1071.

NICHOLSON, L.B. and KUCHROO, V.K. (1996). Manipulation of the T_{H1}/T_{H2} balance in autoimmune disease. *Curr. Opin. Immunol.* 8, 837–842.

NICKERSON, P., STEURER, W., STEIGER, J., ZHENG, X., STEELE, A.W., and STROM, T.B. (1994). Cytokines and the T_H1/T_H2 paradigm in transplantation. *Curr. Opin. Immunol.* **6**, 757–764.

NIEMINEN, A.L. and LEMASTERS, J.J. (1996). Hepatic injury by metal accumulation, in Chang, L.W. (Ed.) *Toxicology of Metals*, pp. 887–899, Boca Raton, FL: CRC Lewis Publishers.

OBEL, N., HANSEN, B., CHRISTENSEN, M.M., NIELSEN, S.L., and RUNGBY, J. (1993). Methyl mercury, mercuric chloride, and silver lactate decrease superoxide anion formation and chemotaxis in human polymorphonuclear leucocytes. *Human Exp. Toxicol.* **12**, 361–364.

OCHEL, M., VOHR, H.W., PFEIFFER, C., and GLEICHMANN, E. (1991). IL-4 is required for the IgE and IgG$_1$ increase and IgG$_1$ autoantibody formation in mice treated with mercuric chloride. *J. Immunol.* **146**, 3006–3011.

OHSAWA, M. and KIMURA, M. (1979). Enhancement of β_2-microglobulin formation induced by phytohemagglutinin and mercuric ion in cultured human leucocytes. *Biochem. Biophys. Res. Commun.* **91**, 569–574.

OLIVEIRA, D.B., WOLFREYS, K., and COLEMAN, J.W. (1994). Compounds that induce T_{H2}-driven autoimmunity in Brown Norway rats sensitize their mast cells *in vitro*. *Human Exp. Toxicol.* **13**, 615.

OLSSON, S. and BERGMAN, M. (1992). Daily dose calculations from measurements of intra-oral mercury vapor. *J. Dent. Res.* **71**, 414–423.

PALLOTTINO, M. (1984). *Etruscologia*. Milan, Italy: Editore Ulrico Hoepli.

PELLETIER, L. and DRUET, P. (1995). Immunotoxicology of metals, in Goyer, R.A., and Cherian, M.G. (Eds) *Toxicology of Metals*, pp. 77–92, Berlin: Springer-Verlag.

157

PELLETIER, L., GALCERAN, M., PASQUIER, R., RONCO, P., VERROUST, P., BARIETY, J., and DRUET, P. (1987a). Down modulation of Heymann's nephritis by mercuric chloride. *Kidney Int.* **32**, 227–232.

PELLETIER, L., PASQUIER, R., ROSSERT, J., and DRUET, P. (1987b). HgCl$_2$ induces nonspecific immunosuppression in Lewis rats. *Eur. J. Immunol.* **17**, 49–54.

PELLETIER, L., PASQUIER, R., GUETTIER, C., VIAL, M.C., MANDET, C., NOCHY, D., BAZIN, H., and DRUET, P. (1988a). HgCl$_2$ induces T- and B-cells to proliferate and differentiate in BN rats. *Clin. Exp. Immunol.* **71**, 336–342.

PELLETIER, L., ROSSERT, J., PASQUIER, R., VILLAROYA, H., BELAIR, M.F., VIAL, M.C., ORIOL, R., and DRUET, P. (1988b). Effect of HgCl$_2$ on experimental allergic encephalomyelitis in Lewis rats. HgCl$_2$-induced down-modulation of the disease. *Eur. J. Immunol.* **18**, 243–247.

PELLETIER, L., ROSSERT, J., PASQUIER, R., VIAL, M.C., and DRUET, P. (1990). Role of CD8[+] T-cells in mercury-induced autoimmunity or immunosuppression in the rat. *Scand. J. Immunol.* **31**, 65–74.

PELLETIER, L., ROSSERT, J., PASQUIER, R., VILLAROYA, H., ORIOL, R., and DRUET, P. (1991). HgCl$_2$-induced perturbation of the T-cell network in experimental allergic encephalomyelitis. II. *In vivo* demonstration of the role of T-suppressor and contrasuppressor cells. *Cell. Immunol.* **137**, 379–388.

PELLETIER, L., BELLON, B., TOURNADE, H., DUBEY, C., GUERY, J.C., SAOUDI, A., HIRSCH, F., and DRUET, P. (1992). Chemical-induced autoimmunity, in Bona, C.A. and Kaushik, A.K. (Eds) *Molecular Immunobiology of Self-Reactivity*, pp. 315–353, New York: Marcel Dekker.

PESZKOWSKI, M.J., WARFVINGE, G., and LARSSON, Å. (1993). HgCl$_2$-induced glandular pathosis in the Brown Norway rat. *Clin. Immunol. Immunopathol.* **69**, 272–277.

PRIGENT, P., PONCET, P., ATEN, J., BLANPIED, C., CHAND, A., FÉVRIER, M., DRUET, P., and HIRSCH, F. (1995). Mercuric chloride-induced programmed cell death of a murine hybridoma. II. Opposite effect of interleukin-2 and interleukin-4. *Cell. Immunol.* **161**, 107–111.

PROUVOST-DANON, A., ABADIE, A., SAPIN, C., BAZIN, H., and DRUET, P. (1981). Induction of IgE synthesis and potentiation of anti-ovalbumin IgE antibody response by HgCl$_2$ in the rat. *J. Immunol.* **126**, 699–702.

QASIM, F.J., MATHIESON, P.W., THIRU, S., OLIVEIRA, D.B., and LOCKWOOD, C.M. (1993). Further characterization of an animal model of systemic vasculitis, in Gross, W.L. (Ed.) *ANCA-Associated Vasculitides: Immunological and Clinical Aspects*, pp. 133–137, New York: Plenum Press.

RAHMAN, S.M., PU, M., HAMAGUCHI, M., IWAMOTOT, T., ISOBE, K., and NAKASHIMA, I. (1993). Redox-linked ligand-independent cell surface triggering for extensive protein tyrosine phosphorylation. *FEBS Lett.* **317**, 35–38.

REIMER, G. (1990). Autoantibodies against nuclear, nucleolar, and mitochondrial antigens in systemic sclerosis (scleroderma). *Rheum. Dis. Clin. North Am.* **16**, 169–183.

REUTER, R., TESSARS, G., VOHR, H.W., GLEICHMANN, E., and LÜHRMANN, R. (1989). Mercuric chloride induces autoantibodies to small nuclear ribonucleoprotein in susceptible mice. *Proc. Natl Acad. Sci. USA* **86**, 237–241.

RÖCKEN, M., RACKE, M., and SHEVACH, E.M. (1996). IL-4-induced immune deviation as antigen-specific therapy for inflammatory autoimmune disease. *Immunol. Today* **17**, 225–231.

RÖGER, J., ZILLIKENS, D., BURG, G., and GLEICHMANN, E. (1992). Systemic autoimmune disease in a patient with long-standing exposure to mercury. *Eur. J. Dermatol.* **2**, 168–170.

ROLINK, A.G., STRASSER, A., and MELCHERS, F. (1990). Autoimmune diseases induced by graft-vs-host disease, in Burakoff, S.J., Deeg, H.J., Ferrara, J., and Atkinson, K. (Eds) *Graft-vs-Host Disease*, pp. 161–175, New York: Marcel Dekker.

ROSSERT, J., PELLETIER, L., PASQUIER, R., VILLAROYA, H., ORIOL, R., and DRUET, P. (1991). HgCl₂-induced perturbation of the T-cell network in experimental allergic encephalomyelitis. I. *In vitro* characterization of T-cells involved. *Cell. Immunol.* **137**, 367–378.

ROSSINI, A.A., GREINER, D.L., FRIEDMAN, H.P., and MORDES, J.P. (1993). Immunopathogenesis of diabetes mellitus. *Diabetes Rev.* **1**, 43–75.

SAEGUSA, J., YAMAMOTO, S., IWAI, H., and UEDA, K. (1990). Antinucleolar autoantibody induced in mice by mercuric chloride. *Ind. Health* **28**, 21–30.

SAOUDI, A., BELLON, B., DE KOZAK, Y., KUHN, J., VIAL, M.C., THILLAYE, B., and DRUET, P. (1991). Prevention of experimental autoimmune uveoretinitis and experimental autoimmune pinealitis in (Lewis × Brown-Norway)F₁ rats by HgCl₂ injections. *Immunology* **74**, 348–354.

SAOUDI, A., KUHN, J., HUYGEN, K., DE KOZAK, Y., VELU, T., GOLDMAN, M., DRUET, P., and BELLON, B. (1993). T_H2 activated cells prevent experimental autoimmune uveoretinitis, a T_H1-dependent autoimmune disease. *Eur. J. Immunol.* **23**, 3096–3103.

SCHOPF, E., SCHULZ, K.H., and GROMM, M. (1967). Transformationen und Mitosen von Lymphocyten *in vitro* durch Quecksilber(II)-chlorid. *Naturwissensch.* **54**, 568–569.

SCHOPF, E., SCHULZ, K.H., and ISENSEE, I. (1969). Investigations on lymphocyte transformation in mercury sensitivity. Nonspecific transformation due to mercury compounds. *Arch. Klin. Exp. Dermatol.* **234**, 420–433.

SCHRALLHAMMER-BENKLER, K., RING, J., PRZYBILLA, B., and LANDTHALER, M. (1992). Acute mercury intoxication with lichenoid drug eruption followed by mercury contact allergy and development of antinuclear antibodies. *Acta Derm. Venereol. (Stockholm)* **72**, 294–296.

SHARMA, R.P. and DUGYALA, R.R. (1996). Effects of metals on cell-mediated immunity and biological response modulators, in Chang, L.W. (Ed.) *Toxicology of Metals*, pp. 785–796, Boca Raton, FL: CRC Lewis Publishers.

SHENKER, B.J., ROONEY, C., VITALE, L., and SHAPIRO, I.M. (1992). Immunotoxic effects of mercuric compounds on human lymphocytes and monocytes. I. Suppression of T-cell activation. *Immunopharmacol. Immunotoxicol.* **14**, 539–553.

SHENKER, B.J., BERTHOLD, P., ROONEY, C., VITALE, L., DeBOLT, K., and SHAPIRO, I.M. (1993). Immunotoxic effects of mercuric compounds on human lymphocytes and monocytes. III. Alterations in B-cell function and viability. *Immunopharmacol. Immunotoxicol.* **15**, 87–112.

SHULMAN, H.M. (1990). Pathology of chronic graft-vs-host disease, in Burakoff, S.J., Deeg, H.J., Ferrara, J., and Atkinson, K. (Eds) *Graft-vs-Host Disease*, pp. 587–614, New York: Marcel Dekker.

SNOVER, D.C. (1990). The pathology of acute graft-vs-host disease, in Burakoff, S.J., Deeg, H.J., Ferrara, J. and Atkinson, K. (Eds) *Graft-vs-Host Disease*, pp. 337–353, New York: Marcel Dekker.

SOLEZ, K. (1983). Acute renal failure ('acute tubular necrosis', infarction, and cortical necrosis), in Heptinstall, R.H. (Ed.) *Pathology of the Kidney*, pp. 1069–1148, Boston: Little, Brown and Company.

STROM, T.B., ROY-CHAUDHURY, P., MANFRO, R., ZHENG, X.X., NICKERSON, P.W., WOOD, K., and BUSHELL, A. (1996). The T_H1/T_H2 paradigm and the allograft response. *Curr. Opin. Immunol.* **8**, 688–693.

TAN, X., TANG, C., CASTOLDI, A.F., MANZO, L., and COSTA, L.G. (1993). Effects of inorganic and organic mercury on intracellular calcium levels in rat T-lymphocytes. *J. Toxicol. Environ. Health* **38**, 159–170.

VAN DER MEIDE, P.H., DE LABIE, M.C., BOTMAN, C.A., VAN BENNEKOM, W.P., OLSSON, T., ATEN, J., and WEENING, J.J. (1993). Mercuric chloride down-regulates T-cell interferon-γ production in Brown Norway but not in Lewis rats: role of glutathione. *Eur. J. Immunol.* **23**, 675–681.

VAN DER MEIDE, P.H., DE LABIE, M.C., BOTMAN, C.A., ATEN, J., and WEENING, J.J. (1995). Nitric oxide suppresses IFNγ production in the spleen of mercuric chloride-exposed Brown Norway rats. *Cell. Immunol.* **161**, 195–206.

VERITY, M.A. (1995). Nervous system, in Goyer, R.A., Klaassen, C.D., and Waalkes, M.P. (Eds) *Metal Toxicology*, pp. 199–235, San Diego: Academic Press.

VOSSEN, J.M. and HEIDT, P.J. (1990). Gnotobiotic measures for the prevention of acute graft-vs-host disease, in Burakoff, S.J., Deeg, H.J., Ferrara, J., and Atkinson, K. (Eds) *Graft-vs-Host Disease*, pp. 403–413, New York: Marcel Dekker.

WARBRICK, E.V., THOMAS, A.L., and COLEMAN, J.W. (1995). The effects of mercuric chloride on growth, cytokine and MHC Class II gene expression in a human leukemic mast cell line. *Toxicology* **104**, 179–186.

WARNER, G.L. and LAWRENCE, D.A. (1986). Stimulation of murine lymphocyte responses by cations. *Cell. Immunol.* **101**, 425–439.

WEENING, J.J., FLEUREN, G.J., and HOEDEMAEKER, P.J. (1978). Demonstration of antinuclear antibodies in mercuric chloride-induced glomerulopathy in the rat. *Lab. Invest.* **39**, 405–411.

WEENING, J.J., GROND, J., VAN DER TOP, D., and HOEDEMAEKER, P.J. (1980). Identification of the nuclear antigen involved in mercury-induced glomerulopathy in the rat. *Invest. Cell Pathol.* **3**, 129–134.

WEENING, J.J., HOEDEMAEKER, P.J., and BAKKER, W.W. (1981). Immunoregulation and anti-nuclear antibodies in mercury-induced glomerulopathy in the rat. *Clin. Exp. Immunol.* **45**, 64–71.

WEINER, J.A. and NYLANDER, M. (1996). Aspects on health risks of mercury from dental amalgams, in Chang, L.W. (Ed.) *Toxicology of Metals*, pp. 469–486, Boca Raton, FL: CRC Lewis Publishers.

WHALEN, B.J., GREINER, D.L., MORDES, J.P., and ROSSINI, A.A. (1994). Adoptive transfer of autoimmune diabetes mellitus to athymic rats: synergy of CD4[+] and CD8[+] T-cells and prevention by RT6[+] T-cells. *J. Autoimmun.* **7**, 819–831.

WHO (World Health Organization) (1989). Environmental Health Criteria No. 86. *Mercury – Environmental Aspects*. World Health Organization, Geneva.

    (1990). Environmental Health Criteria No. 101. *Methylmercury*. World Health Organization, Geneva.

    (1991). Environmental Health Criteria No. 118. *Inorganic Mercury*. World Health Organization, Geneva.

ZDOLSEK, J.M., SODER, O., and HULTMAN, P. (1994). Mercury induces *in vivo* and *in vitro* secretion of interleukin-1 in mice. *Immunopharmacology* **28**, 201–208.

ZELIKOFF, J.T. and COHEN, M.D. (1996). Immunotoxicology of inorganic metal compounds, in Smialowicz, R.J. and Holsapple, M.P. (Eds) *Experimental Immunotoxicology*, pp. 189–228, Boca Raton, FL: CRC Press.

ZELIKOFF, J.T. and SMIALOWICZ, R.J. (1996). Metal-induced alterations in innate immunity, in Chang, L.W. (Ed.) *Toxicology of Metals*, pp. 811–826, Boca Raton, FL: CRC Lewis Publishers.

ZIPRIS, D., GREINER, D.L., MALKANI, S., WHALEN, B., MORDES, J.P., and ROSSINI, A.A. (1996). Cytokine gene expression in islets and thyroids of BB rats. IFNγ and IL-12 p40 mRNA increase with age in both diabetic and insulin-treated nondiabetic BB rats. *J. Immunol.* **156**, 1315–1321.

8

Nickel

RALPH J. SMIALOWICZ

National Health and Environmental Effects Research Laboratory, US Environmental Protection Agency, Research Triangle Park, NC 27711, USA

8.1 History

Nickel (Ni) is a transition metal with atomic number 28 in group VIII of the Periodic Table following iron and cobalt. Metallic nickel is a relatively hard and strong, lustrous, silver-white malleable metal that is moderately reactive. It has typical metallic properties, is magnetic, is a good conductor of electricity and heat, and unlike iron is resistant to corrosion. Nickel is the 24th element in order of natural abundance, making up approximately 0.008 percent of the Earth's crust and 0.01 percent of igneous rocks. There are two generic types of naturally occurring nickel ores, the nickel–iron sulfides and the oxide or silicate laterites. Nickel is present in combination with iron in certain types of meteorites, and in manganese oxide nodules, which contain nickel, copper, and cobalt, and are found on the ocean bottom. Large amounts of nickel ores are also believed to exit in the Earth's core (Cooley, 1987). Nickel is usually dipositive in compounds, but it can also exist in oxidation states of 0, +1, +2, +3, and +4, with +2 being the most common and important oxidation state under environmental conditions. In addition to simple compounds or salts, nickel also forms a variety of coordinate compounds or complexes (da Silva and Williams, 1991; ATSDR, 1993). The green or blue color of nickel compounds is due to its hydration or its binding to other ligands (Cooley, 1987).

Because of their properties of strength, malleability, toughness, and corrosion resistance, naturally occurring nickel ores and alloys were used in ancient times long before nickel was isolated. Copper–nickel coinage was produced in Bactria as early as 235 BC, and in ancient China, ores containing nickel, copper, and zinc were smelted producing 'white copper', which was used to make weapons (Boldt, 1967; Cooley, 1987). In the 15th and 16th centuries, the European metal industry recognized the existence of nickel in ores and alloys

163

due to problems associated with refining copper. Miners in the Hartz mountains of Germany called the ore *Kupfernickel* or 'Old Nick's copper', since it gave off toxic fumes when heated as if Old Nick or the devil was at work (Morgan and Usher, 1994). The Swedish chemist Alex Cronstedt first isolated nickel as an element in 1751, from niccolite (i.e. *Kupfernickel*), and in 1804 H.T. Richter prepared it in a relatively pure form (Illis, 1987).

Nickel became commercially important when large deposits were discovered in the 1850s in New Caledonia off the east coast of Australia and at Sudbury in Ontario, Canada. During this time the world's naval powers adopted nickel-bearing armor. The market for nickel was driven primarily by military requirements until about 1920; however, after World War I there was an increase in the industrial application of nickel (Illis, 1987). Today, annual world consumption of nickel, excluding the former Soviet Union and Eastern bloc countries, is over 670 kilotons (Morgan and Usher, 1994). It is used primarily in the production of stainless and other steels, nickel alloys of which there are more than 3000, electroplating, high-temperature and electrical-resistance alloys, cast irons, and miscellaneous applications such as catalysts, magnets, ceramics, batteries, and pigments (ATSDR, 1993).

The first investigation of the biological effects of nickel were reported by C.G. Gmelin, a professor of chemistry at the University of Tubingen, in 1826 (NAS, 1975). In this work a number of elements were administered to rabbits and dogs by gastric intubation. Exposure to nickel sulfate resulted in cachexia and conjunctivitis at sublethal doses and severe gastritis and fatal convulsions at higher doses (Gmelin, 1826). During the period 1853 to 1885 several clinical studies were performed to investigate the potential therapeutic use of nickel sulfate ($NiSO_4$) in severe diarrhea and nickel bromide ($NiBr_2$) as an anti-epileptic drug. These clinical studies are of scientific value since they represent the only documented pharmacologic effects of nickel in humans (NAS, 1975). Over time, nickel therapy was abandoned after the acute and chronic toxicity of nickel salts in animals was documented (Dzergowsky *et al.*, 1906–1907; Lehmann, 1908–1909).

The toxic volatile liquid nickel carbonyl ($Ni(CO)_4$), which is an intermediate product in the Mond process for nickel refining of nickel sulfide (NiS) ores, was discovered in 1890 by Mond (Mond *et al.*, 1890). Within a few years after the beginning of industrial operations at the Mond nickel refinery in Clydach, Wales, severe cases of $Ni(CO)_4$ poisoning in workers were observed (NAS, 1975). Workers with $Ni(CO)_4$ poisoning had initial symptoms which included headache, vertigo, nausea, vomiting, and gastric pain. This was followed by chest pain, coughing, hypernea, cyanosis, and profound weakness. Terminally ill patients usually became delirious just prior to death (Sunderman, 1970; Vuopala *et al.*, 1970), with death attributed to pneumonitis and cerebral edema and hemorrhage (NAS, 1975; Sunderman and Oskarsson, 1991).

At the turn of the century, a number of investigators reported on the acute toxicity of $Ni(CO)_4$ in humans and animals. During this same period, Herxheimer (1912) described contact dermatitis in industrial workers exposed to nickel (NAS, 1975). More recent studies have examined the acute and chronic toxicity of nickel salts, nickel complexes, and nickel compounds in experimental animals.

8.2 Occurrence

Nickel is widely distributed in the environment as a result of natural and anthropogenic activities. Consequently, exposure of the general population to nickel and nickel compounds occurs via air, water, and food. Nickel and nickel compounds which occur naturally in the Earth's crust are released into the atmosphere as a result of natural processes such as wind-blown dust and volcanic eruptions. Release into the atmosphere also results from anthropogenic activities such as combustion of fossil fuels, nickel mining, smelting and refining operations, steel and nickel alloy production, and waste incineration (NAS, 1975; Sunderman and Oskarsson, 1991; ATSDR, 1993). Approximately 8.5 million kg of nickel are estimated to be emitted into the atmosphere from natural sources each year (ATSDR, 1993). A much greater amount of nickel (i.e. as much as 43 million kg) is estimated to be released from anthropogenic sources, with approximately 62 percent due to burning of residual and fuel oil, followed by nickel metal refining, municipal incineration, steel production, and coal combustion (Schmidt and Andren, 1980; Bennett, 1984; ATSDR, 1993). The average nickel intake by inhalation per individual in the general population has been estimated at 0.1 to 1.0 µg Ni/day. Even considering the highest ambient nickel levels reported, the daily inhalation intake is estimated to be approximately 18 µg (ATSDR, 1993).

Natural weathering of soil as well as discharges and runoff from anthropogenic sources result in the transport of nickel into streams and waterways. Nriagu and Pacyna (1988) estimated that the amount of nickel from anthropogenic sources that entered the world's waterways during 1983 was between 33 and 194 million kg Ni/year. Fresh surface water nickel concentrations are low, with levels reported to average between 15 to 20 µg Ni/L (NAS, 1975; Grandjean, 1984; ATSDR, 1993). Levels in ground water appear to be similar to those of surface water, with median nickel levels of 3.0 µg Ni/L reported from both sources (Page, 1981). Since drinking water generally contains less than 10 µg Ni/L (NAS, 1975; Sunderman, 1986), and if one assumes a consumption rate of 2 L water/day, then the daily intake of nickel from water would be less than 20 µg Ni.

Nickel occurs in plants and animals. The content of nickel in plants is associated with exchangeable nickel in the soil, which is affected by physical, chemical, and biological factors. For example, uptake of nickel by plants is reduced by raising the soil pH (Giashuddin and Cornfield, 1979). In most natural vegetation, nickel content ranges from 0.05 to 5 µg Ni/g dry weight (NAS, 1975; McIlveen and Negusanti, 1994). Edible plants with nickel concentrations of greater than 1.0 mg Ni/kg include certain nuts, oatmeal, wheat bran,

165

beans, cocoa, soya products, sunflower seeds, and certain spices (Sunderman and Oskarsson, 1991; McIlveen and Negusanti, 1994). Aquatic species such as shellfish and crustaceans generally contain higher concentrations of nickel in their edible flesh than do fin fish, while nickel concentrations in animal tissue are in general similar to those found in plants. The content of nickel in food items in the average Danish diet has been reported to be 0.01 to 0.1 ppm in dairy products; 0.02 to 0.11 ppm in meat, fish, and eggs; 0.04 to 0.52 ppm in vegetables; 0.01 to 0.31 ppm in fruits; and 0.1 to 1.76 ppm in grains (Nielsen and Flyvholm, 1984). The average daily dietary intake of nickel is approximately 150 µg, which indicates that the general human population obtains the greatest amount of nickel through food consumption (Nielsen and Flyvholm, 1984; IARC, 1990; ATSDR, 1993). Humans are also exposed to nickel in nickel alloys and nickel-plated materials by dermal contact with steel, coins, fasteners, and jewelry. Certain individuals are exposed to nickel from nickel alloys used in surgical and dental prostheses, and clips, pins, and screws used for fractured bones. Also, residual nickel may be present in soaps, and edible fats and oils (e.g. margarine) hydrogenated with nickel catalysts (Sunderman, 1986; Sunderman and Oskarsson, 1991; ATSDR, 1993).

Since high concentrations of nickel are found in certain occupations, workers in these industrial settings are exposed to higher levels of nickel than the general public. Individuals involved in the production, processing, and use of nickel are among those exposed to the highest levels. Other industries which have a high potential for workers to be exposed to nickel include welding, electroplating, nickel–cadmium battery production, electronics, waste disposal, and jewelry manufacturing. As many as 730 000 US workers were potentially exposed to nickel metal, alloys, dust and fumes, nickel salts, or inorganic nickel compounds according to a 1990 NIOSH study (NIOSH, 1990), with occupational exposure occurring primarily via dermal contact and inhalation (NAS, 1975; ATSDR, 1993).

Based on the mammalian toxicity of nickel and nickel compounds (described below), various national and international regulatory bodies have promulgated guidelines and regulations relative to nickel exposure. For example, nickel compounds have been classified as group 1 (human carcinogens) and metallic nickel as group 2B (possible human carcinogens) by the World Health Organization (IARC, 1990). The US Environmental Protection Agency (EPA) has verified an oral reference dose (RfD) of 0.02 mg/kg per day for soluble salts of nickel (IRIS, 1992), based upon a chronic feeding study in rats (Ambrose *et al.*, 1976). The Occupational Safety and Health Administration (OSHA) and the American Conference of Governmental Industrial Hygienists (ACGIH) have both set an occupational air exposure limit for an 8-h day, 40-h working week, or time-weighted average (TWA) value, at 1 mg Ni/m^3 for nickel metal and nickel insoluble compounds and at 0.1 mg/m^3 for soluble nickel compounds (OSHA, 1989; ACGIH, 1990). A limit value of 7 µg/m^3 for nickel carbonyl has been adopted by many countries (Sunderman and Oskarsson, 1991). Nickel and its compounds are also regulated by the Clean

Water Effluent Guidelines for industrial point sources for a variety of industries which include, but are not limited to, the manufacture of iron, steel, non-ferrous metals, and batteries, as well as electroplating, metal finishing, ore and mineral mining, and paint and ink formulation (EPA, 1988; ATSDR, 1993).

8.3 Essentiality

Dietary nickel deficiency in goats, chickens, mini-pigs and rats has been reported to result in decreased growth (Anke *et al.*, 1983; Nielsen, 1984; Spears, 1984; Fishbein, 1987). In goats, nickel deficiency during pregnancy resulted in significantly decreased birth weights (Anke *et al.*, 1983). In addition to growth reduction, rats fed nickel-deficient diets were anemic as demonstrated by reduced hemoglobin concentrations, erythrocyte counts, and hematocrit (Schnegg and Kirchgessner, 1975a,b). The anemia observed in rats fed a nickel-deficient diet was apparently mediated by impaired iron absorption (Schnegg and Kirchgessner, 1976). Nickel-deficient diets in animals have also been associated with disturbances in calcium and zinc metabolism as manifested by decreased bone calcium concentrations and parakeratosis-like skin and hair diseases, respectively (Anke *et al.*, 1983). High concentrations of nickel have been found in human and animal fetuses and newborns, suggesting that it may be of biological importance during early development (Anke *et al.*, 1983).

While nickel has been shown to be necessary to animal health, no specific physiological function of nickel has been established in animals (Sunderman and Oskarsson, 1991). Similarly, nickel may be necessary for human health, but there are no reports of nickel deficiency in humans (ATSDR, 1993). No nickel-dependent enzymes are known to exist in mammals, while the nickel-dependent enzymes urease and carbon monoxide dehydrogenase are found in certain bacteria and plants (Dixon *et al.*, 1975; NAS, 1975; Dieckert *et al.*, 1979; Sunderman and Oskarsson, 1991). Taken together, the evidence indicates that nickel partially satisfies the criteria (Mertz, 1970) used to identify essential trace elements (NAS, 1975).

Studies of the absorption of nickel by fasting mice or rats, using radiolabeled Ni^{2+}, indicate that approximately 3 to 6 percent is absorbed from the intestinal tract following oral administration (Ho and Furst, 1973; Sunderman and Oskarsson, 1991). Approximately 35 percent of inhaled nickel is absorbed from the respiratory tract of humans (Bennett, 1984; Grandjean, 1984; Sunderman and Oskarsson, 1991). Nickel is transported in plasma bound to a metalloprotein (nickeloplasmin) that does not contain other detectable trace metals (Himmelhoch *et al.*, 1966; Asato *et al.*, 1975). Nickeloplasmin is an α_2-macroglobulin that does not readily exchange its bound nickel with exogenous Ni^{2+}, and as such does not appear to play an important role in extracellular transport of nickel (Sunderman and Oskarsson, 1991). Nickel that is ingested in food, for the most part, remains unabsorbed within the

167

gastrointestinal tract and is excreted in the feces; however, absorbed nickel is eliminated in the urine (NAS, 1975; Sunderman and Oskarsson, 1991). In contrast, inhaled nickel carbonyl, because of its lipid solubility, is rapidly absorbed and is excreted primarily in the urine and to a minor extent in the feces (Tedeschi and Sunderman, 1957; Sunderman and Selin, 1968).

8.4 General Toxicology

Nickel toxicity is dependent on the solubility of the nickel compound and on the route of exposure. Only soluble nickel is absorbed, although low solubility nickel particles such as crystalline, but not amorphous, nickel are phagocytosed and gradually dissolved in the cytosol (Abbracchio *et al.*, 1982). The toxicity of nickel is ultimately associated with the toxic action of Ni^{2+} ions (Hansen and Stern, 1984). The major toxic route associated with nickel exposure is pulmonary absorption, with gastrointestinal and dermal absorption being significantly less. The distribution of nickel in the body is a function of route of exposure and the time subsequent to exposure, with the lungs and the kidneys being the primary organs for the accumulation of nickel (Coogan *et al.*, 1989).

The occupational risks of lung and nasal sinus cancers in nickel refinery workers during the early part of this century have been well documented (NAS, 1975; Doll *et al.*, 1977, 1990; Chovil *et al.*, 1981; ATSDR, 1993). Increased risk was initially attributed to $Ni(CO)_4$ gas; however, while metallic nickel and $Ni(CO)_4$ are carcinogenic under experimental conditions, the epidemiological evidence implicated exposure to dust from preliminary nickel processes such as roasting and smelting. The identification of the specific nickel compounds which induce the cancers is uncertain, although airborne oxidic (e.g. nickel oxide (NiO), nickel–copper oxide, nickel silicate oxides, and complex oxides), sulfidic (e.g. nickel subsulfide (Ni_3S_2) and nickel monosulfide (NiS)), and soluble nickel (e.g. $NiSO_4$ and nickel chloride ($NiCl_2$)) are suspected to be the principal carcinogens (NAS, 1975; Doll *et al.*, 1977; Sunderman and Oskarsson, 1991; ATSDR, 1993; Morgan and Usher, 1994; Hughes *et al.*, 1995).

Inhalation is the major toxic route of exposure to nickel and while the carcinogenic effects of nickel are of primary concern, other pulmonary involvement has been observed. Occupational exposure to inhaled nickel has been associated with the induction of asthma, which can be a result of either a primary irritation or an allergic response (McConnell *et al.*, 1973; Block and Yeung, 1982; Novey, *et al.*, 1983; Dolovich *et al.*, 1984; Shirakawa *et al.*, 1990). Bronchoalveolar hyperplasia and granular pneumocyte proliferation in the lungs of rats and rabbits, respectively, have been reported following chronic inhalation exposure to nickel dust (Camner *et al.*, 1984; Benson *et al.*, 1986).

Tumors have been induced in rats following inhalation exposure of rats to $Ni(CO)_4$ (Sunderman and Donnelly, 1965) and Ni_3S_2 (Ottolenghi *et al.*, 1975).

168

In addition, parenteral administration of nickel results in tumors. For example, intravenous injection of $Ni(CO)_4$ in rats results in sarcomas and carcinomas of the liver and kidney (Lau *et al.*, 1972). The intramuscular or intrarenal injection of Ni_3S_2 in rats results in the development of primarily rhab-domyosarcomas at the injection site (Gilman, 1962; Gilman and Ruckerbauer, 1962) and renal carcinomas (Jasmin and Riopelle, 1976), respectively. An interesting observation is that the development of cancers following parenteral injection of Ni_3S_2 in rats is antagonized by the simultaneous injec-tion of manganese but not of other metal dusts (Sunderman *et al.*, 1974, 1976, 1979). These observations, as well as others, suggest that nickel toxicity and carcinogenicity may be due to interference with the normal physiological roles of other essential metal cations such as magnesium, manganese, calcium, and zinc (Nieboer *et al.*, 1984a; Kasprzak *et al.*, 1986).

Studies have revealed that while certain particulate nickel compounds are potent carcinogens in experimental animals, water-soluble nickel salts are not. Cells phagocytose particulate nickel and the particles undergo dissolution in the cells, leading to the release of nickel ions which interact with protein-rich nuclear heterochromatin. The formation of covalent cross-links of proteins and amino acids to DNA cause deletions either by altering the DNA template and/or by inhibiting DNA polymerase when it attempts to copy the region of the DNA lesion. These lesions appear to be important in the carcinogenicity of nickel because they are not readily repaired (Costa, 1991; Costa *et al.*, 1994).

Nickel is a developmental and reproductive toxicant in rodents, however there is insufficient evidence to suggest that occupational or environmental exposure to nickel causes developmental or reproductive effects in humans (ATSDR, 1993). Oral exposure of mice and rats to nickel results in decreased fetal weight, increased pup mortality, and decreased pup weight (Ambrose *et al.*, 1976; RTI, 1988a,b). Parenteral administration of nickel to rodents results in similar developmental effects (Sunderman *et al.*, 1978; Lu *et al.*, 1979; Chernoff and Kavlock, 1982). Mice and rats display testicular degeneration following inhalation exposure to $NiSO_4$ or Ni_3S_2 (Benson *et al.*, 1987, 1988), while parenteral exposure of rats to $NiSO_4$ also results in testicular damage (Hoey, 1966).

Other toxic effects associated with nickel exposure include nephrotoxicity, hepatotoxicity, and cardiovascular toxicity (Coogan *et al.*, 1989). Nickel-induced nephrotoxicity in experimental animals is manifested by amin-oaciduria and proteinuria following parenteral administration of nickel (Gitlitz *et al.*, 1975) or inhalation exposure to $Ni(CO)_4$ (Horak and Sunderman, 1980); renal tubular lesions (Foulkes and Blank, 1984); and nickel binding to anionic glycosaminoglycan sites of glomerular basement mem-branes resulting in ionic blocking (Templeton, 1987). Acute hepatotoxicity following parenteral exposure of rats to $NiCl_2$ is characterized by enhanced lipid peroxidation, microvesicular steatosis, inflammation, and increased se-rum alanine and aspartate aminotransferase (Donskoy *et al.*, 1986; Athar *et al.*, 1987). Intravenous administration of $NiCl_2$ to dogs results in acute coronary vasoconstriction (Rubanyi *et al.*, 1984), while the addition of low concentrations of $NiCl_2$ to isolated perfused rat hearts also results in the induction of coronary vasoconstriction (Rubanyi *et al.*, 1981).

8.5 Immunotoxicology

Immunotoxicity has been defined as 'the study of injury to the immune system that can result from occupational, inadvertent or therapeutic exposure to a variety of environmental chemicals or biological materials...' and '...involves studies of the suppression of immunity...' and '...studies of enhanced or excessive immune response, such as allergy [hypersensitivity] or autoimmunity' (NRC, 1992). Nickel is an interesting immunotoxicant in that it induces immunosuppression and hypersensitivity (i.e. allergy and asthma) (Zelikoff and Cohen, 1995). Evidence for nickel's ability to suppress immune function has come from numerous experimental animal studies, while nickel hypersensitivity has been documented in clinical and epidemiological studies. From a public health standpoint, hypersensitivity to nickel in humans is of far more concern than the potential for this metal to suppress immune function. Nickel in all of its forms can induce multiple allergic/hypersensitivity reactions in humans. These include cell-mediated delayed-type contact dermatitis and antibody-mediated immediate-type urticaria and asthma which can result from percutaneous or systemic (i.e. oral or inhalation) nickel exposure (Menne and Maibach, 1987).

8.5.1 *Hypersensitivity/allergy*

A few cases of occupational asthma have been reported in nickel-sensitive workers. For example, Block and Yeung (1982) described a patient who suffered from asthma due to occupational exposure to nickel. This individual had an immediate reaction to nickel sulfate in a bronchoprovocation test and in a skin prick test. Other studies also identified individuals with nickel asthma following bronchial challenge tests with $NiSO_4$ or work dust (Malo *et al.*, 1982, 1985; Novey *et al.*, 1983; Davies, 1986; Shirakawa *et al.*, 1990). Using radioimmunoassays, nickel-specific IgE antibodies have been detected in workers with nickel asthma, which along with bronchial challenge and skin test results, suggest that a type I hypersensitivity reaction may be involved in nickel-induced asthma (Novey *et al.*, 1983; Shirakawa *et al.*, 1990). Support for a type I reaction also comes from studies which provided evidence for the production of antibodies to nickel–albumin in a patient with occupational nickel asthma. The antigenicity of nickel was found to result from the binding of nickel to the primary copper binding site of human serum albumin (Dolovich *et al.*, 1984; Niebor *et al.*, 1984b).

Occupational and non-occupational exposure to nickel can result in a type IV cell-mediated delayed-type hypersensitivity (DTH) reaction presenting as an allergic contact dermatitis. In non-occupational exposures, the primary cause of nickel allergic contact dermatitis is dermal exposure to nickel-containing metals (Peltonen, 1979), with up to 10 percent of the Caucasian population testing positive for nickel sensitivity (von Blomberg *et al.*, 1991; Menne, 1994). About a 5 to 10-fold difference in the incidence of nickel-associated dermatitis has been observed for females compared with males (Prystowsky *et al.*, 1979; Nielsen and Menne, 1992). The primary reason for this gender difference is due to the presence of nickel in jewelry and accessories more frequently worn by women, which come in prolonged direct and intimate contact with the skin. Nickel sulfate is one of the most common causes of contact dermatitis (Kanerva *et al.*, 1994), with the Ni^{2+} ion being the ultimate immunogen (Wahlberg, 1976). The fact that nickel is easily oxidized to its allergenic ionic form is a primary factor in nickel-induced contact allergy. The sodium chloride content of human sweat is sufficient to leach out significant amounts of nickel from nickel-containing metal alloys. The level of nickel released from a nickel alloy coin taped to the skin of a sensitized individual is sufficient to elicit a positive reaction (Hostynek *et al.*, 1993). Results from a clinical study which examined the risk for sensitization from different types of nickel coating and alloys indicated that the release of more than $1\,\mu g$ Ni/cm^2 per week caused the majority of cases of sensitization (Menne *et al.*, 1987; Menne, 1994). While nickel dermatitis usually results from dermal exposure to nickel-containing alloys in such items as jewelry and coinage, oral intake of soluble nickel salts also can elicit allergic reactions in sensitive individuals (Christensen and Moller, 1975).

Nickel has been classified as a medium-to-strong contact sensitizer in humans (Kligman, 1966; Vandenberg and Epstein, 1963). However, because nickel is such a ubiquitous allergen, sensitized individuals have one of the worst possible prognoses of allergic individuals due to the great potential for re-exposure to this metal (Hostynek *et al.*, 1993). Treatment of nickel-sensitive individuals includes the use of topical corticoid creams over prolonged (i.e. weeks to months) periods or short-term systemic corticoid treatment in severe cases. However, avoidance of future exposure is the primary goal in the management of individuals with nickel sensitivity (Avnstorp *et al.*, 1990).

Several attempts have been made to determine if genetic markers can be linked to the development of sensitivity to nickel in humans. However, no consistent associations have been found between HLA class I and II genes and disease susceptibility (Menne and Holm, 1989). For example, a study in a twin and nickel-sensitive patient population failed to demonstrate an association between nickel allergy and HLA-A or -B antigens (Hansen *et al.*, 1982). However, in another study HLA typing of nickel-sensitive patients revealed a significant increase in HLA-DRw6 antigen (Mozzanica *et al.*, 1990). Although an association was found between nickel allergy and HLA-DQATaqI using restriction fragment length polymorphism (Olerup and Emtestam, 1988), this was not confirmed using more precise polymerase chain reaction and sequence-specific oligonucleotides for the HLA-DQA1 or -DQB1 loci (Ikaheimo *et al.*, 1993). While there is some evidence for an association between sensitivity to nickel and HLA antigens, more work remains to be performed.

Most cases of sensitization to nickel have been associated with ear piercing and the use of cheap costume jewelry and wrist watches (Avnstorp *et al.*, 1990). Recent Swedish (Larsson-Stymne and Wildstrom, 1985), Finnish (Peltonen and Terho, 1989), Norwegian (Dotterud and Falk, 1994), and Danish (Nielsen and Menne, 1992) epidemiological studies have identified ear piercing as the principal inducer of nickel hypersensitivity in female schoolchildren. Ten to fifteen percent of young girls who have their ears pierced develop nickel allergy (Avnstorp *et al.*, 1990). The sensitization of children takes on added significance because these individuals are subject to indefinite nickel dermatitis (Fisher, 1991). Concern for the consequences of nickel exposure led to the promulgation of a statute in Denmark in 1991 that banned the sale of objects that come in contact with the skin that release more than $0.5\,\mu g$ Ni/cm^2 per week (Kanerva *et al.*, 1994).

While nickel is a relatively strong sensitizer in humans, it has been surprisingly difficult to induce contact allergy consistently in experimental animals. This has led to the implication that nickel is not an inherently potent contact allergen (Avnstorp *et al.*, 1990). More than 25 different methods have been employed in attempts to induce nickel contact sensitivity in experimental animals (Wahlberg, 1989). The methods include epicutaneous, intradermal, or intramuscular administration of several different nickel salts, with or without the use of adjuvants, in several species, but primarily in guinea pigs (Wahlberg, 1989). The intradermal administration of nickel salts, salts mixed with adjuvant, or adjuvant administered by separate injection has achieved limited success in sensitizing guinea pigs (Maurer *et al.*, 1979) and rabbits (Merritt and Brown, 1980); however, the majority of the intradermal methods that have been employed have failed to induce sensitization to nickel. Using the epicutaneous technique, Lammintausta *et al.* (1985) obtained a sensitization rate of 50 percent. However, in a subsequent study using the guinea pig maximization test (GPMT), a sensitization rate of 93 to 100 percent was achieved (Lammintausta *et al.*, 1986). Nevertheless, the rate of success from various laboratories that have employed the GPMT is very inconsistent, with success varying from 0 to 100 percent (Wahlberg, 1989). Attempts to sensitize mice using a variety of techniques, with the exception of repeated and prolonged ear painting, failed to induce contact sensitivity to nickel (Moller, 1984). More recently Wahlberg and Boman (1992) have demonstrated that guinea pigs can be consistently sensitized to nickel. Also, recent reports from two laboratories indicate progress in establishing mouse models for nickel-induced allergy. van Hoogstraten *et al.* (1993) reared mice for at least two generations under a nickel-free environment (i.e. maintained on nickel-free feed and water in metal-free cages). A highly reproducible sensitization to nickel was established by a single intradermal injection of $NiSO_4$ in adjuvant into both flanks. In another model, the continuous epicutaneous exposure of mice, over 7 days, to $NiSO_4$ on the occluded clipped flank also resulted in consistent and reproducible sensitization to nickel (Ishii *et al.*, 1995).

Over the past two decades a tremendous amount of information about the mechanisms underlying allergic contact dermatitis has been gained as a result of studies using experimental animal models. However, until just recently, the lack of reliable animal models for nickel-induced hypersensitivity has dictated that *in vivo* and *in vitro* clinical studies be employed for mechanistic investigations of allergic nickel contact sensitivity. Nickel sensitivity has been found to follow the two phases (i.e. sensitization and elicitation) associated with classic T-lymphocyte-dependent type IV DTH responses (Scheper *et al.*, 1989), which can be briefly described as follows. During the sensitization phase, topical allergen (e.g. nickel) enters the epidermis where it is processed by epidermal dendritic cells (i.e. Langerhans cells, LC). The LC migrate through the dermal lymphatics to regional lymph nodes, where they become highly efficient antigen-presenting cells (APC). In the lymph nodes, the antigen, which is complexed to class II MHC molecules present on LC/dendritic cells, is presented to T-lymphocytes. The T-lymphocytes undergo activation and proliferation, and differentiate into antigen-specific memory T-lymphocytes. The elicitation phase occurs following subsequent epicutaneous challenge with the allergen resulting in its processing and presentation by LC and its recognition by the antigen-specific memory T-lymphocytes. These T-lymphocytes are activated and produce soluble factors, such as cytokine, which induce local inflammation and dermatitis (Barker, 1992; Kimber, 1995).

In the case of nickel contact dermatitis, nickel cations, which may be generated as a result of oxidation of nickel alloys, penetrate the epidermis and bind as haptens on serum or cellular proteins (Scheper *et al.*, 1989). Evidence for the association of nickel with LC/APC comes from studies which have demonstrated that proliferation of nickel-specific human T-lymphocytes requires the presence of APC. Work from Braathen's laboratory has shown that both peripheral blood monocytes and skin-derived epidermal cells are capable of functioning as APC, however, epidermal cells were more potent than monocytes in inducing T-lymphocyte proliferation (Braathen, 1980; Braathen and Thorsby, 1983). Res *et al.* (1987) tested the antigen-presenting capacity of FcR^+ monocytes and FcR^- circulating dendritic cells and found that the critical APC was the dendritic cell. Furthermore, using highly purified $T6^+$ (CD-1) LC and $T6^-$ epidermal cells, it was found that the critical APC within the epidermal cell population were the LC (Res *et al.*, 1987).

Work with nickel-specific T-lymphocyte clones, which were isolated from skin lesion biopsies or peripheral blood of patients with allergic nickel contact dermatitis, demonstrated that nickel-induced activation, proliferation, and lymphokine production by these T-lymphocytes also require the presence of APC and were restricted to HLA class II molecules (Sinigaglia *et al.*, 1985; Kapsenberg *et al.*, 1987). Isolated T-lymphocyte clones were found to be

OKT3$^+$ (CD3), OKT4$^+$ (CD4), OKT8$^-$ (CD8), and to produce interleukin-2 (IL-2) and interferon-γ (IFN-γ) (Sinigaglia *et al.*, 1985). Other work also demonstrates that the response of nickel-specific T-cell clones is restricted to determinants presented on HLA-DR and HLA-DQ molecules (Kapsenberg, *et al.*, 1988; Pistoor *et al.*, 1995). These results indicate that nickel is recognized by T cells in the context of self-MHC class II gene products similar to those of peptide antigens. Other studies have demonstrated that most nickel-specific T-lymphocyte clones isolated from peripheral blood secrete tumor necrosis factor-α (TNF-α), granulocyte–macrophage colony-stimulating factor (GM-CSF), IL-2, and high levels of IFN-γ, but low or undetectable levels of IL-4 and IL-5 (Kapsenberg *et al.*, 1991, 1992). Since this pattern of cytokine production resembles that of the mouse CD4$^+$ T$_H$1-lymphocyte subtype (Mosmann and Coffman, 1989), it was postulated that nickel ions induce a T$_H$1-type response (Kapsenberg *et al.*, 1992). However, in a recent study, T-lymphocyte clones isolated from a skin biopsy of a patient with nickel-induced contact dermatitis produced high levels of IL-4 and IL-5 with only minor amounts of IFN-γ (Probst *et al.*, 1995). This cytokine profile suggests that nickel-specific cells obtained from the actual site of inflammation express the T$_H$2 phenotype. It was suggested that T$_H$2 cytokines may be involved in the immunopathogenetic reaction of nickel-induced contact dermatitis by initiating the inflammatory process during the elicitation phase of the hypersensitivity response (Probst *et al.*, 1995).

Work from Sinigaglia (1994) suggests that T-lymphocytes recognize metal–hapten-modified MHC–peptide complexes on APC. A competition assay, which is based on the proliferative response of a specific T-lymphocyte clone to a specific peptide (Pink and Sinigaglia, 1989), was used to determine if nickel can inhibit the response by either directly modifying the specific MHC molecule or by altering the antigenic peptide before or after its association with the MHC protein. The data indicated that nickel binding to the MHC-associated peptide modified the structure of the complex and altered it in such a way that it was no longer recognized by the peptide-specific T-lymphocyte clone (Sinigaglia, 1994). It was determined that nickel inhibited the peptide-specific T-lymphocyte reaction if the antigenic peptide contained histidine, which is known to be a typical nickel-complexing amino acid (Romagnoli *et al.*, 1991). It is not clear from this work if such nickel–peptide complexes form the determinants that are recognized by nickel-specific T-lymphocytes. However, recent work by Weltzien *et al.* (1996) demonstrated that certain nickel-specific human T-lymphocyte clones, which do not require the constant presence of excess nickel with APC, can trigger these T-lymphocytes to proliferate in the presence of nickel-pulsed APC. This is the first direct evidence that the Ni^{2+} ion participates as a determinant in T-lymphocyte activation (Weltzien *et al.*, 1996).

Epidermal keratinocytes also participate in nickel-specific allergic contact dermatitis. Garioch *et al.* (1991) reported increased expression of intercellular adhesion molecule-1 (ICAM-1) on keratinocytes obtained from nickel-

174

sensitive individuals. In addition, increased numbers of lymphocytes expressing leukocyte function-associated antigen (LFA-1), which is the ligand for
ICAM-1, were observed in the dermis and epidermis. The LFA-1 cells were
found in close proximity to keratinocytes expressing ICAM-1 in tissue biopsies, which suggests the possible attachment of T-lymphocytes to keratinocytes
via LFA-1/ICAM-1 molecules leading to subsequent production of inflammatory cytokines (Garioch *et al.*, 1991). Using normal human keratinocytes cultured in the presence of nickel sulfate, Gueniche *et al.* (1994a,b) demonstrated
that keratinocytes were induced to produce IL-1α, IL-1β and TNF-α, and to
express ICAM-1. These data confirm the existence of direct interactions between nickel and keratinocytes, which generate cytokines that play an important role in the pathophysiology of allergic contact dermatitis (Gueniche *et al.*,
1994a,b).

Vascular endothelial cells are involved in a number of physiological processes including the development of immune responses such as inflammation
and delayed hypersensitivity reactions. It is not surprising, therefore, that
these cells would be expected to play a role in nickel-induced contact dermatitis. Studies employing normal human umbilical vein endothelial cells
(HUVEC) cultured in the presence of nickel demonstrate the upregulated
expression of ICAM-1, vascular cell adhesion molecule-1 (VCAM-1), and
E-selectin (previously known as endothelial leukocyte adhesion molecule 1 or
ELAM-1) (Wildner *et al.*, 1992; Goebeler *et al.*, 1993). These adherence molecules have been demonstrated to mediate leukocyte recruitment from the
vascular compartment to sites of inflammation, including the epidermis in
allergic contact dermatitis (Barker, 1992). As such, their upregulation by
nickel may represent an adjuvant mechanism that promotes the sensitization
and elicitation events associated with contact hypersensitivity to this metal
(Goebeler *et al.*, 1993). More recently, Goebeler *et al.* (1995) studied the
capacity of nickel to activate nuclear factor (NF)-κB, a transcription factor
involved in inducible expression of the adhesion molecules ICAM-1, VCAM-
1, and E-selectin. A strong increase of NF-κB DNA binding was detected after
stimulation of HUVEC with nickel, which was associated with increased adhesion molecule expression. In addition, nickel induced mRNA production and
protein secretion of the NF-κB-controlled proinflammatory cytokine IL-6.
Goebeler *et al.* (1995) concluded that the data suggest that distinct allergens,
such as metals, represent a new class of agents that are capable of inducing NF-
κB binding activity which subsequently modulates transcription of cytokines
and adhesion molecule genes.

In summary, a good deal of information has accrued regarding the mechanisms underlying nickel-induced allergic hypersensitivity. This information
serves not only to delineate the interactions between cells and cell products
which are involved in the development of nickel-specific hypersensitivity, but
also sheds light on the possible mechanisms of contact dermatitis induced by
other agents. Taken together, the results of the studies cited above indicate
that the pathological consequences of contact hypersensitivity to nickel

involve complex and integrated cooperation of APC, antigen-specific T cells, keratinocytes, and endothelial cells. This knowledge will hopefully lead to new and more effective therapies for allergic contact dermatitis.

8.5.2 *Immunosuppression*

Evidence from experimental animal studies indicates that nickel is an immunosuppressant as demonstrated by its ability to alter the functional activity of leukocytes and lymphocytes. These studies have employed *in vivo* exposure to nickel compounds via inhalation, pulmonary instillation, parenteral injection, or oral exposure (Table 8.1). *In vitro* studies, in which nickel has been added to leukocytes and lymphocytes obtained from humans and animals, have also been performed. In several cases, nickel-specific effects observed *in vivo* in animals have been corroborated using *in vitro* techniques.

Animals exposed to nickel via inhalation or intratracheal instillation display altered respiratory system host defense mechanisms and increased susceptibility to infectious agents. For example, exposure of hamsters to aerosols of $NiCl_2$ at $100\,\mu g/m^3$ resulted in depression of the normal ciliary activity of hamster trachea, as determined by an *in vitro* model system using isolated tracheal rings (Adalis *et al.*, 1978). This nickel-induced *in vivo* alteration in the ciliated epithelium of the trachea was corroborated using normal hamster tracheal rings exposed to $NiCl_2$ at a concentration of $0.011\,mM$ *in vitro*. Rabbits exposed to metallic nickel dust aerosols (0.13 to $2\,mg/m^3$), for periods ranging from 1 to 6 months, displayed pathological conditions characterized by increases in lung weight, phospholipid concentration, surfactant activity, and volume and density of epithelial alveolar type II cells, as well as effects on alveolar macrophages (Camner *et al.*, 1978; Jarstrand *et al.*, 1978; Johansson and Camner, 1980; Johansson *et al.*, 1980, 1981, 1983; Curstedt *et al.*, 1983). Exposure of rats for 4 months to nickel oxide (NiO) aerosols at $0.025\,mg/m^3$ resulted in decreased numbers of macrophages obtained in lung lavage fluid (Spiegelberg *et al.*, 1984). A 2-month exposure of mice to $0.45\,mg\ Ni_3S_2/m^3$, $0.45\,mg\ NiSO_4/m^3$, or $0.45\,mg\ NiO/m^3$ resulted in decreased phagocytosis of opsonized sheep red blood cells (SRBC) by alveolar macrophages (Haley *et al.*, 1990). Work by Sunderman *et al.* (1989) suggests that alveolar macrophages are a cellular target for nickel-induced toxicity, regardless of the route of exposure. The subcutaneous administration of $NiCl_2$ to rats initially caused nickel uptake into, and activation of, alveolar macrophages. This was followed within 24 to 48 h by reduced phagocytic activity and increased lipid peroxidation, respectively (Sunderman *et al.*, 1989). Subcutaneous injection of $NiCl_2$ also produced thymic involution in the rat, with marked degenerative changes in the cortical thymocytes (Knight *et al.*, 1987).

Studies using *in vitro* systems demonstrated that $NiCl_2$ at a concentration of $5\,mM$ was cytotoxic to rabbit alveolar macrophages (Waters *et al.*, 1975), and that nickel inhibited alveolar macrophage phagocytosis of latex particles

Table 8.1 Immunosuppression in experimental animals exposed to nickel

Route	Species	Immune function effects	References
Pulmonary	Hamster	↓ Alveolar macrophage phagocytosis	Haley *et al.*, 1990
		↑ Mortality to virus infection	Port *et al.*, 1975
	Rabbit	↓ Alveolar macrophage and lung lysozyme activity	Lundborg and Camner, 1984
		↓ Alveolar macrophage phagocytosis	Wiernik *et al.*, 1983
	Mouse	↓ Bacterial lung clearance and ↑ mortality	Adkins *et al.*, 1979
		↓ Antibody response to SRBC	Graham *et al.*, 1978
	Rat	↓ Alveolar macrophage phagocytosis	Adkins *et al.*, 1979
		↓ Antibody response to SRBC	Spiegelberg *et al.*, 1984
Parenteral	Mouse	↓ Antibody response to SRBC	Graham *et al.*, 1975a Smialowicz *et al.*, 1984a
		↓ Lymphocyte responses to T-cell mitogens and ↓ number of splenic T lymphocytes	Smialowicz *et al.*, 1984a
		↓ NK cell activity	Smialowicz *et al.*, 1984a, 1985a, and 1986a
		↓ Clearance of tumor cells and ↑ tumor burdens in lungs	Smialowicz *et al.*, 1984a, 1985a, and 1987b
		↓ Virus-augmented NK activity and ↑ mortality from MCMV infection	Daniels *et al.*, 1987
	Rat	↓ NK cell activity and ↑ mortality due to tumor burdens	Smialowicz *et al.*, 1987a
		↓ NK cell activity and ↑ development of spontaneous rhabdomyosarcomas	Judde *et al.*, 1987
Oral	Mouse	↓ Antibody response to KLH	Schiffer *et al.*, 1991

KLH, keyhole limpet hemocyanin; MCMV, murine cytomegalovirus; NK, natural killer; SRBC, sheep red blood cell.

(Graham *et al.*, 1975a). Similarly, pretreatment of rat peritoneal macrophages with crystalline NiS ($5\mu g/ml$) inhibited phagocytosis of opsonized SRBC (Jaramillo and Sonnenfeld, 1993). Oxidative metabolic activity of rat alveolar macrophages was maximally inhibited at a concentration of $10\,mM$ $NiCl_2$ (Castranova *et al.*, 1980).

Since the alveolar macrophage is one of the primary cells involved in defense against infectious agents at the alveolar level in the lung, its integrity and activity following nickel exposure has been further scrutinized. Alveolar macrophages from rabbits exposed to $NiCl_2$ aerosols for 1 month at $0.3\,mg$

$NiCl_2/m^3$ displayed decreased *ex vivo* phagocytic activity against opsonized yeast (Wiernik *et al.*, 1983). Furthermore, rabbits similarly exposed to aerosols of 0.2 to 0.6 mg $NiCl_2/m^3$ had decreased lysozyme levels in lung lavage fluid and in alveolar macrophages (Lundborg and Camner, 1984). Macrophages obtained from these same rabbits, when cultured *in vitro*, produced decreased levels of lysozyme activity. Confirmation of this *in vivo*-generated decrease in macrophage lysozyme activity with nickel exposure was obtained in an *in vitro* study (Lundborg *et al.*, 1987). Alveolar macrophages isolated from normal rabbits were cultured in the presence of $NiCl_2$ at concentrations ranging from 0 to 24 µg $NiCl_2$/ml for 2 days. A nickel dose-related suppression of lysozyme activity, as determined by activity in culture supernates, was observed (Lundborg *et al.*, 1987). These data suggest that the decreased lysozyme activity observed *in vivo* is due to a direct effect of Ni^{2+} ions on macrophages. Lysozyme is an important enzyme in alveolar macrophages, which have high levels of this enzyme, because it is active against mucopeptide which is responsible for the structural integrity of the bacterial cell wall. As such, this enzyme activity in alveolar macrophages is related to the capacity of these cells to destroy some of the microorganisms which commonly invade the lower respiratory tract (Myrvik *et al.*, 1961).

Evidence of the consequences of nickel-induced alterations in alveolar macrophage function and in other respiratory tract immunoprotective functions was revealed in a study reported by Adkins *et al.* (1979). Mice were exposed for 2 h to $NiCl_2$ or $NiSO_4$ aerosols (at $0.455 mg/m^3$) and then challenged with an aerosol of viable *Streptococcus pyogenes*. Increased mortality was observed for mice exposed to nickel compared with controls, while nickel-exposed mice demonstrated a reduced ability to clear inhaled streptococci. Alveolar macrophages obtained from rats similarly exposed to nickel aerosols displayed decreased phagocytic activity as measured *in vitro*. Hamsters exposed to 5 mg of NiO via intratracheal instillation following challenge with influenza virus had increased mortality (Port *et al.*, 1975). In addition, the surviving hamsters displayed mild to severe acute interstitial infiltrates of polymorphonuclear cells and macrophages 1 to 2 weeks later.

Nickel also affects various lymphocytes which play important roles in immunosurveillance. The primary antibody response to SRBC in mice, as measured by the splenocyte plaque-forming cell (PFC) assay, was suppressed following a 2-h inhalation exposure to 250 µg $NiCl_2/m^3$ (Graham *et al.*, 1978). Long-term (4 month) inhalation exposure of rats to NiO particles was required to suppress serum anti-SRBC titers (Spiegelberg *et al.*, 1984). A single intramuscular injection of $NiCl_2$ also suppressed the PFC response in mice (Graham *et al.*, 1975b). Suppression of the PFC response to the T-lymphocyte-dependent antigen SRBC, but not the T-lymphocyte-independent antigen polyvinyl-pyrrolidone, following a single intramuscular injection of $NiCl_2$ at 18.3 mg $NiCl_2$/kg, was also reported (Smialowicz *et al.*, 1984a). In this same study, the mitogen-stimulated response of splenic lymphocytes to the T-lymphocyte mitogens, phytohemagglutinin (PHA) and concanavalin-A

(ConA), and the B- and T-lymphocyte mitogen, pokeweed mitogen (PWM), was suppressed in nickel-injected mice. Furthermore, splenic natural killer (NK) cell activity was suppressed, thymus weight was reduced in the absence of decreased body weight, and the percentage of θ-positive splenic T-lymphocytes was reduced in nickel compared with saline injected mice. No alterations, however, were observed in the lymphoproliferative response to the B-lymphocyte mitogen, lipopolysaccharide (LPS), in the percentage of Ig-positive splenic B-lymphocytes, nor in the phagocytic activity of resident peritoneal macrophages for opsonized and non-opsonized chicken erythrocytes. These data indicated that a single intramuscular injection with $NiCl_2$ predominantly affects T-lymphocyte-dependent immune function and NK cell activity in the mouse (Smialowicz *et al.*, 1984a).

Exposure to nickel via the oral route also results in suppression of humoral immunity. However, this route of exposure appears to target the B-lymphocyte rather than the T-lymphocyte following parenteral injection, as well as failing to alter NK cell activity. Mice exposed to $NiSO_4$ orally at 1 to 10 g/L for 6 months displayed dose-related reductions in thymus weight, bone marrow cellularity, and granulocyte–macrophage and pluripotent stem-cell proliferative responses; however, splenic NK cell activity was not altered (Dieter *et al.*, 1988). In addition, the lymphoproliferative response to the B-lymphocyte mitogen LPS was reduced in splenocytes, which was attributed to being secondary to the primary effect of $NiSO_4$ on the myeloid system. Dietary exposure of mice to $NiSO_4$-supplemented diets for 4 weeks resulted in decreased serum IgM- and IgG-specific antibody titers to keyhole limpet hemocyanin (KLH) and lymphoproliferative responses to LPS (Schiffer *et al.*, 1991). In addition, splenocytes obtained from KLH-immunized mice maintained on the nickel-supplemented diet responded poorly to KLH *in vitro*. NK cell activity was not evaluated in this study (Schiffer *et al.*, 1991). The underlying reason for the disparities between the results of oral compared with parenteral exposure studies are not known. However, the route and duration of exposure may be important in so far as these parameters will ultimately influence the nickel concentrations that are delivered to the ultimate effector cell target(s).

NK cells play an important role in anti-tumor and anti-viral immunity as well as participating in the regulation of lymphoid and other hematopoietic cell populations. As such, the perturbation of these cells by exogenous agents, such as nickel, may compromise these functions resulting in increased morbidity and mortality (Smialowicz, 1997). Consequently, other studies have examined the effects of parenteral or inhalation nickel exposure on NK cell function in experimental animals. As indicated above, a single intramuscular injection of $NiCl_2$ in mice suppresses splenic NK cell activity, as determined by the *in vitro* ^{51}Cr-release assay (Smialowicz *et al.*, 1984a, 1985a). In addition, suppression of NK activity by nickel has also been demonstrated by reductions in the clearance of NK-sensitive, radiolabeled, tumor cells from the lungs and spleen, as well as by increased lung tumor burdens following challenge with syngeneic, NK-sensitive, B16F10 melanoma cells (Smialowicz *et al.*, 1984a, 1985a, 1987b).

Young adult (i.e. 8 to 10 weeks old) mice which were exposed to $NiCl_2$ *in utero* via osmotic minipumps implanted subcutaneously, from gestation day 5 to birth, also had reduced NK activity (Smialowicz *et al.*, 1986a). However, these mice did not display increased susceptibility to B16F10 tumors. In another study, mice infected with murine cytomegalovirus (MCMV) and subsequently given a single intramuscular injection of $NiCl_2$ had reduced virus-augmented NK activity and increased mortality to MCMV infection (Daniels *et al.*, 1987). Rat NK activity is also suppressed by $NiCl_2$ exposure via a single intramuscular injection. In addition to suppression of splenic NK cell activity, nickel-injected rats had decreased resistance to challenge with syngeneic, NK-sensitive, MADB106 rat mammary adenocarcinoma cells (Smialowicz *et al.*, 1987a). While NK activity was suppressed in rats following nickel injection, no effects were observed in thymus weights, in the PFC response to SRBC, or in the mitogen-stimulated lymphoproliferative response as were observed in mice (Smialowicz *et al.*, 1984a). *In vitro* exposure of human peripheral blood mononuclear cells to Ni_3S_2 and $NiSO_4$ resulted in a decline in CD56 NK cells, while only Ni_3S_2 reduced $CD4^+$ T cells (Zeromski *et al.*, 1995). Both nickel salts, however, reduced NK cell activity as determined by the ^{51}Cr-release assay.

While a consistent suppression of NK activity has been observed following parenteral administration of nickel, inhalation or lung deposition of nickel compounds has been reported to result in variable changes in NK activity. Inhalation exposure of mice to nickel prior to challenge with MCMV did not affect host susceptibility to this virus nor suppress NK cell activity (Daniels *et al.*, 1987). Splenic NK activity of mice was not altered following exposure to aerosols of Ni_3S_2 for 6h/day for 12 days (Benson *et al.*, 1987, 1988). How-ever, more prolonged exposure of mice to this nickel compound or $NiSO_4$ (i.e. 6h/day, 5 days/week for 65 days) resulted in either decreased splenic NK activity or increased lung tumor burdens following challenge with NK-sensitive B16F10 melanoma cells, respectively (Haley *et al.*, 1990). Unfor-tunately, alveolar NK activity was not determined in any of these three inha-lation studies. On the other hand, lung-associated NK activity was evaluated in cynomolgus monkeys which had been previously immunized and repeatedly challenged with SRBC in specific lung lobes followed by instillation of Ni_3S_2 (Haley *et al.*, 1987). In this study, NK cell activity was enhanced in all lung lobes examined regardless of SRBC or nickel exposure in all nickel-exposed monkeys. While Ni is undoubtedly capable of modulating NK cell activity both *in vivo* and *in vitro*, the biological relevance of Ni-induced NK modulation is debatable given that *in vivo* suppression occurs following parenteral but not inhalation exposure. Future work should focus on determining if long-term inhalation exposure to Ni alters lung-associated NK activity, and host resist-ance models which are dependent upon intact lung NK function.

In contrast to nickel-induced suppression of NK activity, a single intramus-cular injection of manganese chloride in mice resulted in enhanced splenic NK activity, as determined by the *in vitro* ^{51}Cr-release assay, by increased *in vivo* clearance of radiolabeled syngeneic tumor cells, and by decreased B16F10 lung tumor burdens (Rogers *et al.*, 1983; Smialowicz *et al.*, 1984b, 1988). Manganese also enhanced macrophage phagocytic function as well as tumor cytostatic and cytolytic activity in mice (Smialowicz *et al.*, 1985b). Manganese-induced enhancement of NK activity was found to be mediated via the induction of IFN both *in vivo* (Smialowicz *et al.*, 1984b, 1988) and *in vitro* (Smialowicz *et al.*, 1986b). Finally, the co-administration of $NiCl_2$ and manganese chloride in mice resulted in the reversal of suppressed NK activity in mice injected with $NiCl_2$ alone (Smialowicz *et al.*, 1987b). These results are interesting given that while the intramuscular injection of Ni_3S_2 in rats results in the induction of rhabdomyosarcomas (Gilman, 1962; Gilman and Ruckerbauter, 1962), the co-administration of Ni_3S_2 plus manganese dust inhibits the induction of these tumors by nickel (Sunderman *et al.*, 1974, 1976). Taken together, these results suggest that nickel-induced carcinogenesis and its inhibition by manganese may be related to the opposite effects of these two metals on NK cell activity. In fact, results from a study with rats provides supportive evidence for this hypothesis (Judde *et al.*, 1987). Rats given a single intramuscular injection of either metallic powder or nickel subsulfide had reduced NK cell activity in peripheral blood lymphocytes and those rats which had persistent depressed NK activity subsequently developed nickel-induced rhabdomyosarcomas. The inclusion of manganese dust with nickel injected intramuscularly inhibited the development of local tumors and also prevented the depression of NK cell activity produced by nickel alone (Judde *et al.* 1987). Since IFN is an important regulator of NK activity (Smialowicz, 1996), the fact that manganese enhances IFN production (Smialowicz *et al.*, 1984a, 1986b, 1988), while nickel inhibits it (Treagan and Frust, 1970; Gainer, 1977; Sonnenfeld *et al.*, 1983; Jaramillo and Sonnenfeld, 1989) suggests that one possible mechanism of nickel-induced suppression of NK activity may be due to inhibition of basal IFN production. More data, however, are necessary to confirm and support this hypothesis.

8.5.3   *Concluding remarks*

Experimental studies, in which animals have been exposed to different nickel compounds, have revealed variable suppression of natural (i.e. innate) and/or specific (i.e. cell-mediated and humoral) immunity. Nickel-induced depressed immune function has been correlated with decreased host resistance to infection and tumors in some studies. The immunosuppression elicited is dependent not only on the form of nickel, but also on the dose, route, and duration of exposure, as well as on the host species examined. In some cases, similar immune suppression has been observed using environmentally relevant (i.e. inhalation and oral) and non-relevant routes (i.e. subcutaneous and intramuscular) of exposure. The determining factor in whether nickel exposure results in immunosuppression may be dependent upon achieving a nickel concentration at the target tissue(s) or cell(s) that is sufficiently toxic to elicit dysfunction. While under certain conditions animals are immunosuppressed by nickel, there is no clinical evidence for immunosuppression in humans. Nevertheless, the results of studies in animals suggest that the potential exists for nickel-induced immunosuppression in humans, particularly in individuals exposed to high levels. Allergic hypersensitivity, however, remains the most significant clinical immune alteration precipitated by nickel in humans.

Disclaimer

This chapter has been reviewed by the Environmental Protection Agency's Office of Research and Development, and approved for publication. Approval does not signify that the contents necessarily reflect the views and policies of the Agency nor does mention of trade names or commercial products constitute endorsement or recommendation for use.

References

ABBRACCHIO, M.P., SIMMONS-HANSEN, J., and COSTA, M. (1982). Cytoplasmic dissolution of phagocytosed crystalline nickel sulfide particles: a prerequisite for nuclear uptake of nickel. *J. Toxicol. Environ. Health* **9**, 663–676.

ACGIH (American Conference of Governmental Industrial Hygienists) (1990). Threshold limit values for chemical substances and physical agents and biological exposure indices, Cincinnati, OH: American Conference of Government Industrial Hygienists.

ADALIS, D., GARDNER, D.E., and MILLER, F.J. (1978). Cytotoxic effects of nickel on ciliated epithelium. *Amer. Rev. Resp. Dis.* **118**, 347–354.

ADKINS, B., JR, RICHARDS, J.H., and GARDNER, D.E. (1979). Enhancement of experimental respiratory infection following nickel inhalation. *Environ. Res.* **20**, 33–42.

AMBROSE, A.M., LARSON, P.S., BORZELLECA, J.F., and HENNIGAR, G.R., JR (1976). Long term toxicologic assessment of nickel in rats and dogs. *J. Food Sci. Technol.* **13**, 181–187.

ANKE, M., GRUN, M., GROPPEL, B., and KRONEMANN, H. (1983). Nutritional requirements of nickel, in Sarkar, B. (Ed.) *Biological Aspects of Metals and Metal-Related Diseases*, pp. 89–105, New York: Raven Press.

ASATO, N., VAN SOESTBERGEN, M., and SUNDERMAN, F.W., JR (1975). Binding of ^{63}Ni(II) to ultrafiltrable constituents of rabbit serum in vivo and in vitro. *Clin. Chem.* **21**, 521–527.

ATHAR, M., HASAN, S.K., and SRIVASTAVA, R.C. (1987). Evidence for the involvement of hydroxyl radicals in nickel mediated enhancement of lipid peroxidation: implications for nickel carcinogenesis. *Biochem. Biophys. Res. Commun.* **147**, 1276–1281.

ATSDR (Agency for Toxic Substances and Disease Registry) (1993). Toxicological profile for nickel (update). Agency for Toxic Substances and Disease Registry, US

Department of Health and Human Services, Public Health Service, ATSDR/TP-92/14.

AVNSTORP, C., MENNE, T., and MAIBACH, H. (1990). Contact allergy to chromium and nickel, in Dayan, A.D., Hertel, R.F., Heseltine, E., Kazantzis, G., Smith, E.M., and Van der Venne, M.T. (Eds) *Immunotoxicity of Metals and Immunotoxicology*, pp. 83–91, New York: Plenum Press.

BARKER, J.N. (1992). Role of keratinocytes in allergic contact dermatitis. *Contact Dermatitis* **26**, 145–148.

BENNETT, B.G. (1984). Environmental nickel pathways in man, in Sunderman, F.W., JR (Ed.) *Nickel in the Human Environment. Proceedings of a Joint Symposium*, IARC Scientific Publication No. 53, pp. 489–495, Lyon, France: International Agency for Research on Cancer.

BENSON, J.M., HENDERSON, R.F., McCLELLAN, R.O., HANSON, R.L., and REBAR, A.H. (1986). Comparative acute toxicity of four nickel compounds to F344 rat lung. *Fund. Appl. Toxicol.* **7**, 340–347.

BENSON, J.M., CARPENTER, R.L., HAHN, F.F., HALEY, P.J., HANSON, R.L., HOBBS, C.H., PICKRELL, J.A., and DUNNICK, J.K. (1987). Comparative inhalation toxicity of nickel subsulfide to F344/N rats and $B_6C_3F_1$ mice exposed for twelve days. *Fundam. Appl. Toxicol.* **9**, 251–265.

BENSON, J.M., BURT, D.G., and CARPENTER, R.L. (1988). Comparative inhalation toxicity of nickel sulfate to F344/N rats and $B_6C_3F_1$ mice exposed for twelve days. *Fundam. Appl. Toxicol.* **10**, 164–178.

BLOCK, G.T. and YEUNG, M.B. (1982). Asthma induced by nickel. *J. Am. Med. Assoc.* **247**, 1600–1602.

BOLDT, J.R., JR (1967). *The Winning of Nickel. Its Geology, Mining and Extractive Metallurgy*, pp. 83–87, Toronto: D. Van Nostrand Company.

BRAATHEN, L.W. (1980). Studies on human epidermal Langerhans cells III: induction of T-lymphocyte response to nickel sulphate in sensitized individuals. *Br. J. Dermatol.* **103**, 517–526.

BRAATHEN, L.W. and THORSBY, E. (1983). Human epidermal Langerhans cells are more potent than blood monocytes in inducing some antigen-specific T-cell responses. *Br. J. Dermatol.* **108**, 139–146.

CAMNER, P., JOHANSSON, A., and LUNDBORG, M. (1978). Alveolar macrophages in rabbits exposed to nickel dust. Ultrastructural changes and effect on phagocytosis. *Environ. Res.* **16**, 226–235.

CAMNER, P., CASARETT-BRUCE, M., CURSTEDT, T., JARSTRAND, C., WIERNIK, A., JOHANSSON, A., LUNDBORG, M., and ROBERTSON, B. (1984). Toxicology of nickel, in Sunderman, F.W., JR (Ed.) *Nickel in the Human Environment*, IARC Scientific Publication No. 53, pp. 267–276, Lyon, France: International Agency for Research on Cancer.

CASTRANOVA, V., BOWMAN, L., REASOR, M.J., and MILES, P.R. (1980). Effects of heavy metal ions on selected oxidative metabolic processes in rat alveolar macrophages. *Toxicol. Appl. Pharmacol.* **53**, 14–23.

CHERNOFF, N. and KAVLOCK, R.J. (1982). An *in vivo* teratology screen utilizing pregnant mice. *J. Toxicol. Environ. Health* **10**, 541–550.

CHOVIL, A., SUTHERLAND, R.B., and HALLIDAY, M. (1981). Respiratory cancer in a cohort of nickel sinter plant workers. *Br. J. Ind. Med.* **38**, 327–333.

CHRISTENSEN, B. and MOLLER, H. (1975). External and internal exposure to the antigen in the hand eczema of nickel allergy. *Contact Dermatitis* **1**, 134–141.

COOGAN, T.P., LATTA, D.M., SNOW, E.T., and COSTA, M. (1989). Toxicity and carcinogenicity of nickel compounds. *CRC Crit. Rev. Toxicol.* **19**, 341–384.

COOLEY, W.E. (1987). Nickel, in *McGraw-Hill Encyclopedia of Science and Technology*, Vol. 11, 6th Edn, pp. 654–656, New York: McGraw-Hill.

COSTA, M. (1991). Molecular mechanisms of nickel carcinogenesis. *Annu. Rev. Pharmacol. Toxicol.* **31**, 321–337.

COSTA, M., SALNIKOW, K., COSENTINO, S., KLEIN, C.B., HUANG, X., and ZHUANG, Z. (1994). Molecular mechanisms of nickel carcinogenesis. *Environ. Health Perspect.* **102**(Suppl. 3), 127–130.

CURSTEDT, T., HAGMAN, M., ROBERTSON, B., and CAMNER, P. (1983). Rabbit lungs after long-term exposure to low nickel dust concentration. I. Effects on phospholipid concentration and surfactant activity. *Environ. Res.* **30**, 89–94.

DANIELS, M.J., MENACHE, M.G., BURLESON, G.R., GRAHAM, J.A., and SELGRADE, M.J. (1987). Effects of $NiCl_2$ and $CdCl_2$ on susceptibility to murine cytomegalovirus and virus-augmented natural killer cell and interferon responses. *Fundam. Appl. Toxicol.* **8**, 443–453.

DA SILVA, J.J. and WILLIAMS, R.J. (1991). *The Biological Chemistry of the Elements. The Inorganic Chemistry of Life*, pp. 33–45, Oxford: Clarendon Press.

DAVIES, J.E. (1986). Occupational asthma caused by nickel salts. *J. Soc. Occup. Med.* **36**, 29–31.

DIECKERT, G.B., GRAF, E.G., and THAUER, R.K. (1979). Nickel requirements for carbon monoxide dehydrogenase formation in *Clostridium pasteurianum. Arch. Microbiol.* **12**, 117–120.

DIETER, M.P., JAMESON, C.W., TUCKER, A.N., LUSTER, M.I., FRENCH, J.E., HONG, H.L., and BOORMAN, G.A. (1988). Evaluation of tissue disposition, myelopoietic, and immunologic responses in mice after long-term exposure to nickel sulfate in the drinking water. *J. Toxicol. Environ. Health* **24**, 357–372.

DIXON, N.E., GAZZOLA, C., BLAKELEY, R.L., and ZERNER, B. (1975). Jack bean urease, a metalloenzyme: a simple biological role for nickel? *J. Am. Chem. Soc.* **97**, 4131–4133.

DOLL, R., MATHEWS, J.D., and MORGAN, L.G. (1977). Cancers of the lung and nasal sinuses in nickel workers: a reassessment of the period of risk. *Br. J. Ind. Med.* **34**, 102–105.

DOLL, R., ANDERSEN, A., COOPER, W.C., COSMATOS, I., CRAGLE, D.L., EASTON, D., ENTERLINE, P., GOLDBERG, M., METCALFE, L., NORSETH, T., PETO, J., RIGAUT, J.P., ROBERTS, R., SEILKOP, S.K., SHANNON, H., SPEIZER, F., SUNDERMAN, F.W., JR, THORNHILL, P., WARNER, J.S., WEGLO, J., and WRIGHT, M. (1990). Report of the international committee on nickel carcinogenesis in man. *Scand. J. Work Environ. Health* **16**, 1–82.

DOLOVICH, J., EVANS, S.L., and NIEBOER, E. (1984). Occupational asthma from nickel sensitivity: I. Human serum albumin in the antigenic determinant. *Br. J. Ind. Med.* **41**, 51–55.

DONSKOY, E., DONSKOY, M., FOROUHAR, F., GILLIES, C.G., MARZOUK, A., REID, M.C., ZAHARIA, O., and SUNDERMAN, F.W., JR (1986). Hepatic toxicity of nickel chloride in rats. *Ann. Clin. Lab. Sci.* **16**, 108–117.

DOTTERUD, L.K. and FALK, E.S. (1994). Metal allergy in north Norwegian schoolchildren and its relationship with ear piercing and atopy. *Contact Dermatitis* **31**, 308–313.

184

DZERGOWSKY, W.S., DZERGOWSKY, S.K., and SHUMOFF-SIEBER, N.O. (1906–1907). Die Wirkung von Nikelsalzer auf den tierischen Organismus. *Biochem. Z.* **2**, 190–218.

EPA (Environmental Protection Agency) (1988). Environmental Protection Agency, 40 CFR Parts 400–475. Analysis of clean water act effluent guidelines: pollutants. Summary of the chemicals regulated by industrial point source category. Federal Register.

FISHBEIN, L. (1987). Trace and ultra-trace elements in nutrition. *Toxicol. Environ. Chem.* **14**, 73–99.

FISHER, A.A. (1991). Nickel dermatitis in children. *Current Contact News* **47**, 19–20.

FOULKES, E.C. and BLANK, S. (1984). The selective action of nickel on tubule function in rabbit kidneys. *Toxicology* **33**, 245–249.

GAINER, J.H. (1977). Effects of heavy metals and/or of deficiency of zinc on mortality rates in mice infected with encephalomyocarditis virus. *Amer. J. Vet. Res.* **38**, 869–872.

GARIOCH, J.J., MACKIE, R.M., CAMPBELL, I., and FORSYTH, A. (1991). Keratinocyte expression of intercellular adhesion molecule 1 (ICAM-1) correlated with infiltration of lymphocyte function associated antigen 1 (LFA-1) positive cells in evolving allergic contact dermatitis reactions. *Histopathology* **19**, 351–354.

GIASHUDDIN, M. and CORNFIELD, A.H. (1979). Effects of adding nickel (as oxide) to soil on nitrogen and carbon mineralization at different pH values. *Environ. Pollut.* **19**, 67–70.

GILMAN, J.P.W. (1962). Metal carcinogenesis. II. A study on the carcinogenic activity of cobalt, copper, iron, and nickel compounds. *Cancer Res.* **22**, 158–162.

GILMAN, J.P.W. and RUCKERBAUER, G.M. (1962). Metal carcinogenesis. I. Observations on the carcinogenicity of a refiner dust, cobalt oxide, and colloidal thorium dioxide. *Cancer Res.* **22**, 152–157.

GITLITZ, P.H., SUNDERMAN, F.W., JR, and GOLDBLATT, P.J. (1975). Aminoaciduria and proteinuria in rats after a single intraperitoneal injection of Ni (II). *Toxicol. Appl. Pharmacol.* **34**, 430–440.

GMELIN, C.G. (1826). Experiences sur l'action de la baryte, de la strontiane, du chrome, du molybdene, du tungstene, du tellure, de l'osmium, du platine, de l'iridium, du rhodium, du palladium, du nickel, du cobalt, de l'urane, du cerium, dur fer et du manganese sur l'organisme animal. *Bull. Sci. Med.* **7**, 110–117.

GOEBELER, M., MEINARDUS-HAGER, G., ROTH, J., GOERDT, S., and SORG, C. (1993). Nickel chloride and cobalt chloride, two common contact sensitizers, directly induce expression of intercellular adhesion molecule-1 (ICAM-1), vascular cell adhesion molecule-1 (VCAM-1), and endothelial leukocyte adhesion molecule (ELAM-1) by endothelial cells. *J. Invest. Dermatol.* **100**, 759–765.

GOEBELER, M., ROTH, J., BROCKER, E.B., SORG, C., and SCHULZE-OSTHOFF, K. (1995). Activation of nuclear factor-κB and gene expression in human endothelial cells by the common haptens nickel and cobalt. *J. Immunol.* **155**, 2459–2467.

GRAHAM, J.A., GARDNER, D.E., WATERS, M.D., and COFFIN, D.L. (1975a). Effects of trace metals on phagocytosis by alveolar macrophages. *Infect. Immun.* **11**, 1278–1283.

GRAHAM, J.A., GARDNER, D.E., MILLER, F.J., DANIELS, M.J., and COFFIN, D.L. (1975b). Effect of nickel chloride on primary antibody production in the spleen. *Environ. Res.* **16**, 77–87.

185

GRAHAM, J.A., MILLER, F.J., DANIELS, M.J., PAYNE, E.A., and GARDNER, D.E. (1978). Influence of cadmium, nickel, and chromium on primary immunity in mice. *Environ. Res.* **16**, 77–87.

GRANDJEAN, P. (1984). Human exposure to nickel, in Sunderman, F.W. Jr (Ed.) *Nickel in the Human Environment. Proceedings of a Joint Symposium*, IARC Scientific Publication No. 53, pp. 469–485, Lyon, France: International Agency for Research on Cancer.

GUENICHE, A., VIAC, J., LIZARD, G., CHARVERON, M., and SCHMITT, D. (1994a). Effect of nickel on the activation state of normal human keratinocytes through interleukin 1 and intercellular adhesion molecule 1 expression. *Br. J. Dermatol.* **131**, 250–256.

—— (1994b). Effect of various metals on intercellular adhesion molecule-1 expression and tumour necrosis factor alpha production by normal human keratinocytes. *Arch. Dermatol. Res.* **286**, 466–470.

HALEY, P.J., BICE, D.E., MUGGENBURG, B.A., HAHN, F.F., and BENJAMIN, S.A. (1987). Immunopathologic effects of nickel subsulfide on the primate pulmonary immune system. *Toxicol. Appl. Pharmacol.* **88**, 1–12.

HALEY, P.J., SHOPP, G.M., BENSON, J.M., CHENG, Y.S., BICE, D.E., LUSTER, M.I., DUNNICK, J.K., and HOBBS, C.H. (1990). The immunotoxicity of three nickel compounds following 13-week inhalation exposure in the mouse. *Fund. Appl. Toxicol.* **15**, 476–487.

HANSEN, K. and STERN, R.M. (1984). Toxicity and transformation of nickel compounds *in vitro*, in Sunderman, F.W. (Ed.) *Nickel in the Human Environment. Proceedings of a Joint Symposium*, IARC Scientific Publication No. 53, pp. 193–200, Lyon, France: International Agency for Research on Cancer.

HANSEN, H.E., MENNE, T., and OLESEN, L.S. (1982). HLA antigens in nickel sensitive females: based on a twin and a patient population. *Tissue Antigens* **19**, 306–310.

HERXHEIMER, K. (1912). Ueber die gewerblichen erkrankungen der haut. *Dtch. Med. Wochenschr.* **38**, 18–22.

HIMMELHOCH, S.R., SOBER, H.A., VALLEE, B.L., PETERSON, E.A., and FUWA, K. (1966). Spectrographic and chromatographic resolution of metalloproteins in human serum. *Biochemistry* **5**, 2523–2530.

HO, W. and FRUST, A. (1973). Nickel excretion by rats following a single treatment. *Proc. West. Pharmacol. Soc.* **16**, 245–248.

HOEY, M.J. (1966). The effects of metallic salts on the histology and function of the rat testis. *J. Reprod. Fertil.* **12**, 461–471.

HORAK, E. and SUNDERMAN, F.W., JR (1980). Nephrotoxicity of nickel carbonyl in rats. *Ann. Clin. Lab. Sci.* **10**, 425–431.

HOSTYNEK, J.J., HINZ, R.S., LORENCE, C.R., PRICE, M., and GUY, R.H. (1993). Metals and skin. *CRC Crit. Rev. Toxicol.* **23**, 171–235.

HUGHES, K., MEEK, M.E., NEWHOOK, R., and CHAN, P.K.L. (1995). Speciation in health risk assessments of metals: evaluation of effects associated with forms present in the environment. *Regul. Toxicol. Pharmacol.* **22**, 213–220.

IARC (International Agency for Research on Cancer) (1990). *IARC Monographs on Evaluation of Carcinogenic Risks to Humans.* Vol. 49, *Chromium, Nickel, and Welding*, pp. 257–445, Lyon, France: IARC Scientific Publications, International Agency for Research on Cancer.

IKAHEIMO, I., TIILIKAINEN, A., KARVONEN, J., and SILVENNOINEN-KASSINEN, S. (1993). HLA-DQA1 and -DQB1 loci in nickel allergy patients. *Int. Arch. Allergy Immunol.* **100**, 248–250.

ILLIS, A. (1987). Nickel metallurgy, in *McGraw-Hill Encyclopedia of Science and Technology*, Vol. 11, 6th Edn, pp. 658–661, New York: McGraw-Hill.

IRIS (1992). *Integrated Risk Information System.* Office of Health and Environmental Assessment, Environmental Criteria and Assessment Office, US Environmental Protection Agency, Cincinnati, OH.

ISHII, N., SUGITA, Y., NAKAJIMA, H., TANAKA, S.I., and ASKENASE, P.W. (1995). Elicitation of nickel sulfate ($NiSO_4$)-specific delayed-type hypersensitivity requires early-occurring and early-acting $NiSO_4$-specific DTH-initiating cells with an unusual mixed phenotype for an antigen-specific cell. *Cell. Immunol.* **161**, 244–255.

JARAMILLO, A. and SONNENFELD, G. (1989). Effects of amorphous and crystalline nickel sulfide on induction of interferons-α/β and -γ and interleukin-2. *Environ. Res.* **48**, 275–286.

(1993). Effects of nickel sulfide on induction of interleukin-1 and phagocytic activity. *Environ. Res.* **63**, 16–25.

JARSTRAND, C., LUNDBORG, M., WIERNIK, A., and CAMNER, P. (1978). Alveolar macrophage function in nickel dust-exposed rabbits. *Toxicology* **11**, 353–359.

JASMIN, G. and RIOPELLE, J.L. (1976). Renal carcinomas and erythrocytosis in rats following intrarenal injection of nickel subsulfide. *Lab. Invest.* **35**, 71–78.

JOHANSSON, A. and CAMNER, P. (1980). Effects of nickel dust on rabbit alveolar epithelium. *Envir. Res.* **22**, 510–516.

JOHANSSON, A., CAMNER, P., JARSTRAND, C., and WIERNIK, A. (1980). Morphology and function of alveolar macrophages after long-term nickel exposure. *Environ. Res.* **23**, 170–180.

JOHANSSON, A., CAMNER, P., and ROBERTSON, B. (1981). Effects of long-term nickel dust exposure on rabbit alveolar epithelium. *Environ. Res.* **25**, 391–403.

JOHANSSON, A., CAMNER, P., JARSTRAND, C., and WIERNIK, A. (1983). Rabbit lungs after long-term exposure to low nickel dust concentrations. II. Effects on morphology and function. *Environ. Res.* **30**, 142–151.

JUDDE, J.G., BREILLOUT, F., CLEENCEAU, C., POUPON, M.F., and JASMIN, C. (1987). Inhibition of rat natural killer cell function by carcinogenic nickel compounds: preventive action of manganese. *J. Natl Cancer Inst.* **78**, 1185–1190.

KANERVA, L., SIPILAINEN-MALM, T., ESTLANDER, T., ZITTING, A., JOLANKI, R., and TARVAINEN, K. (1994). Nickel release from metals, and a case of allergic contact dermatitis from stainless steel. *Contact Dermatitis* **31**, 299–303.

KAPSENBERG, M.L., RES, P., BOS, J.D., SCHOOTEMIJER, A., TEUNISSEN, M.B., and SCHOOTEN, W.V. (1987). Nickel-specific T-lymphocyte clones derived from allergic nickel-contact dermatitis lesions in man: heterogeneity based on requirement of dendritic antigen-presenting cell subsets. *Eur. J. Immunol.* **17**, 861–865.

KAPSENBERG, M.L., VAN DER POUW-FRANN, T., STIEKEMA, F.E., SCHOOTEMIJER, A., and BOS, J.D. (1988). Direct and indirect nickel-specific stimulation of T-lymphocytes from patients with allergic contact dermatitis to nickel. *Eur. J. Immunol.* **18**, 977–982.

KAPSENBERG, M.L., WIERENGA, E.A., BOS, J.D., and JANSEN, H.M. (1991). Functional subsets of allergen-reactive human $CD4^+$ T-cells. *Immunol. Today* **12**, 392–395.

KAPSENBERG, M.L., WIERENGA, E.A., STEIKEMA, F.E., ANKE, M.B., TIGGELMAN, A.M., and BOS, J.D. (1992). T_{H1} lymphokine production profiles of nickel-specific CD4$^+$ T-lymphocyte clones from nickel contact allergic and non-allergic individuals. *Invest. Dermatol.* **98**, 59–63.

KASPRZAK, K.S., WAALKES, M.P., and POIRIER, L.A. (1986). Antagonism by essential divalent metals and amino acids of nickel (II)-DNA binding *in vitro*. *Toxicol. Appl. Pharmacol.* **82**, 336–343.

KIMBER, I. (1995). Chemical-induced hypersensitivity, in Smialowicz, R.J. and Holsapple, M.P. (Eds), *Experimental Immunotoxicology*, pp. 391–417, Boca Raton, FL: CRC Press.

KLIGMAN, A.M. (1966). Identification of contact allergies by human assay. III. Maximization test: a procedure for screening and rating contact sensitizers. *J. Invest. Dermatol.* **47**, 393–409.

KNIGHT, J.A., REZUKE, W.N., WONG, S.H., HOPFER, S.M., ZAHARIA, O., and SUNDERMAN, F.W., JR (1987). Acute thymic involution and increased lipoperoxides in thymus of nickel chloride-treated rats. *Res. Commun. Chem. Pathol. Pharmacol.* **55**, 101–109.

LAMMINTAUSTA, K., KALIMO, K., and JANSEN, C.T. (1985). Experimental nickel sensitization in the guinea pig: comparison of different protocols. *Contact Dermatitis* **12**, 258–262.

LAMMINTAUSTA, K., KORHONEN, K., and JANSEN, C.T. (1986). Methods of sensitization determines if UVB irradiation inhibits the development of delayed type hypersensitivity to nickel in guinea pigs. *Photodermatology* **3**, 102–103.

LARSSON-STYMNE, B. and WILDSTROM, L. (1985). Ear piercing: a cause of nickel allergy in schoolgirls. *Contact Dermatitis* **13**, 289–293.

LAU, T.J., HACKETT, R.L., and SUNDERMAN, F.W., JR (1972). The carcinogenicity of intravenous nickel carbonyl in rats. *Cancer Res.* **32**, 2253–2258.

LEHMANN, K.B. (1908–1909). Hygienische Studien uber Nickel. *Arch. Hyg.* **68**, 421–465.

LU, C.C., MATSUMOTO, N., and IIJIMI, S. (1979). Teratogenic effects of nickel chloride on embryonic mice and its transfer to embryonic mice. *Teratology* **19**, 137–142.

LUNDBORG, M. and CAMNER, P. (1984). Lysozyme levels in rabbit lung after inhalation of nickel, cadmium, cobalt, and copper chlorides. *Environ. Res.* **34**, 335–342.

LUNDBORG, M., JOHNSSON, A., and CAMNER, P. (1987). Morphology and release of lysozyme following exposure of rabbit lung macrophages to nickel or cadmium in vitro. *Toxicology* **46**, 191–202.

MALO, J.L., CARTIER, A., DOEPNER, M., NIEBOER, E., EVANS, S., and DOLOVICH, J. (1982). Occupational asthma caused by nickel sulfate. *J. Allergy Clin. Immunol.* **69**, 55–59.

MALO, J.L., CARTIER, A., GAGNON, G., EVANS, S., and DOLOVICH, J. (1985). Isolated late asthmatic reaction due to nickel antibody. *Clin. Allergy* **15**, 95–99.

MAURER, T., THOMANN, P., WEIRICH, E.G., and HESS, R. (1979). Predictive evaluation in animals of the contact allergenic potential of medically important substances. II. Comparison of different methods of cutaneous sensitization with 'weak' allergens. *Contact Dermatitis* **5**, 1–10.

MCCONNELL, L.H., FINK, J.N., SCHLUETER, D.P., and SCHMIDT, M.G., JR (1973). Asthma caused by nickel sensitivity. *Ann. Intern. Med.* **78**, 888–890.

MCILVEEN, W.D. and NEGUSANTI, J.J. (1994). Nickel in the terrestrial environment. *Sci. Total Environ.* **148**, 109–138.

MENNE, T. (1994). Quantitative aspects of nickel dermatitis. Sensitization and elicited threshold concentrations. *Sci. Total Environ.* **148**, 275–281.

MENNE, T. and HOLM, N.V. (1989). Genetic aspects of nickel sensitization, in Maibach, H.I. and Menne, T. (Eds) *Nickel and Skin: Immunology and Toxicology*, pp. 101–107, Boca Raton, FL: CRC Press.

MENNE, T. and MAIBACH, H.I. (1987). Systemic contact allergy reactions. *Semin. Dermatol.* **6**, 108–118.

MENNE, T., BRANDUP, F., THESTRUP-PEDERSEN, K., VEIEN, N.K., ANDERSEN, J.R., YDING, F., and VALEUR, G. (1987). Patch test reactivity to nickel alloys. *Contact Dermatitis* **16**, 255–259.

MERRITT, K. and BROWN, S.A. (1980). Tissue reaction and metal sensitivity. An animal study. *Acta Orthop. Scand.* **51**, 403–411.

MERTZ, W. (1970). Some aspects of nutritional trace element research. *Fed. Proc.* **29**, 1482–1488.

MOLLER, H. (1984). Attempts to induce contact allergy to nickel in the mouse. *Contact Dermatitis* **10**, 65–68.

MOND, L., LANGER, C., and QUINCKE, F. (1890). Action of carbon monoxide on nickel. *J. Chem. Soc.* **67**, 749–753.

MORGAN, L.G. and USHER, V. (1994). Health problems associated with nickel refining and use. *Ann. Occup. Hyg.* **38**, 189–198.

MOSMANN, T.R. and COFFMAN, R.L. (1989). T_{H1} and T_{H2} cells: Different patterns of lymphokine secretion lead to different functional properties. *Ann. Rev. Immunol.* **7**, 145–173.

MOZZANICA, N., RIZZOLO, L., VENERONI, G., DIOTTI, R., HEPEISEN, S., and FINZI, A.F. (1990). HLA-A,B,C and DR antigens in nickel contact sensitivity. *Br. J. Dermatol.* **122**, 309–313.

MYRVIK, Q.N., LEAKE, E.S., and FARISS, B. (1961). Lysozyme content of alveolar and peritoneal macrophages from the rabbit. *J. Immunol.* **86**, 133–136.

NAS (National Academy of Sciences) (1975). *Nickel.* Washington, DC: National Academy of Sciences, Printing and Publishing Office of the National Academy of Sciences.

NIEBOER, E., MAXWELL, R.I., and STAFFORD, A.R. (1984a). Chemical and biological reactivity of insoluble nickel compounds and the bioinorganic chemistry of nickel, in Sunderman, F.W. JR (Ed.) *Nickel in the Human Environment. Proceedings of a Joint Symposium*, IARC Scientific Publication No. 53, pp. 439–468, Lyon France: International Agency for Research on Cancer.

NIEBOER, E., EVANS, S.L., and DOLOVICH, J. (1984b). Occupational asthma from nickel sensitivity. II. Factors influencing the interaction of Ni^{2+}, HSA, and serum antibodies with nickel related specificity. *Br. J. Ind. Med.* **41**, 56–63.

NIELSEN, F.H. (1984). Ultratrace elements in nutrition. *Ann. Res. Nutr.* **4**, 22–41.

NIELSEN, F.H. and FLYVHOLM, M. (1984). Risks of high nickel intake with diet, in Sunderman, F.W., JR (Ed.) *Nickel in the Human Environment. Proceedings of a Joint Symposium*, IARC Scientific Publication No. 53, pp. 333–338, Lyon, France: International Agency for Research on Cancer.

NIELSEN, N.H. and MENNE, T. (1992). Allergic contact sensitization in an unselected Danish population. The Glostrup allergy study, Denmark. *Acta. Dermatol. Venereol.* **72**, 456–460.

NIOSH (National Institute for Occupational Safety and Health) (1990). *National Occupational Exposure Survey* (*NOES*), Cincinnati, OH: US Department of Health and Human Services, Public Health Service, Centers for Disease Control, National Institute for Occupational Safety and Health.

NOVEY, H.S., HABIB, M., and WELLS, I.D. (1983). Asthma and IgE antibodies induced by chromium and nickel salts. *J. Allergy Clin. Immunol.* **72**, 407–412.

NRC (National Research Council) (1992). *Biological Markers in Immunotoxicology*, Washington, DC: National Research Council. National Academic Press.

NRIAGU, J.O. and PACYNA, J.M. (1988). Quantitative assessment of worldwide contamination of air, water and soils by trace metals. *Nature* **333**, 134–139.

OLERUP, O. and EMTESTAM, L. (1988). Allergic contact dermatitis to nickel is associated with Taq I HLA-DQA allelic restriction fragment. *Immunogenetics* **28**, 310–313.

OSHA (Occupational Safety and Health Administration) (1989). US Department of Labor. Occupational Safety and Health Administration. Air contaminant: Final rule. Code of Federal Regulations. 29 CFR 1910, 2946.

OTTOLENGHI, A.D., HASEMAN, J.K., PAYNE, W.W., FALK, H.L., and MACFARLAND, H.N. (1975). Inhalation studies of nickel sulfide in pulmonary carcinogenesis of rats. *J. Natl Cancer Inst.* **54**, 1165–1172.

PAGE, G.W. (1981). Comparison of groundwater and surface water for patterns and levels of contamination by toxic substances. *Environ. Sci. Technol.* **15**, 1475–1481.

PELTONEN, L. (1979). Nickel sensitivity in the general population. *Contact Dermatitis* **5**, 27–32.

PELTONEN, L. and TERHO, P. (1989). Nickel sensitivity in school children in Finland, in Frosch, P.J. and Dooms-Goosens, A. (Eds) *Current Topics in Contact Dermatitis*, pp. 184–187, Berlin: Springer-Verlag.

PINK, J.R. and SINIGAGLIA, F. (1989). Characterizing T-cell epitopes in vaccine candidates. *Immunol. Today* **10**, 408–409.

PISTOOR, F.H., KAPSENBERG, M.L., BOS, J.D., MEINARDI, M.M., VON BLOMBERG, B.M., and SCHEPER, R.J. (1995). Cross-reactivity of human nickel-reactive T-lymphocyte clones with copper and palladium. *J. Invest. Dermatol.* **105**, 92–95.

PORT, C.D., FENTERS, J.D., EHRLICH, R., COFFIN, D.L., and GARDNER, D. (1975). Interaction of nickel oxide and influenza infection in the hamster. *Environ. Health Perspect.* **10**, 268.

PROBST, P., KUNTZLIN, D., and FLEISCHER, B. (1995). T_{H2}-type infiltrating T-cells in nickel-induced contact dermatitis. *Cell. Immunol.* **165**, 134–140.

PRYSTOWSKY, S.D., ALLEN, A.M., SMITH, R.W., NONOMURA, J.H., ODOM, R.B., and AKERS, W.A. (1979). Allergic contact hypersensitivity to nickel, neomycin, ethylenediamine, and benzocaine. Relationship between age, sex, history of exposure and reactivity of standard patch tests and use tests in a general population. *Arch. Dermatol.* **115**, 959–962.

RES, P., KAPSENBERG, M.L., BOS, J.D., and STIEKEMA, F. (1987). The critical role of human dendritic antigen-presenting cell subsets in nickel-specific T-cell proliferation. *J. Invest. Dermatol.* **88**, 550–554.

ROGERS, R.R., GARNER, R.J., RIDDLE, M.M., LUEBKE, R.W., and SMIALOWICZ, R.J. (1983). Augmentation of murine natural killer cell activity by manganese chloride. *Toxicol. Appl. Pharmacol.* **70**, 7–17.

ROMAGNOLI, P., LABHARDT, A.M., and SINIGAGLIA, F. (1991). Selective interaction of Ni with an MHC-bound peptide. *EMBO J.* **10**, 1303–1306.

RTI (Research Triangle Institute) (1988a). Two-generation reproductive and fertility study of nickel chloride administered to CD rats in the drinking water: Fertility and reproductive performance of the P_0 generation. Final Study report (2 of 3). Report to the Office of Solid Waste Management, US Environmental Protection Agency by Research Triangle Institute, Research Triangle Park, NC.

(1988b). Two-generation reproductive and fertility study of nickel chloride administered to CD rats in the drinking water: Fertility and reproductive performance of the F_1 generation. Final Study report (2 of 3). Report to the Office of Solid Waste Management, US Environmental Protection Agency by Research Triangle Institute, Research Triangle Park, NC.

RUBANYI, G., LIGETI, L., and KOLLER, A. (1981). Nickel is released from the ischemic myocardium and contracts coronary vessels by a Ca-dependent mechanism. *J. Mol. Cell Cardiol.* **13**, 1023–1026.

RUBANYI, G., LIGETI, L., KOLLER, A., and KOVACH, A.G. (1984). Possible role of nickel ions in the pathogenesis of ischemic coronary vasoconstriction in the dog heart. *J. Mol. Cell Cardiol.* **16**, 533–546.

SCHEPER, R.J., VON BLOMBERG, M., VREEBURG, K.J., and VAN HOOGSTRATEN, I.M. (1989). Recent advances in immunology of nickel sensitization, in Maibach H.I. and Menne, T. (Eds) *Nickel and the Skin: Immunology and Toxicology*, pp. 55–63, Boca Raton, FL: CRC Press.

SCHIFFER, R.B., SUNDERMAN, F.W., JR, BAGGS, R.B., and MOYNIHAN, J.A. (1991). The effects of exposure to dietary nickel and zinc upon humoral and cellular immunity in SJL mice. *J. Neuroimmunol.* **43**, 229–239.

SCHMIDT, J.A. and ANDREN, A.W. (1980). The atmospheric chemistry of nickel, in Nriagu, J.O. (Ed.) *Nickel in the Environment*, pp. 93–135, New York: John Wiley & Sons.

SCHNEGG, A. and KIRCHGESSNER, M. (1975a). Essentiality of nickel for the growth of animals. *Z. Tierphysiol. Tierernahr. Futtermittelkd.* **36**, 63–74.

(1975b). Changes in the hemoglobin content, erythrocyte count and hematocrit in nickel deficiency. *Nutr. Metab.* **19**, 263–278.

(1976). Absorption and metabolic efficiency of iron during nickel deficiency. *Int. J. Vitamin Nutr. Res.* **46**, 96–99.

SHIRAKAWA, T., KUSAKA, Y., FUJIMURA, N., KATO, M., HEKI, S., and MORIMOTO, K. (1990). Hard metal asthma: cross immunological and respiratory reactivity between cobalt and nickel? *Thorax* **45**, 267–271.

SINIGAGLIA, F. (1994). The molecular basis of metal recognition by T-cells. *J. Invest. Dermatol.* **102**, 398–401.

SINIGAGLIA, F., SCHEIDEGGER, D., GAROTTA, G., SCHEPER, R., PLETSCHER, M., and LANZAVECCHIA, A. (1985). Isolation and characterization of Ni-specific T-cell clones from patients with Ni-contact dermatitis. *J. Immunol.* **135**, 3929–3932.

SMIALOWICZ, R.J. (1997). Natural Killer (NK) Cells, in Sipes, I.G., McQueen, C.A., and Gandolfi, A.J. (Eds) *Comprehensive Toxicology, Volume 5: Toxicology of the Immune System*, pp. 77–96, Oxford: Pergamon Press.

SMIALOWICZ, R.J., ROGERS, R.R., RIDDLE, M.M., and STOTT, G.A. (1984a). Immunologic effects of nickel. I. Suppression of cellular and humoral immunity. *Environ. Res.* **33**, 413–427.

SMIALOWICZ, R.J., ROGERS, R.R., RIDDLE, M.M., LUEBKE, R.W., ROWE, D.G., and GARNER, R.J. (1984b). Manganese chloride enhances murine cell-mediated cytotoxicity: effects on natural killer cells. *Immunopharmacology* **6**, 1–23.

SMIALOWICZ, R.J., ROGERS, R.R., RIDDLE, M.M., GARNER, R.J., ROWE, D.G., and LUEBKE, R.W. (1985a). Immunologic effects of nickel. II. Suppression of natural killer (NK) cell activity. *Environ. Res.* **36**, 56–66.

SMIALOWICZ, R.J., LUEBKE, R.W., ROGERS, R.R., RIDDLE, M.M., and ROWE, D.G. (1985b). Manganese chloride enhances natural cell-mediated immune effector cell function: effects on macrophages. *Immunopharmacology* **9**, 1–11 .

SMIALOWICZ, R.J., ROGERS, R.R., RIDDLE, M.M., ROWE, D.G., and LUEBKE, R.W. (1986a). Immunological studies in mice following *in utero* exposure to NiCl₂. *Toxicology* **38**, 293–303.

(1986b). *In vitro* augmentation of natural killer cell activity by manganese chloride. *J. Toxicol. Environ. Health* **19**, 243–254.

SMIALOWICZ, R.J., ROGERS, R.R., ROWE, D.G., RIDDLE, M.M., and LUEBKE, R.W. (1987a). The effect of nickel on immune function in the rat. *Toxicology* **44**, 271–281.

SMIALOWICZ, R.J., ROGERS, R.R., RIDDLE, M.M., LUEBKE, R.W., FOGELSON, L.D., and ROWE, D.G. (1987b). Effects of manganese, calcium, magnesium, and zinc on nickel-induced suppression of murine natural killer cell activity. *J. Toxicol. Environ. Health* **20**, 67–80.

SMIALOWICZ, R.J., RIDDLE, M.M., ROGERS, R.R., LUEBKE, R.W., and BURLESON, G.R. (1988). Enhancement of natural killer cell activity and interferon production by manganese in young mice. *Immunopharmac. Immunotoxic.* **10**, 93–107.

SONNENFELD, G., STREIPS, U.N., and COSTA, M. (1983). Differential effects of amorphous and crystalline nickel sulfide on murine α/β interferon production. *Environ. Res.* **32**, 474–479.

SPEARS, J.W. (1984). Nickel as a 'newer trace element' in the nutrition of domestic animals. *J. Anim. Sci.* **59**, 823–834.

SPIEGELBERG, T., KORDEL, W., and HOCHRAINER, D. (1984). Effects of NiO inhalation on alveolar macrophages and the humoral immune system of rats. *Ecotox. Environ. Saf.* **8**, 516–525.

SUNDERMAN, F.W. (1970). Nickel poisoning, in Sunderman, F.W. and Sunderman, F.W. Jr (Eds) *Laboratory Diagnosis of Diseases Caused by Toxic Agents*, pp. 387–396, St Louis: Warren H. Green.

SUNDERMAN, F.W. and DONNELLY, A.J. (1965). Studies of nickel carcinogenesis. Metastasizing pulmonary tumors in rats induced by the inhalation of nickel carbonyl. *Am. J. Pathol.* **46**, 1027–1041.

SUNDERMAN, F.W., JR (1986). Sources of exposure and biological effects of nickel, in O'Neill, K., Schuller, P., and Fishbein, L. (Eds) *Environmental Carcinogens Selected Methods of Analysis. Vol. 8: Some Metals: As, Be, Cd, Cr, Ni, Pb, Se, Zn*, IARC Scientific Publication No. 70, pp. 79–92, Lyon, France: International Agency for Research on Cancer.

SUNDERMAN, F.W., JR and SELIN, C.E. (1968). The metabolism of nickel-63 carbonyl. *Toxicol. Appl. Pharmacol.* **12**, 207–218.

SUNDERMAN, F.W., JR and OSKARSSON, A. (1991). Nickel, in Merian, E. (Ed.) *Metals and Their Compounds in the Environment, Occurrence, Analysis, and Biological Relevance*, pp. 1101–1126, New York: VCH.

SUNDERMAN, F.W., JR, LAU, T.J., and CRALLEY, L.J. (1974). Inhibitory effect of manganese upon muscle tumorigenesis by nickel subsulfide. *Cancer Res.* **34**, 92–95.

192

SUNDERMAN, F.W., JR, KASPRZAK, K.S., LAU, T.J., MINGHETTI, P.P., MAENZA, R.M., BECKER, N., ONKELINX, C., and GOLDBLATT, P.J. (1976). Effects of manganese on carcinogenicity and metabolism of nickel subsulfide. *Cancer Res.* **36**, 1790–1800.

SUNDERMAN, F.W., JR, SHEN, S.K., MITCHELL, J.M., ALLPASS, P.R., and DAMJANOV, I. (1978). Embryotoxicity and fetal toxicity of nickel in rats. *Toxicol. Appl. Pharmacol.* **43**, 381–390.

SUNDERMAN, F.W., JR, MAENZA, R.M., HOPFER, S.M., MITCHELL, J.M., ALLPASS, P.R., and DAMJANOV, I. (1979). Induction of renal cancers in rats by intrarenal injection of nickel subsulfide. *J. Environ. Pathol. Toxicol.* **2**, 1511–1527.

SUNDERMAN, F.W., JR, HOPFER, S.M., LIN, S.M., PLOWMAN, M.C., STOJANOVIC, T., WONG, S.H., ZAHARIA, O., and ZIEBKA, L. (1989). Toxicity to alveolar macrophages in rats following parenteral injection of nickel chloride. *Toxicol. Appl. Pharmacol.* **100**, 107–118.

TEDESCHI, R.E. and SUNDERMAN, F.W. (1957). Nickel poisoning. V. The metabolism of nickel under normal conditions and after exposure to nickel carbonyl. *A.M.A. Arch. Ind. Health* **16**, 486–488.

TEMPLETON, D.M. (1987). Interaction of toxic cations with the glomerulus: binding of Ni to purified glomerular basement membrane. *Toxicology* **43**, 1–15.

TREAGAN, L. and FURST, A. (1970). Inhibition of interferon synthesis in mammalian cell cultures after nickel treatment. *Res. Commun. Chem. Pathol. Pharmacol.* **1**, 395–402.

VANDENBERG, J. and EPSTEIN, W. (1963). Experimental skin contact sensitization in man. *J. Invest. Dermatol.* **41**, 413–416.

VAN HOOGSTRATEN, I.M., BOOS, C., BODEN, D., VON BLOMBERG, M.E., SCHEPER, R.J., and KRAAL, G. (1993). Oral induction of tolerance to nickel sensitization in mice. *J. Invest. Dermatol.* **101**, 26–31.

VON BLOMBERG, M.E., BRUYNZEEL, D.P., and SCHEPER, R.J. (1991). Advances in mechanisms of allergic contact dermatitis: *in vivo* and *in vitro*, in Manzuli, F.N. and Maibach, H.I. (Eds) *Dermatotoxicology*, pp. 255–362, Washington: Hemisphere Publishing Co.

VUOPALA, U., HUHTI, E., TAKKUNEN, J., and HUIKKO, M. (1970). Nickel carbonyl poisoning. Report of 25 cases. *Ann. Clin. Res.* **2**, 214–222.

WAHLBERG, J.E. (1976). Sensitisation of guinea pigs with nickel sulphate. *Dermatologica* **152**, 321–330.

   (1989). Nickel: animal sensitization assays, in Maibach, H.I. and Menne, T. (Eds) *Nickel and the Skin: Immunology and Toxicology*, pp. 65–73, Boca Raton, FL: CRC Press.

WAHLBERG, J.E. and BOMAN, A.S. (1992). Cross-reactivity to palladium and nickel studied in the guinea pig. *Acta Derm. Venereol.* **72**, 95–97.

WATERS, M.D., GARDNER, D.E., ARANYI, C., and COFFIN, D.L. (1975). Metal toxicity for rabbit macrophages *in vitro*. *Environ. Res.* **9**, 32–47.

WELTZIEN, H.U., MOULON, C., MARTIN, S., PADOVAN, E., HARTMANN, U., and KOHLER, J. (1996). T-cell immune responses to haptens. Structural models for allergic and autoimmune reactions. *Toxicology* **107**, 141–151.

WIERNIK, A., JOHANSSON, A., JARSTRAND, C., and CAMNER, P. (1983). Rabbit lung after inhalation of soluble nickel. I. Effects on alveolar macrophages. *Environ. Res.* **30**, 129–141.

WILDNER, O., LIPKOW, T., and KNOP, J. (1992). Increased expression of ICAM-1, E-selectin, and VCAM-1 by cultured human endothelial cells upon exposure to haptens. *Exp. Dermatol.* **1**, 191–198.

ZELIKOFF, J.T. and COHEN, M.D. (1995). Immunotoxicity of inorganic metal compounds, in Smialowicz, R.J. and Holsapple, M.P. (Eds) *Experimental Immunotoxicology*, pp. 189–228, Boca Raton, FL: CRC Press.

ZEROMSKI, J., JEZEWSKA, E., SIKORA, J., and KASPRZAK, K.S. (1995). The effects of nickel compounds on immunophenotype and natural killer cell function of normal human lymphocytes. *Toxicology* **97**, 39–48.

9

Platinum

KATHLEEN RODGERS

Livingston Research Institute, University of Southern California, 1321 North Mission Road, Los Angeles, CA 90033, USA

9.1 Introduction

Platinum (Pt) is a member of Group VIII metals of the Periodic Table. It is a silver-gray, lustrous, malleable, and ductile precious metal that is of high commercial value due to its high resistance to most corrosive agents. Platinum salts are commonly used as catalysts in automotive and chemical industries. Platinum is generally non-toxic in the metallic state, but the soluble halide salts are highly reactive. Exposure to the soluble halide salts of platinum can result in hypersensitivity-allergy reactions in susceptible individuals.

Many complex salts of platinum exist. The most common of the platinum salts are: platinum chloride (platinum tetrachloride, platinic acid, $PtCl_4$); platinum dichloride (platinous chloride, $PtCl_2$); platinum dioxide (platinic oxide, PtO_2); and platinum sulfate (platinic sulfate, $Pt(SO_4)_2$). Other salts which are water soluble include: ammonium and sodium chloroplatinates; sodium, potassium, and ammonium tetrachloroplatinates; and sodium, potassium, and ammonium hexachloro-platinates (Proctor *et al.*, 1988).

The platinoids form an important group of metal alloys. These include (major use for the alloy in parentheses): platinum black (catalyst), platinum–cobalt (permanent magnets), and platinum–rhenium (catalyst) (Sax and Lewis, 1987). In clinical medicine, the two most widely used chemotherapeutic platinum complexes are *cis*-diamminedichloroplatinum(II) (cisplatin, CDDP) and *cis*-diammine(1,1-cyclobutane dicarboxylato)platinum(II) (carboplatin, CBDCA).

The most biologically relevant oxidation states of platinum are +2 (II) and +4 (IV). The coordination chemistry of platinum(II) complexes is square-planar, while that of platinum(IV) complexes is octahedral (McBryde, 1973).

In general, very few studies have been conducted to examine the immunotoxic effects of compounds that contain platinum. The majority of the studies conducted have examined the structure–activity relationship and mechanism of allergenicity associated with exposure to halide salts of platinum and the potential immunomodulation associated with the therapeutic use of the chemotherapeutic CDDP.

9.2 History

Disease has long been associated with occupational exposure to platinum salts. The first report of platinum allergy was in 1804. However, in spite of this, because of the physical and chemical properties described below, platinum has been used in numerous industries including photography, refining, and the manufacturing of fiberglass and high octane fuel.

The therapeutic use of platinum was known as early as 1841, when it was used against syphilis and rheumatism. Since then, platinum compounds have been found to be potent anti-tumor agents with cisplatin as one of the most effective chemotherapeutic agents available. In fact, the implementation of cisplatin in the treatment of malignancies has had a dramatic impact on the prognosis of these cancers.

9.3 Occurrence

Several industries utilize platinum metal and its compounds. These include (percentage of total use in parentheses): automotive (51 percent), electrical (13 percent), petroleum refining (10 percent), chemical (9 percent), ceramics and glass (3 percent), jewelry and arts (3 percent), dentistry (2 percent), and other, including pharmaceutical (9 percent) (Boggs, 1985). Most of the commercial applications exploit the catalytic activities, nobility (resistance to oxidation), and strength of platinum. The chemical industry uses platinum–rhenium catalysts for production of nitric acid and spinnerets for rayon, glass fiber, and Plexiglas manufacture. The pharmaceutical industry produces some drugs and vitamins with platinum catalysts, mostly in hydrogenation or dehydrogenation reactions. Platinum catalysts used in the automotive industry dominate the use of this metal as a result of its being a component of catalytic converters for air pollution abatement (McBryde, 1973; Hawley, 1977; Boggs, 1985). Alloys of platinum–iridium are the most important alloys because they are harder and more resistant to chemical attack than platinum alone. Such alloys are used for jewelry, electrical contacts, fuse wire, and hypodermic needles. Platinum–cobalt alloys have been developed into powerful magnets and are used in hearing aids, self-winding watches, and dental alloys. Other uses for platinum alloys include electroplating and surgical wire (Sax and Lewis, 1987).

9.3.1 *Exposure limits and environmental monitoring*

The current threshold limit values/time weighted averages (TLV-TWA 8 h/day, 40 h/week) for metallic platinum dusts and soluble platinum salts are 1.0 and 0.002 mg/m³, respectively (Sax and Lewis, 1987; ACGIH, 1991). The TLV for platinum salts is set at a level to prevent respiratory effects and is believed to provide protection against sensitization; however, it does not offer protection to a previously sensitized individual.

To monitor the occupational environment for platinum, air is drawn through a cellulose ester filter for approximately 2 h. The filter is treated with acid to dissolve soluble platinum salts, and the resultant solution is analyzed for platinum by inductively coupled plasma atomic emission spectroscopy (ICP-AES).

9.4 General Toxicology

9.4.1 *Platinum uptake and excretion*

Absorption of platinum from most exposure routes is very low, with the exception of inhalation or intravenous injection (Venugopal and Luckey, 1978). Following inhalation, a majority of the platinum dose and its salts is retained in the lungs and respiratory tract. After intravenous injection, most platinoids distribute to soft tissues, mainly kidney, liver, muscle, and spleen. Very little evidence exists demonstrating the long-term accumulation of platinoids. Excretion of platinoid salts after intravenous injection is mainly in urine. Orally administered platinoids are excreted in the feces.

9.4.2 *Acute toxicity*

The majority of data on the acute toxicity of platinum-containing compounds concern mainly its complex coordination compounds, the chloroplatinates and amines. The intravenous LD_{50} for $PtCl_4$ for the rat is approximately 25 mg/kg according to Moore *et al.* (1975). For the complex platinum salts, such as dipotassium hexachlor and tetrachloroplatinate, the intradermal lowest toxic dose (TD_{LO}) for humans is 40 mg/kg (Khan *et al.*, 1975). Cisplatin has been shown to have an intravenous TD_{LO} of 2.5 mg/kg in many species.

After inhalation and dermal exposures, platinum oxides and soluble platinum salts can act as irritants or sensitizers (as will be discussed further below). In albino rabbits, platinum(IV) tetrachloride ($PtCl_4$) was considered to be unsafe for intact or abraded skin contact, while platinum(II) dichloride ($PtCl_2$) and platinum(IV) dioxide (PtO_2) were considered to be only a minor hazard (Campbell *et al.*, 1975).

Cisplatin is a widely used anti-tumor agent that acts to covalently bind DNA and inhibit DNA replication. In early clinical trials, nephrotoxicity was the major dose-limiting effect (Ellenhorn and Barceloux, 1988). The mechanism by which cisplatin induces renal toxicity and the role that the platinum molecule plays in this toxicity are unclear. Administration of transplatin at doses that lead to comparable levels of platinum in the kidney is not nephrotoxic, indicating that the geometry of the complexes is important (Goldstein and Mayor, 1983). Other effects associated with short-term cisplatin use are gastrointestinal disturbances, myelosuppression, allergic reactions, and electrolyte disturbances (Ellenhorn and Barceloux, 1988).

9.4.3 Chronic toxicity

Very few toxicities are uniquely associated with chronic exposure to platinum salts. Chronic occupational exposure to platinum compounds may exacerbate platinum hypersensitivity reactions as will be discussed below. Exposure to platinum salts in an occupational setting is regulated to minimize the occurrence of toxicity.

Several long-term toxic effects have been associated with chronic use of cisplatin. As with acute toxicity, the role of the platinum molecule in chronic toxicity is unclear. Reproductive toxicity, including infertility and embryotoxicity, has been associated with long-term cisplatin use. Cisplatin has also been shown to be toxic to central and peripheral nervous systems (Loehrer and Einhorn, 1984; Calabresi and Parks, 1985). Sensory peripheral neuropathies (paresthesias) are the most common neurotoxicities observed and are reversible after discontinuation of treatment (Loehrer and Einhorn, 1984). Hematologic effects, which occur from 6 to 26 days after initiation of treatment, include hypomagnesemia, leukopenia, and thrombocytopenia (Ellenhorn and Barceloux, 1988). Ototoxicity caused by cisplatin is often irreversible and is manifested by tinnitus and hearing loss in the high frequency range.

9.5 Immunotoxicology

9.5.1 Human immunotoxicology

Several definitive reports have appeared that describe the ability of soluble complex platinum salts to lead to the generation of allergic-like symptoms (platinum allergy, platinum asthma, or platinosis). The syndrome formerly known as 'platinosis' can manifest the following symptomatology: lacrimation, sneezing, rhinorrhea, cough, dyspnea, bronchial asthma (from chloroplatinates), and cyanosis (Plunkett, 1976). This syndrome is now described as 'allergy to platinum compounds containing reactive halogen ligands' (Hughes, 1980; Boggs, 1985; Jacobs, 1987). These symptoms may be mediated via an immediate (type I) hypersensitivity or a delayed (type IV, within 24 h) hypersensitivity reaction. As is characteristic of allergic hypersensitivity, platinum allergy disappears upon removal from exposure, but asthmatic attacks occur immediately upon low-level re-exposure. The common skin lesions are mainly between the fingers and in the antecubital fossae. Platinosis or platinum allergy is caused by either dermal contact or inhalation of chloroplatinic acid or its salts, but not metallic platinum. Ammonium tetrachloroplatinite(II) ($[NH_4]_2PtCl_4$) and ammonium hexachloroplatinate(IV) ($[NH_4]_2PtCl_6$) are the main occupational sensitizing agents. The latency period of sensitization may last for weeks or several months but could take years of working with platinum compounds prior to being sensitized (Seiler and Sigel, 1988).

Allergic rhinitis was first reported in 1804 (Harris, 1975). Hunter *et al.* (1945) described rhinorrhea in 52 of 91 exposed workers employed in four platinum refineries. Air sampling showed platinum levels ranging from 0.9 to $1700 \mu g/m^3$. The severity of the response was greatest in workers crushing platinum salts where airborne levels ranged from 400 to $1700 \mu g/m^3$. Roberts (1951) first used the term 'platinosis' to describe a syndrome among workers in another platinum refinery. Platinosis was considered to be a progressive, allergic reaction to platinum salts. In this study, latency varied from a few months to years. The incidence of reaction to platinum salts was also found to be high (about 70 percent) in a German study of 51 platinum refiners (Massmann and Opitz, 1954). The major form of platinum allergy observed in this study (termed symptomatic form) manifested itself as asthma, eczema, and urticaria, as individual symptoms or in combination. The minor form (termed asymptomatic form) manifested itself as rhinopharyngitis, cough, dyspnea, itching, and conjunctival vasodilation. Similar reports of adverse reactions to platinum salts as described from the British, American, and German literature occur in the Swiss and South African literature (Jordi, 1951; Marshall, 1952; Massman and Opitz, 1954; Freedman and Krupey, 1968; Hunter, 1969; Parrot *et al.*, 1969).

Allergic skin disease in workers exposed to soluble platinum salts has also been described (Parrot *et al.*, 1969; Levene, 1971). It is thought that the dermal allergic hypersensitivity to platinum salts may result from histamine release directly from basophilic cells in response to exposure of workers to hexachloroplatinate intermediates (Saindelle and Ruff, 1969).

More recently, as part of a National Institute for Occupational Safety and Health (NIOSH) health hazards evaluation, workers employed in a precious metal refinery were evaluated for skin test sensitivity to platinum (Biagini *et al.*, 1985). Current (a total of 107 individuals) and former (a total of 30 individuals) workers, who were no longer employed because of platinum-related health problems, were evaluated for sensitivity to ($[NH_4]_2PtCl_6$).

Fourteen percent (current) to 27 percent (former) of the persons evaluated had positive skin tests. Platinum-specific antibodies (as assessed by passive cutaneous anaphylaxis or RAST) were found at higher levels in persons with a positive skin test compared with skin-test negative workers or non-exposed controls.

Further study by Merget *et al.* (1988) examined the anamnestic and immunological data of workers in a platinum refinery. The following groups were examined: (i) workers with work-related symptoms, (ii) workers with symptoms not clearly work-related, (iii) asymptomatic workers, (iv) atopic controls, and (v) non-atopic controls. All persons in groups (i) and (ii), but none in the other groups, showed a positive cutaneous reaction to hexachloroplatinate(IV) ($[PtCl_6]^{2-}$). Total serum IgE was higher in groups (i) and (iv) than groups (ii), (iii), and (v). Histamine release in response to $(PtCl_6)^{2-}$ was observed in all groups but was highest in group (iv). In skin-test positive subjects, cutaneous sensitivity to $(PtCl_6)^{2-}$ was linked to high histamine release with $(PtCl_6)^{2-}$. These types of studies that evaluate the process by which platinum hypersensitivity develops are being continued.

A case of atopic hypersenstivity to cisplatin has been reported by Khan *et al.* (1975). During administration of the eighth dose (8 weeks after the seventh dose), the patient became flushed, vomited, and developed respiratory difficulties within 3 min of the start of infusion. Skin test resulted in a typical wheal-and-flare response which was 4+ for cisplatin and K_2PtCl_6 and 3+ for K_2tCl_4 suggesting that the observed reaction was due to a response to the platinum portion of the molecule. A positive histamine release from leukocytes was obtained. These data are consistent with the potential for anaphylactic immediate hypersensitivity reactions to cisplatin.

In summary, studies in humans suggest that repeated exposure to complex salts of platinum can result in symptoms of immediate and delayed-type hypersensitivities.

9.5.2   Animal studies

Complex platinum salts

Extensive animal experimentation has been conducted to determine the mechanism by which platinum allergy or platinosis occurs after exposure to platinum salts. Initial studies showed that platinum salts can cause a direct release of histamine from basophilic cells (Saindelle and Ruff, 1969). However, expression of symptoms in susceptible individuals after inhalation exposure requires repeated exposure. Symptoms can then occur in response to suboptimal concentrations of antigen upon re-exposure. Therefore, the clinical manifestation seems to be, for the most part, in response to the generation of a specific immune response and not entirely due to a non-specific release of histamine.

In the study of Saindelle and Ruff (1969), it was shown that chloroplatinate ions can cause the release of histamine from the tissues of various animal species. Intravenous administration of sodium chloroplatinate resulted in anaphylactic shock and increased the histamine level in the serum of guinea pigs, dogs, and rats. Intradermal injection of sodium chloroplatinate led to increased capillary permeability as evidenced by the local accumulation of Evans blue dye injected intravenously 10 min prior to intradermal injection of chloroplatinate. Studies were then conducted to evaluate the spatial configuration of the complex platinum salt that can lead to histamine release. Although numerous complex platinum salts were evaluated, only chloroplatinate $(PtCl_6)^{2-}$ with a coordination number of 6 led to the release of histamine from guinea pig tissues. In addition, this action was unique to platinum as other complex ions with a coordination number of 6 (e.g. ammonium chloroiridate and chlorouthenate) had no effect.

Studies were conducted in animals which examined the conditions necessary to generate a specific immune response to various complex platinum salts. After subcutaneous and intravenous injection of $Pt(SO_4)_2$, three times per week for 4 weeks, there was no induction of an allergic state as measured by skin tests (using guinea pigs and rabbits), nor by passive transfer and footpad tests (using mice) (Rosner and Merget, 1990). Attempts to sensitize female hooded Lister rats with $(NH_4)_2[PtCl_4]$ via intraperitoneal, intramuscular, intradermal, subcutaneous, intratracheal, and footpad routes were unsuccessful (Murdock and Pepys, 1984). Cynomolgus monkeys exposed to $Na_2[PtCl_6]$ by nose-only exhalation at $200 \mu g/m^3$, 4 h/day, biweekly for 12 weeks, demonstrated pulmonary deficits as a result of exposure. Exposure to $(NH_4)_2[PtCl_6]$ produced significant skin hypersensitivity and pulmonary hyperreactivity only with concomitant exposure to ozone (Biagini *et al.*, 1982, 1983, 1985, 1986).

The requirements for sensitization to complex salts of platinum were further assessed in the popliteal lymph node (PLN) assay (Schuppe *et al.*, 1992). A single subcutaneous injection of hexachloroplatinates (90 to 180 nM of sodium ammonium salt) resulted in a dose-dependent increase in PLN weight and cellularity; peak reaction was obtained 6 days after administration. A secondary, enhanced response to suboptimal doses of agent was also observed upon subsequent exposures. This secondary response was specific to the initiating agent suggesting a specific immune response. Consistent with the studies described above, the immunogenicity of platinum salts in mice was not confined to hexachloroplatinates, but other compounds, including cisplatin, induced PLN reactions in this model.

An additional study examined the immunopathological reactions to platinum compounds in mice (Schuppe *et al.*, 1991). Mice received subcutaneous injections of 0.4, 2, or 10 μg Na_2PtCl_6 three times per week. Within 14 weeks, 37.5 percent of C57/Bl6, but not B10.S, mice treated with 0.4 or 2 μg Na_2PtCl_6 developed anti-nuclear antibodies. A two- to fivefold increase in serum IgE was observed in C57/Bl6 mice within 10 to 20 weeks when the platinum salt was combined with monthly intraperitoneal injections of aluminum hydroxide (Al(OH)$_3$), but not when given alone.

Recent studies by Dearman *et al.* (1997) showed that ammonium hexachloroplatinate and cisplatin could modify the release of cytokines, which could identify the helper T (T$_H$)-lymphocyte type stimulated by these platinum-containing compounds, by lymph nodes cells isolated from animals exposed to 0.25 to 1 percent platinum compounds. Release of interleukin-4 (IL-4) 12 to 48 h after stimulation with 2 µg/ml concanavalin A (ConA) was elevated by exposure to both complex platinum molecules compared with lymph node cells from animals treated with vehicle alone. In addition, IL-10 release from lymph node cells from platinum salt-treated animals was elevated 72 to 120 h after initiation of culture without exogenous stimulation. The release of interferon-γ (IFN-γ) by lymph node cells after exposure to the complex platinum compounds in dimethylsulfoxide was also examined. In this model, the dimethylsulfoxide vehicle elevated the production of IFN-γ in unstimulated cultures of lymph node cells. Compared with the vehicle control, exposure to cisplatin reduced IFN-γ production 72 to 120 h after initiation of culture. These data suggest that these complex platinum-containing compounds can induce a selective T$_H$2 response.

In summary, exposure of basophilic cells to certain complex platinum salts can result in histamine release. The ability of these compounds to generate such a response is dependent upon several factors including the reactivity and the coordination number of the salt. In addition, in several models, an allergic immune response to platinum-containing compounds could not be generated. However, the PLN assay showed a specific immune response to complex platinum molecules. Finally, cytokine release studies indicate the generation of a selective T$_H$2 response after exposure to complex platinum compounds *in vivo*.

Platinum-containing chemotherapeutic agents

Initial studies of cisplatin and other platinum-containing chemotherapeutic agents showed that intraperitoneal administration of 1 to 3 mg/kg cisplatin to rats for 3 days resulted in a reduction in total reticulocytes and thymic size, and a decrease in circulating leukocytes (Thompson and Gale, 1971). Several subsequent studies have been conducted to evaluate the potential functional changes in immune responsivity which may occur after cisplatin administration. One study examined the effect of intraperitoneal injection of cisplatin (5 mg/ml) on the immune responses of tumor-bearing mice (Andrade-Mena *et al.*, 1985). Administration of cisplatin 4 to 7 days prior to injection of antigen (sheep red blood cells to assess humoral immunity or oxazolone to assess delayed hypersensitivity) elevated the generation of the primary immune response by the mice. On the other hand, administration of cisplatin to tumor-free mice had no effect on the generation of immune responses to these antigens. Studies have also been conducted to assess the ability of cisplatin to modulate the generation of an immune response to allogeneic or syngeneic tumors. In all of these studies, immunization of mice with syngeneic or allogeneic tumors exposed to cisplatin either *in vitro* or *in vivo* resulted in an enhancement of the cellular immune response of splenocytes or peritoneal exudate cells that was generated to that tumor (Bahadur *et al.*, 1984; Bahadur, 1985; Bahadur and Sodhi, 1985; Sodhi *et al.*, 1985).

Further studies indicate that cisplatin augmented the tumoricidal activity of human monocytes and the ability of mice treated with recombinant IL-2 to eliminate tumors (Kleinerman *et al.*, 1980; Bernsen *et al.*, 1993). One mechanism by which this may occur is physical modification of tumor cells rendering them more susceptible to immunologic attack (Gold *et al.*, 1995). Renal carcinoma cells exposed *in vitro* to cisplatin were more easily lysed *in vitro* by autologous peripheral blood mononuclear cells than untreated renal carcinoma cells. In addition, studies to examine the mechanism by which cisplatin stimulates the activity of monocytes–macrophages indicated that tyrosine phosphorylation of several proteins in macrophages was increased upon cisplatin exposure (Kumar *et al.*, 1995). Recently, two new platinum-containing chemotherapeutic agents, oxoplatinum and cycloplatinum, were shown to stimulate the phagocytic activity of peritoneal leukocytes (Brin and Uteshev, 1990).

In contrast, some studies have shown that cisplatin and other platinum-containing chemotherapeutics can have an inhibitory effect on immune function. For example, intraperitoneal administration of higher doses (12 mg/kg) of cisplatin was shown to inhibit the generation of a humoral immune response to sheep red blood cells and trinitrophenol haptenated-keyhole limpet hemocyanin (TNP-KLH), lymphocyte mitogenesis, graft-versus-host reaction, and splenic weight (Berenbaum, 1971; Khan and Hill, 1971a,b, 1972, 1973; Bagasra *et al.*, 1985). In one report, the authors showed that immune enhancement (humoral immune response to SRBC or TNP-KLH) was observed after administration of lower doses of cisplatin (2 mg/kg administered intraperitoneally) (Bagasra *et al.*, 1985). More recently, oxoplatinum was shown to inhibit the generation of a humoral immune response to sheep erythrocytes (Brin *et al.*, 1990). Additionally, platinum-containing chemotherapeutics (carboplatin, cisplatin, oxoplatin, and iproplatin) inhibited the generation of delayed-type hypersensitivity (Jilek *et al.*, 1989). In the latter study, the agents were only effective when administered after sensitization. Therefore, this reduction may be the result of inhibition of the proliferative phase of the immune response.

In summary, these data suggest that cisplatin can have a variety of effects on the immune system depending upon the route, dose, and disease state of the host. Cisplatin can augment the generation of an immune response to a variety of antigens in tumor-bearing and normal mice. Alternatively, these compounds can be immunosuppressive if given at high doses or during the proliferative phase of an immune response.

References

ACGIH (American Conference of Governmental Industrial Hygienists) (1991). *Threshold limit values for chemical substances in the work environment.* Cincinnati, OH: American Conference of Governmental Industrial Hygienists.

ANDRADE-MENA, C.E., ORBACH-ARBOUYS, S., and MATHE, G. (1985). Enhancement of immune responses in tumor-bearing mice after administration of *cis*-diamino-dichloro-platinum (II). *Int. Arch. Allergy Appl. Immunol.* **76**, 341–343.

BAGASRA, O., CURRAO, L., DeSOUZA, L.R., OOSTERHUIS, J.W., and DAMJANOV, I. (1985). Immune response of mice exposed to *cis*-diamminedichloroplatinum. *Cancer Immunol. Immunother.* **19**, 142–147.

BAHADUR, A. (1985). Chemoimmunotherapeutical studies on Dalton's lymphoma with cis-platin. *Pol. J. Pharmacol. Pharm.* **37**, 463–468.

BAHADUR, A. and SODHI, A. (1985). Induction of cell mediated immunity *in vitro* and *in vivo* after cis-platin treatment. *Pol. J. Pharmacol. Pharm.* **37**, 453–461.

BAHADUR, A., SARNA, S., and SODHI, A. (1984). Enhanced cell mediated immunity in mice after cisplatin treatment. *Pol. J. Pharmacol. Pharm.* **36**, 441–448.

BERENBAUM, M.C. (1971). Immunosuppression by platinum diamine. *Br. J. Cancer* **25**, 208–211.

BERNSEN, M.R., VAN BARLINGEN, H.J., VAN DER VELDEN, A.W., DULLENS, H.F., DEN OTTER, W., and HEINTZ, A.P. (1993). Dualistic effects of *cis*-diamminedichloro-platinum on the anti-tumor efficacy of subsequently applied recombinant interleukin-2 therapy: a tumor-dependent phenomenon. *Int. J. Cancer* **54**, 513–517.

BIAGINI, R.E., CLARK, J.C., GALLAGHER, J.S., BERNSTEIN, I.L., and MOORMAN, W.M. (1982). Passive transfer in the monkey of human immediate hypersensitivity to complex salts of platinum and palladium. *Fed. Proc.* **41**, 827.

BIAGINI, R.E., MOORMAN, W.J., SMITH, R.J., LEWIS, T.R., and BERNSTEIN, I.L. (1983). Pulmonary hyperreactivity in cynomolgus monkeys (*Macaca fasicularis*) from nose-only inhalation exposure to disodium hexachloroplatinate, Na_2PtCl_6. *Toxicol. Appl. Pharmacol.* **69**, 377–384.

BIAGINI, R.E., BERNSTEIN, I.L., GALLAGHER, J.S., MOORMAN, W.J., BROOKS, S., and GANN, P.H. (1985). The diversity of reaginic immune responses to platinum and palladium metallic salts. *J. Allergy Clin. Immunol.* **76**, 794–802.

BIAGINI, R.E., MOORMAN, W.J., LEWIS, T.R., and BERNSTEIN, I.L. (1986). Ozone enhancement of platinum asthma in a primate model. *Am. Rev. Respir. Dis.* **134**, 719–725.

BOGGS, P.B. (1985). Platinum allergy. *Cutis* **35**, 318–320.

BRIN, E.V. and UTESHEV, B.S. (1990). The phagocytic activity of the macrophages from mouse peritoneal exudate under the action of platinum preparations. *Biull. Eksper. Biol. I Med.* **110**, 513–515.

BRIN, E.V., RYTENKO, A.N., and UTESHEV, B.S. (1990). The effect of the complex platinum (IV) compound oxoplatinum on the immune system. *Farmakol. I Toksikol.* **53**, 52–54.

CALABRESI, P. and PARKS, R.E., JR (1985). Antiproliferative agents and drugs used for immunosuppression, in Gilman, A.G., Goodman, L.S., Rall, T.W., and Murad, F. (Eds) *The Pharmacological Basis of Therapeutics*, 7th Edn, pp. 1290–1291, New York: Macmillan Press.

CAMPBELL, K.I., GEORGE, E.L., HALL, L.L., and STARA, J.F. (1975). Dermal irritancy of metal compounds. *Arch. Environ. Health* **30**, 168–170.

DEARMAN, R.J., MEEKS, D.A., and KIMBER, I. (1997). Selective induction of type 2 cytokines by topical exposure of mice to platinum salts. *The Toxicologist* **36**, 207.

ELLENHORN, M.J. and BARCELOUX, D.G. (Eds) (1988). *Medical Toxicology – Diagnosis and Treatment of Human Poisoning*, New York: Elsevier.

FREEDMAN, S.O. and KRUPEY, J. (1968). Respiratory allergy caused by platinum salts. *J. Allergy* **42**, 233–237.

GOLD, J.E., MASTERS, T.R., BABBIT, B., FINE, E.M., and WEBER, H.N. (1995). *Ex vivo* activated memory T-lymphocytes as adoptive cellular therapy of human renal cell tumour targets with potentiation by *cis*-diamminedichloroplatinum(II). *Br. J. Urology* **76**, 115–122.

GOLDSTEIN, R.S. and MAYOR, G.H. (1983). Minireview – the nephrotoxicity of cisplatin. *Life Sci.* **32**, 685–690.

HARRIS, S. (1975). Nasal ulceration in workers exposed to ruthenium and platinum salts. *J. Soc. Occup. Med.* **25**, 133–134.

HAWLEY, G.G. (Ed.) (1977). *The Condensed Chemical Dictionary*, 9th Edn, pp. 691–692, New York: Van Nostrand Reinhold.

HUGHES, E.G. (1980). Medical survellience of platinum refinery workers, *J. Soc. Occup. Med.* **30**, 27–30.

HUNTER, D. (Ed.) (1969). *Diseases of Occupations,* 4th Edn, p. 480, London: English University Press.

HUNTER, D., MILTON, R., and PERRY, K.M.A. (1945). Asthma caused by the complex salts of platinum. *Br. J. Ind. Med.* **2**, 99–101.

JACOBS, L. (1987). Platinum salt sensitivity. *Nurs. RSA Verpleging (Republic of South Africa)* **2**, 34–37.

JILEK, P., DVORAKOVA, J., TURECKOVA, J., and PROCHAZKOVA, J. (1989). The effect of platinum cytostatics on delayed-type hypersensitivity in mice. *Neoplasma* **36**, 659–665.

JORDI, A. (1951). Bronchial asthma and allergic manifestations of skin caused by complex salts of platinum: new occupational disease. *Schweiz. Med. Wochenschr.* **81**, 1117.

KHAN, A. and HILL, J.M. (1971a). Inhibition of lymphocyte blastogenesis by platinum compounds. *J. Surg. Oncol.* **3**, 565–567.

(1971b). Immunosuppression with cis-platinum (II) diaminodichloride: Effect on antibody plaque-forming cells. *Infect. Immun.* **4**, 320–321.

(1972). Suppression of graft-versus-host reaction by cis-platinum (II) diaminodichloride. *Transplantation* **13**, 55–57.

(1973). Suppression of lymphocyte blastogenesis in man following *cis*-platinum diaminodichloride administration. *Proc. Soc. Exp. Biol. Med.* **142**, 324–326.

KHAN, A., HILL, J.M., GRATER, W.N., LOEB, E., MACLELLAN, A., and HILL, N. (1975). Atopic hypersensitivity to *cis*-dichlorodiammineplatinum (II) and other platinum complexes. *Cancer Res.* **35**, 2766–2670.

KLEINERMAN, E.S., ZWELLING, L.A., and MUCHMORES, A.V. (1980). Enhancement of naturally occurring human spontaneous monocyte-mediated cytotoxicity by *cis*-diaminodichloroplatinum (II). *Cancer Res.* **40**, 3099–3102.

KUMAR, R., SHRIVASTAVA, A., and SODHI, A. (1995). Cisplatin stimulates protein tyrosine phosphorylation in macrophages. *Biochem. Mol. Biol. Int.* **35**, 541–7.

LEVENE, G.M. (1971). Platinum sensitivity. *Br. J. Derm.* **85**, 590–593.

LOEHRER, P.J. and EINHORN, L.H. (1984). Drugs five years later. Cisplatin. *Ann. Int. Med.* **100**, 704–713.

MARSHALL, J. (1952). Asthma and dermatitis caused by chloroplatinic acid. *S. Afr. Med. J.* **26**, 8–13.

MASSMANN, W. and OPITZ, H. (1954). On platinum allergy. *Zentrabl. Arbeitsmed. Arbeitss.* **4**, 1–17.

McBRYDE, W.A.E. (1973). Platinum and its compounds, in Hampel C.A. and Hawley, G.G. (Eds), *The Encyclopedia of Chemistry*, 3rd Edn, pp. 865–867, New York: Van Nostrand Reinhold.

MERGET, R., SCHULTZE-WERNINGHAUS, G., MUTHORST, T., FRIEDRICH, W., and MEIER-SYDOW, J. (1988). Asthma due to the complex salts of platinum – a cross-sectional survey of workers in a platinum refinery. *Clin. Allergy* **18**, 569–580.

MOORE, W., JR, HYSELL, D., HALL, L., CAMPBELL, K., and STARA, J. (1975). Preliminary studies on the toxicity and metabolism of palladium and platinum. *Environ. Health Perspect.* **10**, 63–71.

MURDOCH, R.D. and PEPYS, J. (1984). Immunological responses to complex salts of platinum. I. Specific IgE antibody production in the rat. *Clin. Exp. Immunol.* **57**, 107–114.

PARROT, J.L., HERBERT, R., SAINDELLE, A., and RUFF, F. (1969). Platinum and platinosis. *Arch. Environ. Health* **19**, 685–691.

PLUNKETT, E.R. (Ed.) (1976). Platinum, in *Handbook of Industrial Toxicology*, pp. 341–342, New York; Chemical Publishing Co.

PROCTOR, N.H., HUGHES, J.P., and FISCHMAN, M. (Eds) (1988). *Chemical Hazards of the Workplace*, pp. 393–436, Philadelphia: J.B. Lippincott Co.

ROBERTS, A.D. (1951). Platinosis. *Arch. Ind. Hyg. Occup. Med.* **4**, 549–559.

ROSNER, G. and MERGET, R. (1990). Allergenic potential of platinum compounds, in Dayan, A.D., Hertel, R.F., Haseltine, E., Dazantzis, G., Smith, E.M., and van der Venne, M.T. (Eds), *Immunotoxicity of Metals and Immunotoxicology*, pp. 93–102, New York: Plenum Press.

SAINDELLE, A. and RUFF, F. (1969). Histamine release by sodium chloroplatinate. *Br. J. Pharmacol.* **35**, 313–321.

SAX, N.I. and LEWIS, R.J. (Eds) (1987). *Hawley's Condensed Chemical Dictionary*, 11th Edn, pp. 926–928, New York: Van Nostrand Reinhold.

SCHUPPE, H.C., HAAS-RAIDA, D., KULIG, J., DICKOPP, U., GLEICHMANN, E., and KIND, P. (1991). Platinum compounds induce pathological immune reactions in mice. *Immunobiology* **181**, 238.

SCHUPPE, H.C., HAAS-RAIDA, D., KULIG, J., BOMER, U., GLEICHMANN, E., and KIND, P. (1992). T-cell-dependent popliteal lymph node reactions to platinum compounds in mice. *Int. Arch. Allergy Immunol.* **97**, 308–14.

SEILER, H.G. and SIGEL, H. (Eds) (1988). *Handbook on Toxicity of Inorganic Compounds*, pp. 341–344 and 501–574, New York: Marcel Dekker.

SODHI, A., TANDON, P., and SARNA, S. (1985). Adoptive transfer of immunity against solid fibrosarcoma in mice with splenocytes and peritoneal exudate cells obtained after *in vitro* sensitization and *in vivo* immunization with *cis*-dichlorodiamine platinum(II) treated fibrosarcoma cells. *Archiv. für Geschwulstforsch.* **55**, 47–61.

THOMPSON, H.S. and GALE, G.R. (1971). *cis*-Dichlorodiammineplatinum(II): hematopoietic effects in rats. *Toxicol. Appl. Pharmacol.* **19**, 602–609.

VENUGOPAL, B. and LUCKEY, T.D. (Eds) (1978). Toxicity of group VIII metals, in *Metal Toxicity in Mammals – 2*, pp. 273–305, New York: Plenum Press.

206

10

Vanadium

MITCHELL D. COHEN

Institute of Environmental Medicine, New York University Medical Center, Long Meadow Road, Tuxedo, NY 10987, USA

10.1 History

Unlike some of the other metals discussed in this book, vanadium does not have a long and glorious history replete with anecdotal stories about its influence on history or on entire civilizations. Vanadium was first discovered by Del Rio in 1801, who, in fact, thought he had only found an ore that resembled lead chromate and which he named erythronium. The element itself was 'rediscovered' and purified as multiple oxides from the smelting products of iron ores by Sefstrom in 1831; because of the mixture of beautifully colored oxides that were obtained, Sefstrom renamed the 'element' vanadin (vanadium) after the North German goddess of beauty, Vanadis. However, it was not until 1927 that vanadium in a pure (99.7 percent) ductile form was isolated by Marden and Rich (Faulkner-Hudson, 1964; Rehder, 1995).

10.2 Occurrence

Vanadium (V) is a Group VB transition element, and as such, can exist in multiple valence states (0, +2, +3, +4, +5) in both anionic and cationic forms. Although the tetravalent and pentavalent forms are the most stable, discrete ions of each do not exist in nature. Most commonly, these ions are bound to oxygen as negatively charged polymeric oxyanions which readily complex with polarizable ligands such as sulfur or phosphorus.

In nature, pentavalent vanadium is most often encountered in the form of vanadium pentoxide (V_2O_5), although ferrovanadium, vanadium carbide, and various forms of vanadates also exist. Colloidal V_2O_5 can liberate vanadate (VO_3^- and VO_4) agents by loss of water, and the resulting monomeric vanadate ions can be further converted into higher polymeric forms (Figure 10.1), much in the way that chromate ions are linked during olation. These conversions, and therefore the distribution of vanadium species, in solution are dependent both on pH and vanadium concentration. As a rule, as the number of vanadate units in the polymer increases, overall toxicity declines; however, even large polymers such as decavanadate can be toxic.

Vanadium is one of the more ubiquitous trace elements found in the environment (Hopkins *et al.*, 1977; Nriagu and Pacyna, 1988). Vanadium is the 10th most abundant of the elements thought to have biological roles, and the 21st among all elements found in the Earth's crust, with levels averaging about 150 ppm V (ATSDR, 1991). The highest natural soil levels (more than 300 ppm V) are often associated with soils rich in clays and shales (Waters, 1977; ATSDR, 1991). Coals often contain upwards of 1 percent vanadium by weight, while petroleum oils often have between 100 and 1400 ppm V, depending on the site of recovery (McTurk *et al.*, 1966; Davies, 1971; Mayotte *et al.*, 1981).

The combustion of fossil fuels and the use of vanadium-bearing ores for steel production and chemical processes represent identifiable sources for both an occupational risk of exposure to vanadium-bearing gases/particles and for the delivery of same into the surrounding environs (Zychlinski, 1980; ATSDR, 1991). Examples of high-risk exposure occupations include: mining and milling of vanadium-bearing ores, oil-fired boiler cleaning, vanadium metal/vanadium oxide production, and vanadium catalyst production. Often in these settings, ambient levels of vanadium can be more than $30\,mg\,V/m^3$ (Faulkner-Hudson, 1964; Vouk, 1979); the established value for immediate danger to life or health (IDLH) is $70\,mg\,V/m^3$ (NIOSH, 1985). The current acceptable limits for both vanadium-bearing dusts and fumes in the workplace are 0.14 to $50\,\mu g\,V/m^3$ per 8- to 24-h period (NIOSH, 1985; ACGIH, 1986; WHO, 1987).

This increased use of vanadium-containing ores/chemicals by several industries, in conjunction with increased combustion of fuel for heating and energy production, has resulted in high ambient levels of respirable vanadium particles in many metropolitan environments (Lee *et al.*, 1972; Schiff and Graham, 1984; Nriagu and Pacyna, 1988). Levels of vanadium have been shown to vary from 0.02 percent by weight in soil-derived aerosols, to 0.02 to 0.20 percent in automobile-derived fumes, and from 0.54 to 0.82 percent in oil combustion-generated aerosols, depending upon the region under study (Thurston and Spengler, 1985; Watson *et al.*, 1994). Though air concentrations of vanadium can vary as a function of both location and season, typical concentrations in rural areas range from 0.25 to $75\,ng\,V/m^3$ (Zoller *et al.*, 1973; Byerrum *et al.*, 1974; WHO, 1988); in urban settings, ambient concentrations are higher, ranging from 60 to $300\,ng\,V/m^3$ (Mamuro *et al.*, 1970; Zoller *et al.*, 1973), but on average are several $\mu g\,V/m^3$ in most major cities. The seasonal variations, wherein winter urban air levels of vanadium are increased sixfold over levels measured in spring and summer months, result from large quantities of

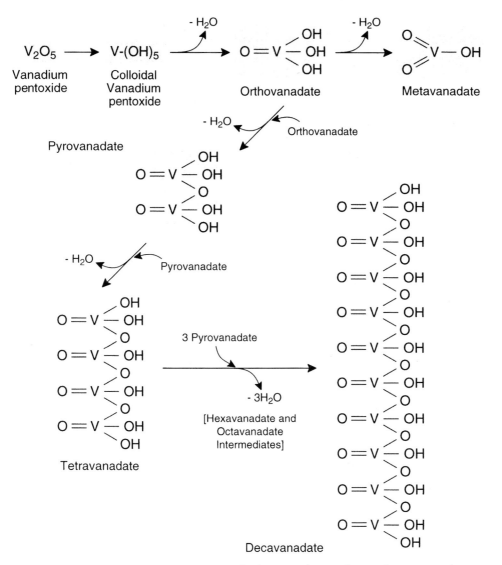

Figure 10.1 Formation of vanadates and other vanadium polymers from parental vanadium pentoxide.

vanadium-bearing oils, shales, and coals being consumed for heat and electricity generation (Lee *et al.*, 1972).

The levels of vanadium in water depend upon geographical considerations and often vary widely as a result of effluents and leachates (from both anthropogenic and natural sources) entering the water table. The levels of vanadium are higher in freshwater than in seawater (0.3 to 200 μg V/L compared with 1 to 3 μg V/L, respectively) (van Zinderen *et al.*, 1980); this is primarily due to saline-induced precipitation in the latter environment. Although average concentrations vary (from 0.2 to more than 100 µg V/L) (Vouk, 1979), the majority of municipal drinking water concentrations is usually no greater than 10 µg V/L (averaging 1.0 to 4.3 µg V/L); thus, drinking water is not considered an important source of vanadium exposure.

Food represents the primary source of non-inhalation vanadium intake in both humans and animals. As vanadium levels in most natural foodstuffs are only several ppb, daily dietary intake is estimated to be from 0.01 to 2 mg V/ day (Shroeder *et al.*, 1963; Byrne and Kosta, 1978; ICRP, 1975). Unlike copper, lead, and tin, whose increased presence in consumable products arises from either purposeful supplementation or from product-induced container leaching, the major contaminating source of foodstuffs by vanadium is soil.

10.3 Essentiality

Although it has been suggested that vanadium is an essential element for certain animals, including the chicken and rat (Schwartz and Milne, 1971; Nielsen and Ollerich, 1973; Hopkins, 1974; Hopkins and Mohr, 1974; Myron *et al.*, 1975; Hill, 1979; Nielsen, 1995), its essentiality in humans and most other higher animals is still not clear (Golden and Golden, 1981). At this point, it appears that only in certain plants (i.e. marine algae), in some bacteria (i.e. nitrogen-fixing *Azotobacter*: Eady, 1995) and fungi (i.e. *Amanita* mushroom species: Bayer, 1995), and in a few lower life forms (i.e. the tunicate ascidians *Ascidia nigra* and *A. ceratodes*: Macara *et al.*, 1979; Oltz *et al.*, 1988; Brand *et al.*, 1989; Smith *et al.*, 1995; and the fan worm *Pseudopotamilla occelata*: Ishii *et al.*, 1995) does vanadium have some demonstrable function in the host's biochemical life processes.

10.4 General Toxicology

At the average level of ambient vanadium commonly encountered in most urban/industrial areas (approximately 50 ng V/m^3), and based on experimental inhalation studies, it has been estimated that about 1 µg V/day enters the average adult human lung (Byrne and Kosta, 1978). The degree of clearance of vanadium is dependent upon the solubility of the vanadium compound inhaled. With the commonly encountered insoluble vanadium pentoxide or more soluble vanadates, initial clearance is fairly rapid, with nearly 40 percent of both chemical classes cleared within 1 h of exposure (Conklin *et al.*, 1982; Sharma *et al.*, 1987). However, significant amounts of the cleared material can enter the systemic circulation and give rise to absorption levels ranging from 50 to 85 percent of an inhaled dose (depending on the compound solubility) (Oberg *et al.*, 1978; Sharma *et al.*, 1987; Edel and Sabbioni, 1988). After 24 h, the two forms diverge in their ability to be cleared, with the insoluble form persisting longer in the lung. As such, total clearance of vanadium is never achieved, with 1 to 3 percent of the original dose persisting as long as 65 days or more (Oberg *et al.*, 1978; Rhoads and Sanders, 1985; Paschoa *et al.*, 1987). As a result, lung vanadium burdens increase with length of employment in these contaminated environments (Tipton and Shafer, 1964); whether a similar effect occurs in exposed non-worker populations is still not clear.

In the event of oral ingestion of vanadium from contaminated water, soils, or foodstuffs, or by swallowing vanadium-containing sputum, absorption from the gastrointestinal tract is quite low (averaging 0.1 to 1.0 percent), irrespective of the parent compound (Curran *et al.*, 1959; Roschin *et al.*, 1980). There appears to be greater intestinal uptake of vanadium in younger animals than in adults, possibly due to a greater non-selective permeability of the immature intestinal barrier (Edel *et al.*, 1984). Oddly enough, exposure via non-inhalation routes (such as oral intake) can still give rise to increased lung vanadium burdens and toxicities (Hopkins and Tilton, 1966; Kacew *et al.*, 1982). Regardless of the route of entry, the vanadium that does finally enter the circulation is preferentially distributed to the kidneys, liver, blood, and bone (ICRP, 1975; Oberg *et al.*, 1978; Sharma *et al.*, 1980; van Zinderen *et al.*, 1980; Rhoads and Sanders, 1985). Though these sites also have their own vanadium clearance mechanisms and kinetics, it appears that the site of long-term vanadium retention is primarily bone. Because bone acts as a repository for vanadium, the possibility of toxic effects is great, not only on bone-related biochemistry/function (Krieger and Tashjian, 1983; Blumenthal and Cosma, 1989; Yamaguchi *et al.*, 1989), but also hematological endpoints (Zaporowska and Wasilewski, 1992) including immune system cell development and function.

At the cellular level, vanadate ions are able to enter cells through the same channels utilized by phosphate and chromate anions; insoluble forms of vanadium can be introduced via pinocytic uptake. Once inside the cell (Figure 10.2), vanadate is rapidly acted upon by cellular reductants (such as NAD(P)H, gluthathione, ascorbate, and catechols) and converted to tetravalent vanadyl species (Heide *et al.*, 1983; Bruech *et al.*, 1984). Unlike other toxic metal oxyanions (such as chromate), the vandayl can then either be bound to proteins *or* readily oxidized back to the vanadate ion. As it is thought that only the pentavalent form of V can exit the cell, this represents a means of detoxification. However, the shuttling back and forth between oxidation states also represents a means for *re*-toxification. Not only are levels of cellular reductants depleted during these shuttling events (Cantley and Aisen, 1979; Bracken and Sharma, 1985; Cohen and Wei, 1988), but other alterations in cell functions are induced. These include: increased generation of reactive oxygen species (Djordjevic *et al.*, 1983; Liochev *et al.*, 1988; Liochev and Fridovich, 1989, 1990; Reif *et al.*, 1989), alterations in the phosphorylative balance of numerous proteins secondary to inhibition of cellular phosphatases (Klarlund, 1985; Bernier *et al.*, 1988; Klarlund *et al.*, 1988; Krivanek and Novakova, 1989; Yang *et al.*, 1989; Heffetz *et al.*, 1990), and modification of the activities of

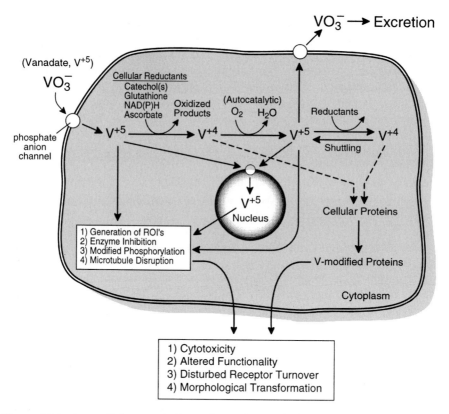

Figure 10.2 Intracellular metabolism of vanadate and selective toxicologic changes. ROI's, reactive oxygen intermediates.

those enzymes involved in both nuclear and cytoplasmic processes, including those used for detoxification of reactive oxygen species and for glycolysis (Tolman *et al.*, 1979; Climent *et al.*, 1981; Singh *et al.*, 1981; Cohen *et al.*, 1987a,b; Stankiewicz, 1989; Serra *et al.*, 1990).

Apart from the redox-related potential for damage to exposed cells and tissues, it is the inhibition of enzyme systems by vanadate ion mimicry of phosphate moieties which has the greatest toxicologic implications. These effects upon normal enzyme functionality may ultimately be contributing factors in: altered intestinal absorption of nutrients (Witkowska *et al.*, 1986; Hajjar *et al.*, 1987, 1989a,b), increased tissue peroxidative damage (Inouye *et al.*, 1980; Donaldson and LaBella, 1983; Donaldson *et al.*, 1985; Elfant and Keen, 1987), and disturbed glucose/glycogen metabolism and cholesterol synthesis (Curran, 1954; Curran *et al.*, 1959; Clausen *et al.*, 1981; Gomez-Foix *et al.*, 1988; Al-Bayati *et al.*, 1990) observed in animals and humans following vanadium exposure.

Although vanadium compounds are one of the most abundant in the natural environment, and are readily encountered within occupational settings, very little is known regarding their carcinogenic/mutagenic effects in humans and animal models; in addition, only a few studies regarding the teratogenic/embryotoxic effects of vanadium exist (Hoffman, 1979; Carlton *et al.*, 1982; Gomez *et al.*, 1992; Llobet *et al.*, 1993). Following life-term feeding studies of rodents with tetravalent vanadium, there was inconclusive evidence for carcinogenicity (Kanisawa and Schroeder, 1967; Schroeder and Balassa, 1967; Schroeder *et al.*, 1970; Schroeder and Mitchener, 1975). This is likely the result of the low levels of gastrointestinal uptake of vanadium, as is the case with many known carcinogenic metals which do not display carcinogenic potentials.

As the most likely route for vanadium intoxication is via inhalation, studies using rodents exposed to atmospheres of vanadium pentoxide indicated a dose-related increase in the incidence of pulmonary/sino-nasal epithelial hyperplasia and metaplasia (NIH, 1993). Epidemiological studies have demonstrated that both acute and chronic exposure to moderate-to-high levels of vanadium pentoxide- or vanadate-bearing dusts/fumes in the workplace resulted in increased localized fibrotic foci and lung weights, as well as an enhanced incidence of lung cancers initiated by other agents (Stocks, 1960; Hickey *et al.*, 1967; Hopkins *et al.*, 1977; Kivuoloto, 1980; Zychlinksi, 1980; Rhoads and Sanders, 1985; Kowalska, 1989). Apparently, like several other metals, vanadium might not act as a direct carcinogen, but instead, exerts secondary toxicities (i.e. immunosuppressive effects) within exposed hosts, thereby allowing initiated cancers to be expressed into neoplasms.

It should be noted that certain vanadium compounds have also been shown to act as anticarcinogens. For example, vanadocene given to rodents bearing Erlich ascites or liver tumors displays cancerostatic activity (Kopf-Maier and Kopf, 1979; Kopf-Maier *et al.*, 1980). In addition, in the form of a dietary supplement, certain vanadium compounds have been shown to block cancer induction by other known carcinogenic agents (Thompson *et al.*, 1984; Kingsnorth *et al.*, 1986). While the mechanisms of the anticarcinogenic activity in these studies are not clear, it should be noted that the effect observed with vanadocene is not unique. Similar results have been obtained using metallocene complexes with other transition elements, including germanium, titanium, molybdenum, and hafnium.

In numerous mammalian and prokaryote assay systems, vanadium has been shown to be a mutagen and a clastogen (Leonard and Gerber, 1994; reviewed in Cohen *et al.*, 1996a). In bacteria, V_2O_5, ammonium metavanadate (NH_4VO_3), and vanadyl dichloride ($VOCl_2$) were initially shown to induce revertants in *Bacillus subtilis* rec⁻ mutants (Kanematsu *et al.*, 1980), but not in *Salmonella typhimurium* or in *Escherichia coli* strains WP₂ and B/r WP₂ (Kanematsu and Kada, 1978; Kada *et al.*, 1980a,b). Subsequent studies found that NH_4VO_3 was mutagenic in *Salmonella* strains TA1537 (in a modified incorporation assay) and TA100 (in a fluctuation test) (Arlauskas *et al.*, 1985). Insoluble V_2O_5 was able to induce reverse mutations with *E. coli* WP₂, WP₂ *uvr*A, and Cm-981, but was still unable to induce mutations in *E. coli* ND160 or MR102, nor with *Salmonella* strains TA1535, TA1537, TA100, or TA98 (Sun, 1987).

In yeast, pentavalent vanadium compounds were capable of increasing both the convertant and revertant mutation frequencies of mitotic gene conversions and reverse point mutations, respectively, in *Saccharomyces cerevisiae* D7, and giving rise to aneuploidy in strain D61M (Bronzetti *et al.*, 1990; Galli *et al.*, 1991). In addition, tetravalent vanadyl sulfate ($VOSO_4$) was able to increase diploid spore formation (Sora *et al.*, 1986).

In cultured mammalian cells, pentavalent vanadium has been shown to modulate DNA synthesis, depending on the concentration of vanadium present in the culture medium (Hori and Oka, 1980; Carpenter, 1981; Jackson and Linskens, 1982; Sabbioni *et al.*, 1983; Smith, 1983; Jones and Reid, 1984). Possible effects upon RNA metabolism have also been suggested (Lindquist *et al.*, 1973; Borah *et al.*, 1985; Krauss and Basch, 1992). Several forms of genetic damage, such as DNA strand breaks, DNA–protein cross-links, and sister chromatid exchanges, have been demonstrated in vanadium-treated human leukocytes and Chinese hamster ovary cells (Birnboim, 1988; Owusu-Yaw *et al.*, 1990; Cohen *et al.*, 1992a). However, as with the procaryote assays, other investigations using similar test systems have not obtained similar results (McLean *et al.*, 1982; Roldan and Altamirano, 1992).

The treatment of human lymphocytes with vanadate resulted in increased numbers of polyploid cells without concurrent increases in mitotic indices (Bracken and Sharma, 1985; Sharma and Talukder, 1987; Roldan and Altamirano, 1992). Similarly, treatment of Chinese hamster V79 lung fibroblasts with V_2O_5 increased the levels of micronucleated cells (Zhong *et al.*, 1994). Vanadium also has the capacity to alter the colony formation capacity of cultured human tumor cell lines (Hanauske *et al.*, 1987; Sheu *et al.*, 1990), and can directly affect cell transformation of cultured cells or after transfection with intact viral DNA/oncogenes (Klarlund, 1985; Dessureault and Weber, 1990; Kowalski *et al.*, 1992).

10.5 Immunotoxicology

As either the pentavalent vanadate or vanadium pentoxide, vanadium has been known to alter immunological responses in humans and experimental animals since the early 1900s (Dutton, 1911; Jackson, 1912; Proescher and Seil, 1917). Its immunosuppressive effects became apparent long after its initial pharmacological use as an immunoenhancer in humans. Many workers exposed to atmospheric vanadium had increased occurrences of prolonged coughing spells, tuberculosis, and general irritation of the respiratory tract. Post-mortem examinations revealed extensive lung damage; often, the primary cause of death was respiratory failure secondary to bacterial infection. Later epidemiological studies demonstrated that acute exposure to high concentrations and/or chronic exposure to moderate levels of workplace vanadium-bearing dusts or fumes resulted in a higher incidence of several pulmonary diseases, including asthma, rhinitis, pharyngitis, chronic productive cough, pneumonia, and bronchitis ('boilermakers bronchitis') (Symanski, 1939; Sjoberg, 1950; Zenz *et al.*, 1962; Roschin, 1967; Zenz and Berg, 1967; Lees, 1980; Musk and Tees, 1982; Levy *et al.*, 1984). More detailed cytological studies with cells from exposed workers demonstrated vanadium-induced disturbances in polymorphonuclear (neutrophil) and plasma cell numbers, immunoglobulin production, and lymphocyte mitogenic responsiveness (Kivuoloto *et al.*, 1979, 1980, 1981; Marini *et al.*, 1987).

The vanadium-induced changes in immunological functions are reproducible in a variety of animal models (reviewed in Zelikoff and Cohen, 1995). Subchronic and acute exposure of rodents to pentavalent vanadium compounds (via various exposure routes) has been shown to alter: mitogen-induced lymphoproliferation (Sharma *et al.*, 1981; Ramanadham and Kern, 1983), alveolar/peritoneal macrophage phagocytosis and lysosomal enzyme activity/release (Waters *et al.*, 1974; Fisher *et al.*, 1978; Labedzka *et al.*, 1989; Vaddi and Wei, 1991a), host resistance to bacterial endotoxin (LPS) and intact microorganisms (Cohen *et al.*, 1986, 1989), lung immune cell populations (Knecht *et al.*, 1985, 1992; McManus *et al.*, 1995; Cohen *et al.*, 1996c), *in situ* induction of interferon-γ (IFN-γ) and interleukin-6 (IL-6) by polyinosinic:polycytidilic acid (Cohen *et al.*, 1997), *ex vivo* formation of tumor necrosis factor-α (TNF-α) and reactive oxygen intermediates (such as superoxide anion and hydrogen peroxide) by lung macrophages (Cohen *et al.*, 1996c), mast cell histamine release (Al-Laith and Pearce, 1989), and both lung macrophage MHC class II antigen expression/induction by IFN-γ and cellular calcium ion balance (Cohen *et al.*, 1996c). Exposures also produced pathological alterations in immune system organs (i.e. Peyer's patches, thymus, and spleen) (Wei *et al.*, 1982; Cohen *et al.*, 1986; Al-Bayati *et al.*, 1992). Studies with either cultured macrophage cell lines or mice exposed to NH_4VO_3 have demonstrated decreased surface levels of F_c receptors for binding differing subclasses of immunoglobulins (Vaddi and Wei, 1991b) and diminished production and activity of TNF-α and IL-1 (Cohen *et al.*, 1992b, 1993). The latter study also showed that, in the absence of exogenous stimuli, vanadium-exposed macrophages spontaneously released significantly greater amounts of inflammatory prostaglandin E_2 than did the untreated controls.

Studies of host resistance to the bacterial pathogen *Listeria monocytogenes* after acute/subchronic vanadate exposures demonstrated that resident peritoneal macrophage function, and consequently cell-mediated immunity, was adversely affected. At the site of infection, bacterial numbers increased rapidly, but no increase in macrophage or neutrophil numbers occurred (Cohen *et al.*, 1989). Macrophages recovered from vanadate-treated mice displayed decreased capacities to phagocytose opsonized *Listeria* or to kill the few organisms that were ingested. These defects were thought to be attributable to vanadium-induced disturbances in superoxide anion formation, glutathione redox cycle activity, and hexose-monophosphate shunt activation

(Cohen and Wei, 1988), events critical to maintaining energy for phagocytosis and intracellular killing.

While the precise mechanisms underlying the immunomodulatory effects of vanadium are not yet clear, *in vivo* and *in vitro* studies with vanadium have begun to yield information to enable hypothetical mechanisms to be proposed. In macrophages (as well as other cell types), vanadate ions have been shown to: disrupt microtubule and microfilament structural integrity (Cande and Wolniak, 1978; Wang and Choppin, 1981; Bennett *et al.*, 1993), induce alterations in local pH due to vanadate polyanion formation (Chasteen, 1983; Rehder, 1991), modify lysosomal enzyme release and activity (Vaddi and Wei, 1991b), alter secretory vesicle fusion to lysosomes (Goren *et al.*, 1984), disrupt cell protein metabolism at both the level of synthesis (Montero *et al.*, 1981; Sabbioni *et al.*, 1983; Kingsnorth *et al.*, 1986) and of catabolism (Seglen and Gordon, 1981), and modulate the inducibility and magnitude of reactive oxygen intermediate formation/release (Cohen *et al.*, 1996b,c).

Apart from these structural/biochemical alterations which could disrupt the functionality of the vanadium-exposed macrophages, changes in the capacity of these cells to interact with, and respond to, signalling agents during an immune response might also underly vanadium-induced immunomodification. Chemical agents (such as vanadium) that are able to disrupt the endocytic delivery of surface receptor–ligand complexes to lysosomes, subsequent complex dissociation, and receptor recycling or *de novo* receptor synthesis can diminish the magnitude of cytokine- and antigen-induced responses by macrophages. Recent studies in this area have indicated that macrophage priming by T-lymphocyte-derived IFN-γ is adversely affected by vanadium exposure. Activation of cellular protein kinases C (PKC) and A, and increases in intracellular calcium ion concentrations, events which can result in a downregulation of IFN-γ receptor expression, have been observed in macrophages harvested from vanadate-treated mice, as well as in J774 murine macrophage cultures treated with vanadium (Vaddi and Wei, 1996). In these studies, cellular calcium burdens were increased fivefold as compared with those levels measured in untreated control cells; these levels were not increased further by treatment with endotoxin or formyl peptide stimulants. In addition, while cytosolic PKC activities were unaffected by vanadate treatment, membrane-bound PKC levels were reduced. Similarly, in mouse WEHI-3 macrophages, the levels of two classes of surface IFN-γ receptors and their binding affinities for IFN-γ were greatly modified by vanadium treatment (Cohen *et al.*, 1996b). In these same treated cells, IFN-γ-inducible responses (including enhanced calcium ion influx, MHC class II antigen expression, and zymosan-inducible reactive oxygen intermediate formation) were diminished secondary to the changes in IFN-γ receptor expression/binding activity.

The mechanism(s) underlying these alterations in surface receptor expression and binding activity in vandium-exposed macrophages are not as yet clear. Studies with lymphocytes (and other non-immune cell types) treated *in vitro* with vanadate have demonstrated an altered affinity for hormones (i.e.

epidermal growth factor and insulin) or cytokines by treated cells (Kadota *et al.*, 1987; Torossian *et al.*, 1988; Evans *et al.*, 1994). Altered surface receptor expression and receptor processing dynamics have been observed in vanadium-exposed hepatocytes (Oka and Weigel, 1991), and may, in part, underly observed modificiations in complement/Fc receptor expression and binding activity in peritoneal macrophages obtained from vanadate-exposed mice (Cohen *et al.*, 1986; Vaddi and Wei, 1991a).

While it has been suggested that vanadium could have the potential to modify directly proteins which consitute the various surface cytokine/antigen/opsonin receptors on macrophages (Vaddi and Wei, 1991a; Cohen *et al.*, 1996b), it has also been hypothesized that modified receptor responses could be related to vanadium-induced changes in cellular protein kinase/phosphorylase enzyme activities (Lopez *et al.*, 1976; Swarup *et al.*, 1982; Nechay, 1984; Nechay *et al.*, 1986; Grinstein *et al.*, 1990; Trudel *et al.*, 1991). In these cases, prolonged phosphorylation of surface receptor proteins or cytokine-induced secondary messenger proteins might then induce false states of cell activation (Pumiglia *et al.*, 1992; Imbert *et al.*, 1994), which could

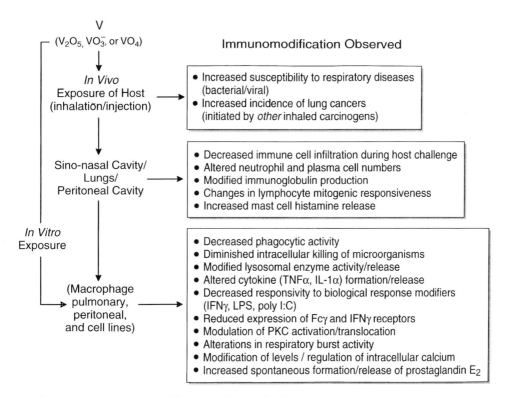

Figure 10.3 Immunomodifications observed following *in vivo* or *in vitro* exposures to vanadium compounds.

subsequently result in modified cytokine receptor expression in these exposed cells. In addition, prolonged phosphorylation of cellular proteins could also lead to the bypass of normal signal transduction pathways and subsequent activation of cytokine DNA response elements (Igarishi *et al.*, 1993), which, in turn, can lead to down-regulation of cytokine receptor expression/ functionality as well (Wietzerbin *et al.*, 1986).

In view of the evidence to indicate that vanadium compounds are immunomodulatory (primarily at the level of the macrophage; Figure 10.3) and the increased concern regarding potential exposures of human populations, vanadium has recently been included as a Superfund target metal. Further studies to elucidate more clearly the mechanisms by which vanadium may alter immunological responsiveness are clearly warranted.

References

ACGIH (American Conference of Government Industrial Hygienists) (1986). *Documentation of the Threshold Limit Values and Biological Exposure Indices*, 5th Edn, Cincinnati, Ohio: American Conference of Government Industrial Hygienists.

AL-BAYATI, M.A., GIRI, S.N., and RABE, O.G. (1990). Time and dose-response study of the effects of vanadate in rats: changes in blood cells, serum enzymes, protein, cholesterol, glucose, calcium, and inorganic phosphate. *J. Environ. Pathol. Toxicol. Oncol.* **9**, 435–455.

AL-BAYATI, M.A., CULBERTSON, M.R., SCHREIDER, J.P., ROSENBLATT, L.S., and RAABE, O.G. (1992). The lymphotoxic action of vanadate. *J. Environ. Path. Toxicol. Oncol.* **11**, 19–27.

AL-LAITH, M. and PEARCE, F.L. (1989). Some further characteristics of histamine secretion from rat mast cells stimulated with sodium orthovanadate. *Agents Action* **27**, 65–67.

ARLAUSKAS, A., BAKER, R.S.U., BONIN, A.M., TANDON, R.K., CRISP, P.T., and ELLIS, J. (1985). Mutagenicity of metal ions in bacteria. *Environ. Res.* **36**, 379–388.

ATSDR (Agency for Toxic Substances and Disease Registry) (1991). *Toxicological Profile for Vanadium and Compounds*. Atlanta, GA: US Department of Health and Human Services, Public Health Service.

BAYER, E. (1995). Amavadin, the vanadium compound of *Amanitae*, in Sigel, H. and Sigel, A. (Eds) V*anadium and Its Role in Life; Metal Ions in Biological Systems*,Vol. 31, pp. 407–422, New York: Marcel Dekker.

BENNETT, P.A., DIXON, R.J., and KELLIE, S. (1993). The phosphotyrosine phosphatase inhibitor vanadyl hydroperoxide induces morphological alterations, cytoskeletal rearrangements, and increased adhesiveness in rat neutrophil leukocytes. *J. Cell Sci.* **106**, 891–901.

BERNIER, M., LAIRD, D.M., and LANE, M.D. (1988). Effect of vanadate on the cellular accumulation of pp15, an apparent product of insulin receptor tyrosine kinase action. *J. Biol. Chem.* **263**, 13626–13634.

BIRNBOIM, H.C. (1988). A superoxide anion-induced DNA strand break metabolic pathway in human leukocytes: effects of vanadate. *Biochem. Cell Biol.* **66**, 374–381.

BLUMENTHAL, N.C. and COSMA, V. (1989). Inhibition of apatite formation by titanium and vanadium ions. *J. Biomed. Mater. Res.: Appl. Biomater.* **23**, 13–22.

BORAH, B., CHEN, C., EGAN, W., MILLER, M., WLODAWER, A., and COHEN, J.S. (1985). Nuclear magnetic resonance and neutron diffraction studies of the complex of ribonuclease A with uridine vanadate, a transition-state analogue. *Biochemistry* **24**, 2058–2067.

BRACKEN, W.M. and SHARMA, R.P. (1985). Cytotoxicity-related alterations of select cellular functions after *in vivo* vanadate exposure. *Biochem. Pharmacol.* **34**, 2465–2470.

BRAND, S.G., HAWKINS, C.J., MARSHALL, A.T., NETTE, G.W., and PARRY, D.L. (1989). Vanadium chemistry of ascidians. *Comp. Biochem. Physiol.* **93B**, 425–436.

BRONZETTI, G., MORICHETTI, E., DELLA CROCE, C., DEL CARRATORE, R., GIROMINI, L., and GALLI, A. (1990). Vanadium: genetical and biochemical investigations. *Mutagenesis* **5**, 293–295.

BRUECH, M., QUINTANILLA, M.E., LEGRUM, W., KOCH, J., NETTER, K.J., and FUHRMANN, G.F. (1984). Effects of vanadate on intracellular reduction equivalents in mouse liver and the fate of vanadium in plasma, erythrocytes, and liver. *Toxicology* **31**, 283–295.

BYERRUM, R.U., ECKHARDT, R.E., and HOPKINS, L.L. (1974). *Vanadium*, Washington, DC: National Academy of Sciences.

BYRNE, A.R. and KOSTA, L. (1978). Vanadium in foods and in human body fluids and tissues. *Sci. Total Environ.* **10**, 17–30.

CANDE, W.Z. and WOLNIAK, S.M. (1978). Chromosome movement in lysed mitotic cells is inhibited by vanadate. *J. Cell Biol.* **79**, 573–580.

CANTLEY, L.C. and AISEN, P. (1979). The fate of cytoplasmic vanadium: implications on (Na,K)-ATPase inhibition. *J. Biol. Chem.* **254**, 1781–1784.

CARLTON, B.D., BENEKE, M.B., and FISHER, G.L. (1982). Assessment of the teratogenicity of ammonium vanadate using Syrian Golden hamsters. *Environ. Res.* **29**, 256–262.

CARPENTER, G. (1981). Vanadate, epidermal growth factor, and the stimulation of DNA synthesis. *Biochem. Biophys. Res. Commun.* **102**, 1115–1121.

CHASTEEN, N.D. (1983). The biochemistry of vanadium. *Struct. Bonding* **53**, 105–138.

CLAUSEN, T., ANDERSEN, T., STURUP-JOHANSEN, M., and PETKOVA, O. (1981). The relationship between the transport of glucose and cations across cell membranes in isolated tissues. XI. The effect of vanadate on ^{45}Ca-efflux and sugar transport in adipose tissue and skeletal muscle. *Biochim. Biophys. Acta* **646**, 261–267.

CLIMENT, F., BARTRONS, R., PONS, G., and CARRERAS, J. (1981). Effect of vanadate on phosphoryl transfer enzymes involved in glucose metabolism. *Biochem. Biophys. Res. Commun.* **101**, 570–576.

COHEN, M.D. and WEI, C.I. (1988). Effects of ammonium metavanadate treatment upon macrophage glutathione redox cycle activity, superoxide production, and intracellular glutathione status. *J. Leukocyte Biol.* **44**, 122–129.

COHEN, M.D., WEI, C.I., TAN, H., and KAO, K.J. (1986). Effect of ammonium metavanadate on the murine immune response. *J. Toxicol. Environ. Health* **19**, 279–298.

COHEN, M.D., SEN, A.C., and WEI, C.I. (1987a). Ammonium metavanadate complexation with glutathione disulfide: a contribution to the inhibition of glutathione reductase. *Inorg. Chim. Acta* **138**, 91–93.

(1987b). Vanadium inhibition of yeast lucose-6-phosphate-dehydrogenase. *Inorg. Chim. Acta* **138**, 179–186.

COHEN, M.D., CHEN, C.M., and WEI, C.I. (1989). Decreased resistance to *Listeria monocytogenes* in mice following vanadate exposure: effects upon the function of macrophages. *Int. J. Immunopharmacol.* **11**, 285–292.

COHEN, M.D., KLEIN, C.B., and COSTA, M. (1992a). Forward mutations and DNA–protein crosslinks induced by ammonium metavanadate in cultured mammalian cells. *Mutat. Res.* **269**, 142–148.

COHEN, M.D., PARSONS, E., SCHLESINGER, R.B., and ZELIKOFF, J.T. (1992b). Immunomodulating effects of vanadium pentoxide (V_2O_5) and ammonium metavanadate (NH_4VO_3) on mouse macrophage (WEHI-3) monokine and prostaglandin release. *The Toxicologist* **12**, 47.

(1993). Immunotoxicity of *in vitro* vanadium exposure: effects on interleukin-1, tumor necrosis factor, and prostaglandin E_2 production by WEHI-3 macrophages. *Int. J. Immunopharmacol.* **15**, 437–446.

COHEN, M.D., BOWSER, D., and COSTA, M. (1996a). The carcinogenicity and genotoxicity of lead, beryllium, and other metals, in Chang, L. (Ed.) *Toxicology of Metals*, pp. 253–284, Boca Raton, FL: CRC Press.

COHEN, M.D., MCMANUS, T.P., YANG, Z., QU, Q., SCHELSINGER, R.B., and ZELIKOFF, J.T. (1996b). Vanadium affects macrophage interferon-γ binding and inducible responses. *Toxicol. Appl. Pharmacol.* **138**, 110–120.

COHEN, M.D., YANG, Z., ZELIKOFF, J.T., and SCHLESINGER, R.B. (1996c). Pulmonary immunotoxicity of inhaled ammonium metavanadate in Fisher 344 rats. *Fundam. Appl. Toxicol.* **33**, 254–263.

COHEN M.D., BECKER, S., DEVLIN, R., SCHLESINGER, R.B., and ZELIKOFF, J.T. (1997). Effects of vanadium upon polyI:C-induced responses in rat lung and alveolar macrophages. *J. Toxicol. Environ. Health* **51**, 591–608.

CONKLIN, A.W., SKINNER, C.S., FELTEN, T.L., and SANDERS, C.L. (1982). Clearance and distribution of intratracheally-instilled vanadium compounds in the rat. *Toxicol. Lett.* **11**, 199–203.

CURRAN, G.L. (1954). Effect of certain transition group elements on hepatic synthesis of cholesterol in the rat. *J. Biol. Chem.* **210**, 765–770.

CURRAN, G.L., AZARNOFF, D.L., and BOHNGER, R.E. (1959). Effect of cholesterol synthesis inhibition in normocholesteremic young men. *J. Clin. Invest.* **38**, 1251–1261.

DAVIES, W.E. (1971). *National Inventory of Sources and Emissions: Vanadium – 1968*, Research Triangle Park, NC: US Environmental Protection Agency, ADTD-1511.

DESSUREAULT, J. and WEBER, J.M. (1990). Retransformation of a revertant cell line with adenovirus E1 oncogenes and vanadate. *J. Cell. Biochem.* **43**, 293–296.

DJORDJEVIC, C., GONSHOR, L.G., and BYNUM, M.A. (1983). Peroxides of vanadium and related metals in biological and medicinal chemistry. *Inorg. Chim. Acta* **79**, 301–302.

DONALDSON, J. and LABELLA, F. (1983). Prooxidant properties of vanadate *in vitro* on catecholamines and on lipid peroxidation by mouse and rat tissues. *J. Toxicol. Environ. Health* **12**, 119–126.

DONALDSON, J., HEMMING, R., and LaBELLA, F. (1985). Vanadium exposure enhances lipid peroxidation in the kidney of rats and mice. *Can. J. Physiol. Pharmacol.* **63**, 196–199.

DUTTON, W.F. (1911). Vanadiumism. *J. Am. Med. Assoc.* **56**, 1648–1658.

EADY, R.R. (1995). Vanadium nitrogenases of *Azobacter*, in Sigel, H. and Sigel, A. (Eds) *Vanadium and Its Role in Life; Metal Ions in Biological Systems*, Vol. 31, pp. 363–406, New York: Marcel Dekker.

EDEL, J. and SABBIONI, E. (1988). Retention of intratracheally-instilled and ingested tetravalent and pentavalent vanadium in the rat. *J. Trace Elem. Electrolytes Health Dis.* **2**, 23–30.

EDEL, J., PIETRA, R., SABBIONI, E., MARAFANTE, E., SPRINGER, A., and UBERTALLI, L. (1984). Disposition of vanadium in rat tissues at different ages. *Chemosphere* **13**, 87–93.

ELFANT, M. and KEEN, C.L. (1987). Sodium vanadate toxicity in adult and developing rats: role of peroxidative damage. *Biol. Trace Elem. Res.* **14**, 193–208.

EVANS, G.A., GARCIA, G.G., ERWIN, R., HOWARD, O.M., and FARRAR, W.L. (1994). Pervanadate simulates the effects of interleukin-2 (IL-2) in human T-cells and provides evidence for the activation of two distinct tyrosine kinase pathways by IL-2. *J. Biol. Chem.* **269**, 23407–23412.

FAULKNER-HUDSON, T.G. (1964). In Browing, E. (Ed.) *Vanadium: Toxicology and Biological Significance*, Elsevier Monographs on Toxic Agents, New York: Elsevier Scientific Publishing.

FISHER, G.L., McNEILL, K.L., WHALEY, C.B., and FONG, J. (1978). Attachment and phagocytosis studies with murine pulmonary alveolar macrophages. *J. Reticuloendothel. Soc.* **24**, 243–252.

GALLI, A., VELLOSI, R., FLORIO, R., DELLA CROCE, C., DEL CARROTORE, R., MORICHETTI, E., GIROMINI, L., ROSELLINI, D., and BRONZETTI, G. (1991). Genotoxicity of vanadium compounds in yeast and cultured mammalian cells. *Teratogen. Carcinogen. Mutagen.* **11**, 175–183.

GOLDEN, M.H. and GOLDEN, B.E. (1981). Trace elements: potential importance in human nutrition with particular reference to zinc and vanadium. *Br. Med. Bull.* **37**, 31–36.

GOMEZ, M., SANCHEZ, D.J., DOMINGO, J.L., and CORBELLA, J. (1992). Embryotoxic and teratogenic effects of intraperitoneally-administered metavanadate in mice. *J. Toxicol. Environ. Health* **37**, 47–56.

GOMEZ-FOIX, A.M., RODRIGUEZ-GIL, J.E., FILLAT, C., and GUINOVART, J.J. (1988). Vanadate raises fructose 2,6-bisphosphate concentrations and activates glycolysis in rat hepatocytes. *Biochem. J.* **255**, 507–512.

GOREN, M.B., SWENDSEN, S.L., FISCUS, J., and MIRANTI, C. (1984). Fluorescent markers for studying phagosome–lysosome fusion. *J. Leukocyte Biol.* **36**, 273–292.

GRINSTEIN, S., FURUYA, W., Lu, D.J., and MILLS, G.B. (1990). Vanadate stimulates oxygen consumption and tyrosine phosphorylation in electropermeabilized human neutrophils. *J. Biol. Chem.* **265**, 318–327.

HAJJAR, J.J., FUCCI, J.C., ROWE, W.A., and TOMICIC, T.K. (1987). Effect of vanadate on amino acid transport in rat jejunum. *Proc. Soc. Exp. Biol. Med.* **184**, 403–409.

HAJJAR, J.J., DOBISH, M.P., and TOMICIC, T.K. (1989a). Effect of chronic vanadate ingestion on amino acid and water absorption in rat intestine. *Arch. Toxicol.* **63**, 29–33.

HAJJAR, J.J., DOBISH, M.P., and TOMICIC, T.K. (1989b). Effect of chronic vanadate ingestion on sugar transport in rat jejunum. *Proc. Soc. Exp. Biol. Med.* **190**, 35–41.

HANAUSKE, U., HANAUSKE, A.R., and MARSHALL, M.H. (1987). Biphasic effect of vanadium salts on *in vitro* tumor colony growth. *Int. J. Cell Cloning* **5**, 170–178.

HEFFETZ, D., BUSHKIN, I., DROR, R., and ZICK, Y. (1990). The insulinomimetic agents H_2O_2 and vanadate stimulate protein tyrosine phosphorylation in intact cells. *J. Biol. Chem.* **265**, 2896–2902.

HEIDE, M., LEGRUM, W., NETTER, K.J., and FUHRMANN, G.F. (1983). Vanadium inhibits oxidative drug demethylation *in vivo* in mice. *Toxicology* **26**, 63–71.

HICKEY, R.J., SCHOFF, E.P., and CLELLAND, R.C. (1967). Relationship between air pollution and certain chronic disease death rates. *Arch. Environ. Health* **15**, 728–739.

HILL, C.H. (1979). Dietary influences on resistance to *Salmonella* infection in chicks. *Fed. Proc.* **38**, 2129.

HOFFMAN, D.J. (1979). Embryotoxic effects of crude oil containing nickel and vanadium in mallards. *Bull. Environ. Contam. Toxicol.* **23**, 203–206.

HOPKINS, L.L. (Ed.) (1974). Vanadium, in *Trace Element Metabolism in Animals*, Vol. 2, pp. 397–410, Baltimore: University Park Press.

HOPKINS, L.L. and MOHR, H.E. (1974). Vanadium as an essential nutrient. *Fed. Proc.* **33**, 1773–1775.

HOPKINS, L.L. and TILTON, B.E. (1966). Metabolism of trace amounts of vanadium[48] in rat organs and liver subcellular particles. *Am. J. Physiol.* **211**, 169–172.

HOPKINS, L.L., CANNON, H.L., NIESCH, A.T., WELCH, R.M., and NIELSEN, F.H. (1977). Vanadium, in *Geochemistry and the Environment, Vol. II. The Relation of Other Selected Trace Elements to Health and Disease*, pp. 93–107, Washington, DC: National Academy of Sciences.

HORI, C. and OKA, T. (1980). Vanadate enhances the stimulatory action of insulin on DNA synthesis in cultured mouse mammary glands. *Biochem. Biophys. Acta* **610**, 235–240.

ICRP (International Committee on Radiological Protection) (1975). *Report of the Task Group on Reference Man*, No. 23, Oxford: Pergamon Press.

IGARISHI, K., DAVID, M., LARNER, A.C., and FINBLOOM, D.S. (1993). *In vitro* activation of a transcription factor by gamma interferon requires a membrane-associated tyrosine kinase and is mimicked by vanadate. *Mol. Cell. Biol.* **13**, 3984–3989.

IMBERT, V., PEYRON, J.F., FAR, D.F., MARI, B., AUBERGER, P., and ROSSI, B. (1994). Induction of tyrosine phosphorylation and T-cell activation by vanadate peroxide, an inhibitor of protein tyrosine phosphatases. *Biochem. J.* **297**, 163–173.

INOUYE, B., MORITA, K., ISHIDA, T., and OGATA, M. (1980). Cooperative effect of sulfite and vanadium compounds on lipid peroxidation. *Toxicol. Appl. Pharmacol.* **53**, 101–107.

ISHII, T., NAKAI, I., and OKOSHI, K. (1995). Biochemical significance of vanadium in a polychaete worm, in Sigel, H. and Sigel, A. (Eds) *Vanadium and Its Role in Life; Metal Ions in Biological Systems*, Vol. 31, pp. 491–510, New York: Marcel Dekker.

JACKSON, D.E. (1912). The pharmacological action of vanadium. *J. Pharm. Exp. Ther.* **3**, 477–514.

JACKSON, J.F. and LINSKENS, H.F. (1982). Metal ion-induced unscheduled DNA synthesis in petunia pollen. *Mol. Gen. Genet.* **187**, 112.

JONES, T.R. and REID, T.W. (1984). Sodium orthovanadate stimulation of DNA synthesis in Nakano mouse lens epithelia cells in serum-free medium. *J. Cell. Physiol.* **121**, 199–205.

KACEW, S., PARULEKAR, M.R., and MERALI, Z. (1982). Effects of parenteral vanadium administration on pulmonary metabolism of rats. *Toxicol. Lett.* **11**, 119–124.

KADA, T., HIRANO, K., and SHIRASU, Y. (1980a). Screening of environmental chemical mutagens by the rec-assay system with *Bacillus subtilis*. *Chem. Mut.* **6**, 149–173.

KADA, T., HIRANO, K., and SHIRASU, Y. (1980b). Screening of environmental chemical mutagens by the rec-assay system with *Bacillus subtilis*, in deSerres, F.J. and Hollander, A. (Eds) *Chemical Mutagens: Principles and Methods for Their Detection*, pp. 149–173, New York: Plenum Press.

KADOTA, S., FANTUS, I.G., DERAGON, G., GUYDA, H.J., and POSNER, B.I. (1987). Stimulation of insulin-like growth factor II receptor binding and insulin receptor kinase activity in rat adipocytes. Effects of vanadate and H_2O_2. *J. Biol. Chem.* **262**, 8252–8256.

KANEMATSU, N. and KADA, T. (1978). Mutagenicity of metal compounds. *Mutat. Res.* **53**, 207–208.

KANEMATSU, N., HARA, M., and KADA, T. (1980). Rec assay and mutagenicity studies on metal compounds. *Mutat. Res.* **77**, 109–116.

KANISAWA, M. and SCHROEDER, H.A. (1967). Life-term studies on the effects of arsenic, germanium, tin, and vanadium on spontaneous tumors in mice and rats. *Cancer Res.* **27**, 1192–1195.

KINGSNORTH, A.N., LAMURAGLIA, G.M., ROSS, J.S., and MALT, R.A. (1986). Vanadate supplements and 1,2-dimethylhydrazine-induced colon cancer in mice: increased thymidine incorporation without enhanced carcinogenesis. *Br. J. Cancer* **53**, 683–686.

KIVUOLOTO, M. (1980). Observations on the lungs of vanadium workers. *Br. J. Ind. Med.* **37**, 363–366.

KIVUOLOTO, M., RASANEN, O., RINNE, A., and RISSANEN, M. (1979). Effects of vanadium on the upper respiratory tract of workers in a vanadium factory. A macroscopic and microscopic study. *Scand. J. Work Environ. Health* **5**, 50–58.

KIVUOLOTO, M., PAKARINEN, A., and PYY, L. (1980). Clinical laboratory results of vanadium-exposed workers. *Arch. Environ. Health* **36**, 109–113.

KIVUOLOTO, M., RASANEN, O., RINNE, A., and RISSANEN, A. (1981). Intracellular immunoglobulin in plasma cells of nasal biopsies taken from vanadium-exposed workers. *Annt. Anz. Jena* **149**, 446–450.

KLARLUND, J.K. (1985). Transformation of cells by an inhibitor of phosphatases acting on phosphorylation in proteins. *Cell* **41**, 707–717.

KLARLUND, J.K., LATINI, S., and FORCHHAMMER, J. (1988). Numerous proteins phosphorylated on tyrosine and enhanced tyrosine kinase activities in vanadate-treated NIH 3T3 fibroblasts. *Biochim. Biophys. Acta* **971**, 112–120.

KNECHT, E.A., MOORMAN, W.J., CLARK, J.C., LYNCH, D.W., and LEWIS, T.R. (1985). Pulmonary effects of acute vanadium pentoxide inhalation in monkeys. *Am. Rev. Resp. Dis.* **132**, 1181–1185.

KNECHT, E.A., MOORMAN, W.J., CLARK, J.C., HULL, R.D., BIAGINI, R.E., LYNCH, D.W., BOYLE, T.J., and SIMON, S.D. (1992). Pulmonary reactivity to vanadium pentoxide following subchronic inhalation exposure in a non-human primate animal model. *J. Appl. Toxicol.* **12**, 427–434.

Immunotoxicology of Environmental and Occupational Metals

KOPF-MAIER, P. and KOPF, H. (1979). Vanadocene dichloride: another anti-tumor agent from the metallocene series. *Z. Naturforsch.* **34B**, 805–807.

KOPF-MAIER, P., HESSE, B., and KOPF, H. (1980). Tumor inhibition by metallocenes: effect of titanocene and hafnocene dichlorides on Erlich ascites tumor in mice. *J. Cancer Res. Clin. Oncol.* **96**, 43–51.

KOWALSKA, M. (1989). Changes in rat lung collagen after life-time treatment with vanadium. *Toxicol. Lett.* **47**, 185–190.

KOWALSKI, L.A., TSANG, S.S., and DAVISON, A.J. (1992). Vanadate enhances transformation of bovine papillomavirus DNA-transfected C3H/10T$_{1/2}$ cells. *Cancer Lett.* **64**, 83–90.

KRAUSS, M. and BASCH, H. (1992). Is the vanadate anion an analogue of the transition state of RNAse A? *J. Am. Chem. Soc.* **14**, 3630–3634.

KRIEGER, N.S. and TASHJIAN, A.H. (1983). Inhibition of stimulated bone resorption by vanadate. *Endocrinology* **113**, 324–328.

KRIVANEK, J. and NOVAKOVA, L. (1989). Inhibition of phosphorylation of the mitochondrial 34 kDa protein: a unique effect of vanadium ions? *Biochem. Pharmacol.* **38**, 2713–2717.

LABEDZKA, M., GULYAS, H., SCHMIDT, N., and GERCKEN, G. (1989). Toxicity of metallic ions and oxides to rabbit alveolar macrophages. *Environ. Res.* **48**, 255–274.

LEE, R.E., GORANSON, S.S., ENRIONE, R.E., and MORGAN, G.B. (1972). National air surveillance cascade impactor network. II. Size distribution measurements of trace element components. *Environ. Sci. Technol.* **6**, 1025–1030.

LEES, R.E.M. (1980). Changes in lung function after exposure to vanadium compounds in fuel oil ash. *Br. J. Ind. Med.* **37**, 253–256.

LEONARD, A. and GERBER, G.B. (1994). Mutagenicity, carcinogenicity, and teratogenicity of vanadium compounds. *Mutat. Res.* **317**, 81–88.

LEVY, B.S., HOFFMAN, L., and GOTTSEGEN, S. (1984). Boilermakers' bronchitis. Respiratory tract irritation associated with vanadium pentoxide during oil-to-coal conversion of a power plant. *J. Occup. Med.* **26**, 567–570.

LINDQUIST, R.N., LYNN, J.L., and LIENHARD, G.E. (1973). Possible transition-state analogs for ribonuclease: the complexes of uridine with oxovanadium(IV) and vanadium(V) ions. *J. Amer. Chem. Soc.* **95**, 8762–8768.

LIOCHEV, S.I. and FRIDOVICH, I. (1989). Vanadate-stimulated oxidation of NAD(P)H. *Free Rad. Biol. Med.* **6**, 617–622.

(1990). Vanadate-stimulated oxidation of NAD(P)H in the presence of biological membranes and other sources of O_2^-. *Arch. Biochem. Biophys.* **279**, 1–7.

LIOCHEV, S., IVANCHEVA, E., and RUSSANOV, E. (1988). Vanadyl- and vanadate-induced lipid peroxidation in mitochondria and in phosphytidylcholine suspensions. *Free Rad. Biol. Med.* **4**, 317–323.

LLOBET, J.M., COLOMINA, M.T., SIRVENT, J.J., DOMINGO, J.L., and CORBELLA, J. (1993). Reproductive toxicity evaluation of vanadium in male mice. *Toxicology* **80**, 199–206.

LOPEZ, V., STEVENS, T., and LINDQWUIST, R.N. (1976). Vanadium ion inhibition of alkaline phosphatase-catalyzed phosphate ester hydrolysis. *Arch. Biochem. Biophys.* **175**, 31–38.

MACARA, I.G., McLEOD, G.C., and KUSTIN, K. (1979). Vanadium in tunicates: oxygen-binding studies. *Comp. Biochem. Physiol. A* **62**, 821–826.

MAMURO, T., MATSUDA, Y., MIZOHATA, A., TAKEUCHI, T., and FUJITA, A. (1970). *Annual Report to the Radiation Center of Osaka Prefecture*, Vol. 11.

MARINI, M., ZUNICA, G., BAGNARA, G.P., and FRANCESCHI, C. (1987). Effect of vanadate on PHA-induced proliferation of human lymphocytes from young and old subjects. *Biochem. Biophys. Res. Commun.* **142**, 836–842.

MAYOTTE, D.H., WONG, J., ST PETER, R.L., LYTLE, F.W., and GREEGOR, R.B. (1981). X-ray absorption spectroscopic investigation of trace vanadium sites in coal. *Science* **214**, 554–556.

MCLEAN, J.R., MCWILLIAMS, R.S., KAPLAN, J.G., and BIRNBOIM, H.C. (1982). Rapid detection of DNA strand breaks in human peripheral blood cells and animal organs following treatment with physical and chemical agents, in Bora, K.C., Douglas, G.R., and Nestmann, E.R. (Eds) *Chemical Mutagenesis, Human Population Monitoring, and Genetic Risk Assessment*, pp. 137–158, New York: Elsevier Scientific Publishing.

MCMANUS, T.P., ZELIKOFF, J.T., SCHLESINGER, R.B., and COHEN, M.D. (1995). Immunotoxic effects in the rat lung from vanadium inhalation. *The Toxicologist* **15**, 257.

MCTURK, L.C., HUIS, C.H., and ECKHARDT, R.E. (1966). Health hazards of vanadium-containing residual oil ash. *Ind. Med. Surg.* **25**, 29–36.

MONTERO, M.R., GUERRI, C., RIBELLES, M., and GRISOIA, S. (1981). Inhibition of protein synthesis in cell cultures by vanadate and in brain homogenates of rats fed vanadate. *Physiol. Chem. Phys.* **13**, 281–287.

MUSK, A.W. and TEES, J.G. (1982). Asthma caused by occupational exposure to vanadium compounds. *Med. J. Aust.* **1**, 183–184.

MYRON, D.R., GIVAND, S.H., HOPKINS, L.L., and NIELSEN, F.H. (1975). Studies on vanadium deficiency in the rat. *Fed. Proc.* **34**, 923.

NECHAY, B.R. (1984). Mechanisms of action of vanadium. *Ann. Rev. Pharmacol. Toxicol.* **24**, 501–524.

NECHAY, B.R., NANNINGA, L.B., NECHAY, P.E., POST, R.L., GRANTHAM, J.J., MACARA, I.G., KUBENA, L. F., PHILLIPS, T.D., and NIELSEN, F.H. (1986). Role of vanadium in biology. *Fed. Proc.* **45**, 123–132.

NIELSEN, F.H. (1995). Vanadium in mammalian physiology and nutrition, in Sigel, H. and Sigel, A. (Eds) *Vanadium and Its Role in Life; Metal Ions in Biological Systems*, Vol. 31, pp. 543–574, New York: Marcel Dekker.

NIELSEN, F.H. and OLLERICH, D.A. (1973). Studies on a vanadium deficiency in chicks. *Fed. Proc.* **32**, 929.

NIH (National Institutes of Health) (1993). *Preliminary Report of the 90-Day Subchronic Inhalation Toxicity Study of Vanadium Pentoxide in F344 Rats and $B_6C_3F_1$ Mice*, Research Triangle Park, NC: National Institutes of Health.

NIOSH (National Institute for Occupational Safety and Health) (1985). *National Institutes for Occupational Safety and Health Pocket Guide to Chemical Hazards*, 5th Edn, pp. 234–235, Washington, DC: US Department of Health and Human Services.

NRIAGU, J.O. and PACYNA, J.M. (1988). Quantitative assessment of worldwide contamination of air, water, and soils by trace metals. *Nature* **333**, 134–139.

OBERG, S.G., PARKER, R.D.R., and SHARMA, R.P. (1978). Distribution and elimination of intratracheally-administered vanadium compound in the rat. *Toxicology* **11**, 315–323.

OKA, J.A. and WEIGEL, P.H. (1991). Vanadate modulates the activity of a subpopulation of asialoglycoprotein receptors on isolated rat hepatocytes: active surface receptors are internalized and replaced by inactive receptors. *Arch. Biochem. Biophys.* **289**, 362–370.

OLTZ, E.M., BRUENING, R.C., SMITH, M.J., KUSTIN, K., and NAKANISHI, K. (1988). The tunichromes: a class of reducing blood pigments from sea squirts – Isolation, structures, and vanadium chemistry. *J. Am. Chem. Soc.* **110**, 6162–6172.

OWUSU-YAW, J., COHEN, M.D., FERNANDO, S.Y., and WEI, C.I. (1990). An assessment of the genotoxicity of vanadium. *Toxicol. Lett.* **50**, 327–336.

PASCHOA, A.S., WRENN, M.E., SINGH, M.P., BRUENGER, F.W., MILLER, S.C., CHOLEWA, M., and JONES, K.W. (1987). Localization of vanadium-containing particles in the lungs of uranium/vanadium miners. *Biol. Trace Elem. Res.* **13**, 275–282.

PROESCHER, F. and SEIL, H.A. (1917). A contribution to the action of vanadium with particular reference to syphilis. *Am. J. Syph. Gonn. Ven. Dis.* **1**, 347–405.

PUMIGLIA, K.M., LAU, L., HUANG, C., BURROUGHS, S., and FEINSTEIN, M.B. (1992). Activation of signal transduction in platelets by the tyrosine phosphatase inhibitor pervanadate (vanadyl hydroperoxide). *Biochem. J.* **286**, 441–449.

RAMANADHAM, M. and KERN, M. (1983). Differential effect of vanadate on DNA synthesis induced by mitogens in T- and B-lymphocytes. *Mol. Cell. Biochem.* **51**, 67–71.

REHDER, D. (1991). The bioinorganic chemistry of vanadium. *Angew. Chem. Int. Ed. Engl.* **30**, 148–167.

——— (1995). Inorganic considerations on the function of vanadium in biological systems, in Sigel, H. and Sigel, A. (Eds) *Vanadium and Its Role in Life; Metal Ions in Biological Systems, Vol. 31*, pp. 1–44, New York: Marcel Dekker.

REIF, D.W., COULOMBE, R.A., and AUST, S.D. (1989). Vanadate-dependent NAD(P)H oxidation by microsomal enzymes. *Arch. Biochem. Biophys.* **270**, 137–143.

RHOADS, K. and SANDERS, C.L. (1985). Lung clearance, translocation, and acute toxicity of arsenic, beryllium, cadmium, cobalt, lead, selenium, vanadium, and ytterbium oxides following deposition in rat lung. *Environ. Res.* **36**, 359–378.

ROLDAN, R.E. and ALTAMIRANO, L.M.A. (1992). Chromosomal aberrations, sister chromatid exchanges, cell-cycle kinetics, and satellite associations in human lymphocyte cultures exposed to vanadium pentoxide. *Mutat. Res.* **245**, 61–65.

ROSCHIN, I.V. (1967). Toxicology of vanadium compounds used in modern industry. *Gig. Sanit.* **32**, 26–32.

ROSCHIN, I.V., ORDZHONIKIDZE, E.K., and SHALGANOVA, I.V. (1980). Vanadium-toxicity, metabolism, and carrier state. *J. Hyg. Epidemiol. Microbiol. Immunol.* **24**, 377–383.

SABBIONI, E., CLERICI, L., and BRAZZELLI, A. (1983). Different effects of vanadium ions on some DNA-metabolizing enzymes. *J. Toxicol. Environ. Health* **12**, 737–748.

SCHIFF, L.J. and GRAHAM, J.A. (1984). Cytotoxic effect of vanadium and oil-fired fly ash on hamster tracheal epithelium. *Environ. Res.* **34**, 390–402.

SCHROEDER, H.A. and BALASSA, J.J. (1967). Arsenic, germanium, tin, and vanadium in mice: effects on growth, survival, and tissue levels. *J. Nutr.* **92**, 245–252.

SCHROEDER, H.A. and MITCHENER, M. (1975). Life-term effects of mercury, methylmercury, and nine other trace metals on mice. *J. Nutr.* **105**, 452–458.

SCHROEDER, H.A., BALASSA, J.J., and TIPTON, I.H. (1963). Abnormal trace metals in man – vanadium. *J. Chron. Dis.* **16**, 1047–1071.

SCHROEDER, H.A., MITCHENER, M., and NASON, A.P. (1970). Zirconium, niobium, antimony, vanadium, and lead in rats: life-term studies. *J. Nutr.* **100**, 59–68.

SCHWARTZ, K. and MILNE, D.B. (1971). Growth effects of vanadium in the rat. *Science* **174**, 426–428.

SEGLEN, P.O. and GORDON, P.B. (1981). Vanadate inhibits protein degradation in isolated rat hepatocytes. *J. Biol. Chem.* **256**, 7699–7703.

SERRA, M.A., SABBIONI, E., MARCHESINI, A., PINTAR, A., and VALOTI, M. (1990). Vanadate as an inhibitor of plant and mammalian peroxidases. *Biol. Trace Elem. Res.* **23**, 151–164.

SHARMA, A. and TALUKDER, G. (1987). Effects of metals on chromosomes of higher organisms. *Environ. Mutagen.* **9**, 191–226.

SHARMA, R.P., OBERG, S.G., and PARKER, R.D. (1980). Vanadium retention in rat tissues following acute exposures to different dose levels. *J. Toxicol. Environ. Health* **6**, 45–54.

SHARMA, R.P., BOURCLER, D.R., BRINKERHOFF, C.R., and CHRISTENSEN, S.A. (1981). Effects of vanadium on immunologic functions. *Am. J. Indust. Med.* **2**, 91–99.

SHARMA, R.P., FLORA, S.J.S., BROWN, D.B., and OBERG, S.G. (1987). Persistence of vanadium compounds in lungs after intratracheal instillation in rats. *Toxicol. Ind. Health* **3**, 321–329.

SHEU, C.W., RODRIGUEZ, I., and LEE, K.W. (1990). Proliferation and morphological alterations of Balb/3T3 cells by vanadate. *Environ. Molec. Mutagen.* **15**(Suppl. 17), 55.

SINGH, J., NORDLIE, R.C., and JORGENSON, R.A. (1981). Vanadate: a potent inhibitor of multifunctional glucose-6-phosphatase. *Biochim. Biophys. Acta* **678**, 477–482.

SJOBERG, S.G. (1950). Vanadium pentoxide dust. A clinical and experimental investigation on its affect after inhalation. *Acta Med. Scand.* **138**, 1–5.

SMITH, J.B. (1983). Vanadium ions stimulate DNA synthesis in Swiss mouse 3T3 and 3T6 cells. *Proc. Natl Acad. Sci. USA* **80**, 6162–6166.

SMITH, M.J., RYAN, D.E., NAKANISHI, K., FRANK, P., and HODGSON, K.O. (1995). Vanadium in ascidians and the chemistry of tunichromes, in Sigel, H. and Sigel, A. (Eds) *Vanadium and Its Role in Life; Metal Ions in Biological Systems*, Vol. 31, pp. 423–490, New York: Marcel Dekker.

SORA, S., CARBONE, M.L., PACCIARINI, M., and MAGNI, G.E. (1986). Disomic and diploid meiotic products induced in *Saccharomyces cerevisiae* by the salts of 27 elements. *Mutagenesis* **1**, 21–28.

STANKIEWICZ, P.J. (1989). Vanadium(V)-stimulated hydrolysis of 2,3-diphosphoglycerate. *Arch. Biochem. Biophys.* **270**, 489–494.

STOCKS, P. (1960). On the relations between atmospheric pollution in urban and rural localities and mortality from cancer, bronchitis, pneumonia, with particular reference to 3,4-benzopyrene, beryllium, molybdenum, vanadium, and arsenic. *Br. J. Cancer* **14**, 397–418.

SUN, M. (Ed.) (1987). *Toxicity of Vanadium and its Environmental Health Standard*, Chengdu, China: West China University Medical Sciences.

SWARUP, G., COHEN, S., and GARBERS, D.L. (1982). Inhibition of membrane phosphotyrosyl-protein phosphatase activity by vanadate. *Biochem. Biophys. Res. Comm.* **107**, 1104–1109.

SYMANSKI, J. (1939). Industrial vanadium poisoning: its origin and symptomatology. *Arch. Gewerbepath. Gewerbehyg.* **9**, 295–301.

THOMPSON, H.J., CHASTEEN, N.D., and MEEKER, L.D. (1984). Dietary vanadyl(IV) sulfate inhibits chemically-induced mammary carcinogenesis. *Carcinogenesis* **5**, 849–851.

THURSTON, G. and SPENGLER, J. (1985). A quantitative assessment of source contributions to inhalable particulate matter pollution in metropolitan Boston. *Atmos. Environ.* **19**, 9–25.

TIPTON, I.H. and SHAFER, J.J. (1964). Statistical analysis of lung trace element levels. *Arch. Environ. Health* **8**, 56–67.

TOLMAN, E.L., BARRIS, E., BURNS, M., PANSINI, A., and PARTRIDGE, R. (1979). Effects of vanadium on glucose metabolism *in vitro*. *Life Sci.* **25**, 1159–1164.

TOROSSIAN, K., FREEDMAN, D., and FANTUS, I.G. (1988). Vanadate down-regulates cell surface insulin and growth hormone receptor and inhibits insulin receptor degradation in cultured human lymphocytes. *J. Biol. Chem.* **263**, 9353–9359.

TRUDEL, S., PAQUET, M.R., and GINSTEIN, S. (1991). Mechanism of vanadate-induced activation of tyrosine phosphorylation and of the respiratory burst in HL60 cells. *Biochem. J.* **276**, 611–617.

VADDI, K. and WEI, C.I. (1991a). Effect of ammonium metavanadate on the mouse peritoneal macrophage lysosomal enzymes. *J. Toxicol. Environ. Health* **33**, 65–78.

—— (1991b). Modulation of Fc receptor expression and function in mouse peritoneal macrophages by ammonium metavanadate. *Int. J. Immunopharmacol.* **13**, 1167–1176.

—— (1996). Modulation of macrophage activation by ammonium metavanadate. *J. Toxicol. Environ. Health* **49**, 631–645.

VAN ZINDEREN, J., BAKKER, L., and JAWORSKI, J.F. (1980). *Effects of Vanadium in the Canadian Environment*, Ottawa: National Research Council of Canada, Associate Committee Scientific Criteria for Environmental Quality.

VOUK, V.B. (1979). Vanadium, in Friberg, L. Nordberg, G.F., and Vouk, V.B. (Eds), *Handbook on the Toxicology of Metals*, pp. 658–674, New York: Elsevier Press.

WANG, E. and CHOPPIN, P.W. (1981). Effect of vanadate on intracellular distribution and function of 10 nm microfilaments. *Proc. Natl Acad. Sci. USA* **78**, 2363–2367.

WATERS, M.D. (1977). Toxicology of vanadium, in Goyer, R.A. and Mehlman, M.A. (Eds) *Toxicology of Trace Elements; Advances in Modern Toxicology*, pp. 147–189, New York: Halsted Press.

WATERS, M.D., GARDNER, D.E., and COFFIN, D.L. (1974). Cytotoxic effects of vanadium on rabbit alveolar macrophages *in vitro*. *Toxicol. Appl. Pharmacol.* **28**, 253–263.

WATSON, J., CHOW, J., LU, Z., FUJITAS, E., LOWENTHAL, D., and LAWSON, D. (1994). Chemical mass balance source appointment of PM_{10} during the Southern California air quality study. *Aerosol. Sci. Technol.* **21**, 1–36.

WEI, C.I., AL-BAYATI, A., CULBERTSON, M.R., ROSENBLATT, L.S., and HANSEN, L.D. (1982). Acute toxicity of ammonium metavanadate in mice. *J. Toxicol. Environ. Health* **10**, 673–687.

WHO (World Health Organization) (1987). *Air quality guidelines for Europe, Copenhagen Regional Office of Europe*, European Series Number 23, Copenhagen: World Health Organization.

—— (1988). *Environmental Health Criteria 81. Vanadium*, Geneva: World Health Organization.

WIETZERBIN, J., GAUDELET, C., AGUET, M., and FALCOFF, E. (1986). Binding and crosslinking of recombinant mouse interferon-γ to receptors in mouse leukemic L1210 cells: interferon-γ internalization and receptor down-regulation. *J. Immunol.* **136**, 2451–2455.

WITKOWSKA, D., OLEDZKA, R., and PIETRZYK, B. (1986). Influence of intoxication with vanadium compounds on the intestinal absorption of calcium in the rat. *Bull. Environ. Contam. Toxicol.* **37**, 899–906.

YAMAGUCHI, M., OISHI, H., and SUKETA, Y. (1989). Effect of vanadium on bone metabolism in weanling rats: zinc prevents the toxic effect of vanadium. *Res. Exp. Med.* **189**, 47–53.

YANG, D., BROWN, A.B., and CHAN, T.M. (1989). Stimulation of tyrosine-specific protein phosphorylation and phosphatidylinositol phosphorylation by orthovanadate in rat liver plasma membrane. *Arch. Biochem. Biophys.* **274**, 659–662.

ZAPOROWSKA, H. and WASILEWSKI, W. (1992). Haematological results of vanadium intoxication in Wistar rats. *Comp. Biochem. Physiol.* **101C**, 57–61.

ZELIKOFF, J.T. and COHEN, M.D. (1995). Immunotoxicity of inorganic metal compounds, in Smialowicz, R.J. and Holsapple, M.P. (Eds) *Experimental Immunotoxicology*, pp. 189–228, Boca Raton, FL: CRC Press.

ZENZ, C. and BERG, B.A. (1967). Human responses to controlled vanadium pentoxide exposure. *Arch. Environ. Health* **14**, 709–712.

ZENZ, C., BARTLETT, J.P. and THIEDE, A.H. (1962). Acute vanadium intoxication. *Arch. Environ. Health* **5**, 542–546.

ZHONG, B., GU, Z., WALLACE, W.E., WHONG, W., and ONG, T. (1994). Genotoxicity of vanadium pentoxide in Chinese hamster V79 cells. *Mutat. Res.* **321**, 35–42.

ZOLLER, W.H., GORDON, G.E., and GLADNEY, E.S. (1973). The sources and distribution of vanadium in the atmosphere, in Kothny, E.L. (Ed.) *Advances in Chemistry (Series No. 123): Trace Elements in the Environment*, pp. 31–41, Washington, DC: American Chemical Society.

ZYCHLINSKI, L. (1980). Toxicological appraisal of work-places exposed to the dust containing vanadium pentoxide. *Bromat. Chem. Toksykol.* **8**, 195–199.

11

Iron, Zinc, and Copper

FELIX OLIMA OMARA[1], PAULINE BROUSSEAU[2], BARRY RAYMOND
BLAKLEY[3], AND MICHEL FOURNIER[1]

[1] TOXEN, Université du Québec à Montréal, Montréal, Québec, Canada, H3C 3P8;
[2] Concordia University, Montréal, Québec, Canada, H3G 1M8;
[3] Department of Veterinary Physiological Sciences, University of Saskatchewan, Saskatoon, Saskatchewan, Canada, S7N 0W0

11.1 Introduction

Macroelements, such as zinc, iron, and copper, although taken daily in very small amounts, are key components necessary to maintain cellular homeostasis. The primary reason for this necessity is associated with the fact that hundreds of metalloenzymes require a metal element, as a cofactor, to be functional. Among the systems that require these elements, is the immune system. In this chapter, we review the existing data regarding the potential immunomodulatory effects of iron, zinc, and copper through epidemiological and animal data associated with natural and acquired immunity, as well as host resistance.

11.2 Iron and the Immune System

11.2.1 Iron deficiency and overload

Maintenance of cellular homeostasis of iron, an essential element, is required for many biological processes, including the regulation of immune function. Immune dysfunction may be associated with deficient or excessive levels of iron in the body. Iron deficiency may result from a lack of iron in the diet, impaired intestinal absorption, increased demands during pregnancy or infancy, chronic blood loss, and/or chronic diseases (Deiss, 1983; Cotran et al., 1989). Iron overload occurs in hereditary hemochromatosis, transfusional iron overload in sickle-cell or refractory anemias (e.g. thalassemia major), and excess dietary iron intake (Powell et al., 1980; Cotran et al., 1989; Bryan, 1991). Animal and human clinical studies have examined the effects of iron deficiency, primarily nutritional iron deficiency, on immune function, but data on excess iron have been deduced mostly from diseases of iron overload and *in vitro* studies. In clinical disease states, confounding variables complicate interpretation of results. Changes in the immune system induced by iron deficiency or overload range from pathological changes in lymphoid organs to alterations in lymphocyte function, specific and innate immune responses, and host resistance. In iron-deficient rats, the size and weight of the thymus were decreased while spleen weight was increased (Luo *et al.*, 1990); in mice, the thymic cortices and medulla were depleted of thymocytes (Kuvibidila *et al.*, 1990). In addition, thymus weights were normal (Dhur *et al.*, 1991) and spleen weights were lower in iron-deficient rat pups (Lockwood and Sherman, 1988).

11.2.2 *Humoral-mediated immunity*

Conflicting data on the influence of iron deficiency and excess iron on the antibody response and other B-lymphocyte functions have been reported. The proportions of B-lymphocytes have been reported to be decreased (Santos and Falcao, 1990) or unaltered (Krantman *et al.*, 1982) during iron deficiency; these differences may be related to species and differences in the degree and duration of the deficiency. Clinical studies have shown that iron deficiency can compromise antibody-mediated immune responses. The antibody response to diphtheria toxoid antigen was compromised in iron-deficient children (MacDougall and Jacobs, 1978). Iron-deficient children with recurrent pulmonary infections have lower levels of serum IgG subclasses and pneumococcal polysaccharide-specific IgG subclass antibodies (Feng *et al.*, 1994). In addition, the levels of non-specific IgA and IgG antibodies were found to be much lower in patients with iron deficiency anemia (Galàn *et al.*, 1988), although antibody responses to the tetanus toxoid antigen were unaltered in iron-deficient children (Chandra and Saraya, 1975).

In rodent models of nutritional iron deficiency, antibody responses to the sheep red blood cell (SRBC) antigen were consistently reduced (Kuvibidila *et al.*, 1982; Kochanowski and Sherman, 1985; Omara and Blakley, 1994a); these results supported those from human clinical studies. In iron-deprived rats, the number of intestinal mucosal secretory IgA and IgM cells were lower, indicating an alteration in mucosal immunity (Perkkio *et al.*, 1987). The effect of iron deficiency on the mitogenic response of B-lymphocytes is unclear as responses have been reported to be both decreased (Kuvibidila and Sarpong, 1990) and unaffected (Blakley and Hamilton, 1988). The T-lymphocyte-independent/macrophage-dependent antibody response against dinitrophenyl (DNP)-Ficoll was normal in iron-deprived mice (Blakley and Hamilton, 1988), suggesting that suppression of antibody responses to the SRBC antigen may be attributed to T-helper (T_H) lymphocyte dysfunction.

The effects of iron overload on the humoral immune response or B-lymphocyte function are less well characterized and often conflicting. For example, mice fed 8000 ppm iron in the diet for 7 weeks had fewer numbers of IgM- (but normal levels of IgG-) plaque-forming cells (PFC) against SRBC (Omara and Blakley, 1994a). However, rats fed 1250 ppm iron for 8.5 weeks had normal IgM and IgG immune responses to SRBC antigen (Sherman, 1990). The differences between the above observations might be related to the severity or duration of iron overload, or to species susceptibility. In human diseases of iron overload, the changes in B-lymphocyte functions are influenced by the number of blood transfusions, splenectomy, and hetero-geneity of the disease condition. Whether the alterations in B-lymphocyte function in these conditions are related to iron overload is unclear. For exam-ple, in patients with thalassemia who have undergone splenectomy, the number of circulating B-lymphocytes and serum immunoglobulin levels are increased (Grady *et al.*, 1985; Dwyer *et al.*, 1987), while responses of B-lymphocytes to mitogens are normal; in patients who have not under-gone splenectomy, the mitogenic responses are diminished (Bryan and Leech, 1983; Bryan *et al.*, 1984, 1991). In patients with sickle-cell anemia who had undergone transfusion, numbers of spontaneously immunoglobulin-secreting B-lymphocytes were increased while the number of PFC induced by pokeweed mitogen (PWM) was lower due to a reduction in the number of T_H-lymphocytes (Nualart *et al.*, 1987). Treatment of B-lymphocytes *in vitro* with both PWM and interferon-α (IFN-α) increased the PFC re-sponse (Nualart *et al.*, 1987), suggesting a defect in cytokine (e.g. IFN-α) production.

11.2.3 Cell-mediated immunity

Iron deficiency and iron overload produce subtle changes in T-lymphocytes and cell-mediated immune functions. In human clinical and animal studies evaluating iron deficiency, weak cell-mediated immune responses, as moni-tored by the delayed-type hypersensitivity (DTH) response to protein antigen (Joynson *et al.*, 1972; MacDougall *et al.*, 1975; Omara and Blakley, 1994b) and delayed cutaneous hypersensitivity (DCH) to 2, 4-dinitrofluorobenzene (DNFB) (Kuvibidila *et al.*, 1981; Omara and Blakley, 1994b), have been con-sistently reported. The capacity of primed splenocytes from iron-deficient mice to transfer the DTH response to naive mice was reduced (Omara and Blakley, 1994a). In addition, T-lymphocyte mitogenic responses (MacDougall *et al.*, 1975; Soyano *et al.*, 1982; Blakley and Hamilton, 1988; Kuvibidila *et al.*, 1990; Kuvibidila and Sarpong, 1990; Omara and Blakley, 1994a), numbers of $CD4^+$ and $CD8^+$ lymphocytes and the CD4/CD8 ratios (Krantman *et al.*, 1982; Kuvibidila *et al.*, 1990; Santos and Falcao, 1990; Dhur *et al.*, 1991; Feng *et al.*, 1994), and T-lymphocyte secretion of IFN-γ (Omara *et al.*, 1994) were de-creased in iron-deficient hosts. These alterations may contribute, in part, to the suppression of T-lymphocyte mitogenic responses, as well as to reductions in cell-mediated and humoral immune responses.

The decreases in activity of the iron-dependent enzyme ribonucleotide reductase (Weiss *et al.*, 1995), as well as in the production of interleukin-2 (IL-2) and IL-6 production (Galàn *et al.*, 1992; Latunde-Dada and Young, 1992; Feng *et al.*, 1994), observed during iron deficiency may further contribute to altered humoral and cell-mediated responses. The selective sensitivity to iron deprivation by T_{H1} lymphocytes compared with that of T_H2 cells (Thorsan *et al.*, 1991) is consistent with the association between iron deficiency and a diminished cell-mediated immune response. However, some studies have reported no alterations in the proportions of T-lymphocyte CD4$^+$ and CD8$^+$ subsets (Mainou-Fowler and Brock, 1985) or in IL-2 production (Omara and Blakley, 1994b) during iron deficiency. This suggests that the degree of iron deficiency may influence the extent of immune alteration as well as reflecting upon the relative cellular or functional susceptibility.

Human clinical and animal studies of iron overload have demonstrated that iron can impair several T-lymphocyte functions. Mice fed 8000 ppm iron for 7 weeks exhibited decreased splenocyte IFN-γ production and delayed skin contact hypersensitivity (DCH) to DNFB, yet unaltered T-lymphocyte mitogenic responses, splenocyte IL-2 production, and DTH responses to SRBC (Omara and Blakley, 1994b). These findings indicate that excess iron differentially modulates contact hypersensitivity and the classic DTH responses. The capacity of splenocytes to generate allospecific cytotoxic T-lymphocytes (T_C) *in vitro* was decreased in mice with iron overload produced by dietary carbonyl iron or iron–dextran injections (Good *et al.*, 1987a). The splenocytes from carbonyl iron-loaded mice (not iron–dextran) produced less IL-2 as a result of decreased numbers of IL-2-secreting cells amongst the spleen cell population (Good *et al.*, 1987a). These observations may be related to the fact that carbonyl iron induces parenchymal cell iron overload, while injected iron–dextran produces iron overload mainly in mononuclear phagocytic cells. Parenteral iron overload induced by iron–dextran in rats was associated with increased lipid peroxidation in splenic homogenates and decreased lymphocyte proliferative responses to concanavalin A (ConA). This suggests a role for oxygen radicals in iron-dependent T-lymphocyte dysfunction (Cardier *et al.*, 1995).

In vitro studies with an iron overload model have shown that iron (iron citrate) can inhibit mixed lymphocyte reactions (Bryan *et al.*, 1981), decrease cloning efficiency of CD4$^+$ T-lymphocytes (Good *et al.*, 1986), and suppress cell proliferation in response to T-lymphocyte-specific mitogens (Bryan and Leech, 1983; Pons *et al.*, 1992). Treatment of human lymphocytes with iron *in vitro* also inhibited the expression of CD2 molecules and E-rosette formation (Carvalho and de Sousa, 1988), as well as generation of allospecific T_C (Good *et al.*, 1987b, 1988). A recent study by Arosa and de Sousa (1995) found that the activity of CD4–lck complexes (but not CD8–lck), expression of CD4 molecules, percentages of CD4$^+$ T cells, and CD4/CD8 ratios were markedly decreased in human lymphocytes treated *in vitro* with iron citrate. The down-modulation of CD4–lck (i.e. CD4–tyrosine kinase p56^{lck}) activity and the decreased percentage of CD4+ T-lymphocytes may contribute to the observed T-lymphocyte dysfunction associated with iron overload (Arosa and de Sousa, 1995).

Clinical studies performed in situations of iron overload have also documented alterations in T-lymphocyte function. Although DTH reactions, levels of circulating T-lymphocytes and CD4$^+$ T-lymphocyte subsets, CD4/CD8 ratios, and T-lymphocyte mitogenic responses were decreased in patients with thalassemia (Munn *et al.*, 1981; Guglielmo *et al.*, 1984; Dwyer *et al.*, 1987; Pardalos *et al.*, 1987), expression of transferrin and IL-2 receptors were found to be normal (Pattanapanyasat, 1990). In patients with sickle-cell anemia who had undergone transfusion, decreased CD4/CD8 ratios and levels of T_H lymphocytes were also observed (Nualart *et al.*, 1987). However, in patients with thalassemia who had undergone splenectomy, although mitogenic responses were normal, levels of circulating CD8+ T-lymphocyte subsets were elevated (Guglielmo *et al.*, 1984; Grady *et al.*, 1985). In patients with hereditary hemochromatosis, numbers of CD4$^+$ and CD8$^+$ lymphocytes and CD4/CD8 ratios vary among patients; these values range from normal (Porto *et al.*, 1994) to high and low numbers of CD8$^+$ T-lymphocytes in untreated and treated patients, respectively. In addition, lymphocyte mitogenic responses were found to be decreased or unaltered in untreated and treated patients, respectively (Bryan *et al.*, 1991). The decrease in CD4/CD8 ratios in patients with thalassemia may be due to a down-regulation in the numbers of CD4+ T-lymphocytes and/or an up-regulation of CD8$^+$ T-lymphocytes (Dwyer *et al.*, 1987; Pardalos *et al.*, 1987). In addition, suppression of T-lymphocyte functions in diseases of iron overload could result from elevated serum levels of ferritin, as ferritin has been shown to inhibit several T-lymphocyte functions (Matzner *et al.*, 1979; Harada *et al.*, 1987).

11.2.4 *Non-specific immunity*

Non-specific immunity mediated by macrophages, neutrophils, and natural killer (NK) cells can be altered by iron deficiency or overload. For example, phagocytosis of yeast cells by peritoneal macrophages and the tumoricidal/cytotoxic activity of splenic and peritoneal macrophages from iron-deficient mice or rats are impaired (Kuvibidila *et al.*, 1983b; Hallquist *et al.*, 1992; Omara *et al.*, 1994). The *in vivo* clearance of radiolabeled particles by fixed macrophages was depressed in rats with iron deficiency (Kuvibidila and Wade, 1987). In both children and mice with iron deficiency, recovered macrophages were found to secrete normal amounts of IL-1 (Bhaskaram *et al.*, 1989; Munoz *et al.*, 1994; Omara *et al.*, 1994), while those from iron-deficient rats secreted reduced amounts of IL-1 (Helyar and Sherman, 1987). In addition, the amount of IFN-α secreted by peritoneal macrophages was decreased in iron-deficient rats (Sherman, 1990). Tumor necrosis factor-α (TNF-α) secretion was unaffected in macrophages from iron-deficient mice (Omara *et al.*,

1994; Hrabinski *et al.*, 1995); in contrast, macrophages/monocytes from children with iron deficiency were found to secrete increased amounts of TNF-α (Munoz *et al.*, 1994). Therefore, the effect of iron deficiency on cytokine production appears to depend upon the type of cytokine analyzed and the affected host model, as well as the severity and duration of the iron deficiency. Lastly, the ability of splenic macrophages to present antigen to T-lymphocytes was not altered in iron-deficient mice (Omara and Blakley, 1994a). This finding suggests that alterations in humoral and cell-mediated immune responses due to iron deficiency may be attributed, at least in part, to T-lymphocyte dysfunction.

In iron-loaded mice and rats, phagocytosis by peritoneal or splenic macrophages was reduced (Omara *et al.*, 1994; Lu and Hayashi, 1995). However, in murine macrophages exposed to iron citrate *in vitro*, phagocytosis and cell maturation were inhibited (Gebran *et al.*, 1993). In clinical diseases of iron overload, diminished phagocytic cell activities have been demonstrated (van Asbeck *et al.*, 1982, 1984). In addition, bactericidal activities of macrophages from human patients with iron overload, and of leukocytes treated *in vitro* with iron salts, were markedly reduced (Ballart *et al.*, 1986); iron-induced inhibition of basic cationic proteins in leukocytes appeared to be responsible (Kaplan *et al.*, 1975). The improved ability of iron-loaded macrophages to kill *Brucella* organisms appeared to be related to increased hydroxyl radical production by macrophages (Jiang and Baldwin, 1993). In contrast, macrophages containing intermediate levels of iron killed *Listeria* organisms more effectively than macrophages with high iron levels (Alford *et al.*, 1991); stimulation of macrophage intracellular killing of certain microorganisms was found to require iron bound to lactoferrin (Lima and Kierszenbaum, 1987). The uptake of iron by macrophages could also modify macrophage production of nitric oxide, as well as that of cytokines required to kill microorganisms and tumor cells (Weiss *et al.*, 1995). For example, macrophages from iron-loaded mice were found to secrete normal amounts of IL-1α, but above average amounts of TNF-α, upon *in vitro* stimulation with lipopolysaccharide (LPS) (Omara *et al.*, 1994).

The expression of class II major histocompatibility complex (MHC) antigens and the antigen-presenting cell (APC) activity of splenic B-lymphocytes and macrophages were not altered in iron-deficient or -loaded mice (Omara *et al.*, 1994). This finding suggests that alterations in the specific immune response in iron overload and deficiency may not be attributed to alterations in macrophage APC function. The lack of a macrophage functional defect with respect to a specific immune response is also consistent with previous studies (Blakley and Hamilton, 1988) in which the murine T-lymphocyte-independent/macrophage-dependent antibody response against DNP-Ficoll was unaffected by iron deficiency. Other studies have also observed no defect in phagocytic activity of neutrophils during clinical iron deficiency (MacDougall *et al.*, 1975; Walter *et al.*, 1986) or overload (van Asbeck *et al.*, 1982, 1984), although impaired phagocytosis was observed in iron-deficient rat pups

(Kochanowski and Sherman, 1984). As noted above, this effect on phagocytosis appeared to be dependent upon the particular target micro-organisms; phagocytosis of bacteria was unaltered in neutrophils from iron-deficient rats, yet uptake of *Candida albicans* was decreased (Moore and Humbert, 1984).

In clinical (Yetgin *et al.*, 1979; Moore and Humbert, 1984; Hasan *et al.*, 1989) and experimentally induced iron deficiency (Moore and Humbert, 1984; Murakawa *et al.*, 1987), the ability of neutrophils to kill bacteria and/or fungi was severely impaired. As the neutrophilic killing of microorganisms involves activation of myeloperoxidase and the cellular oxidative burst, both of which are iron dependent, a reduced bactericidal/fungicidal activity of neutrophils during iron deficiency may be a consequence of reduced myeloperoxidase and oxidative burst activities. This viewpoint is supported by several studies which have demonstrated low myeloperoxidase and oxidative burst activities in neutrophils from individuals deficient in iron (Arbeter *et al.*, 1971; Prasad, 1979; Mackler *et al.*, 1984; Moore and Humbert, 1984; Turgeon-O'Brien *et al.*, 1985; Murakawa *et al.*, 1987). Interestingly, a study by Baggs and Miller (1973) did not detect a defect in neutrophil bactericidal activity in iron-deficient mice; however, the duration and/or severity of the iron deficiency may have accounted for this discrepancy. Iron deficiency also decreased the bactericidal activity of neutrophils against bacteria other than *Streptococcus pneumoniae* (a catalase-negative organism); this suggested poor oxygen radical generation in iron-deficient leukocytes (Moore and Humbert, 1984). As during an intentionally induced iron deficiency, clinical diseases of iron overload (i.e. thalassemia major) are characterized by normal neutrophilic phagocytosis in conjunction with decreased bactericidal, fungicidal, and oxidative burst activities (Cantinieaux *et al.*, 1988; Sen *et al.*, 1989); in these studies, the oxidative burst activities were diminished in proportion to the degree of iron overload.

Natural killer cells are a lymphocyte subpopulation which play a role in the non-specific destruction of tumors and/or virally infected cells *in vitro* and possibly *in vivo*. In iron-deprived rats, NK cytotoxicity against target tumor cells was reduced (Sherman and Lockwood, 1987; Lockwood and Sherman, 1988; Hallquist *et al.*, 1992; Spear and Sherman, 1992; Hrabinski *et al.*, 1995). Furthermore, while *in vitro* treatment of NK cells (obtained from iron-deficient hosts) with IFN-γ improved cytotoxic activity, an earlier study revealed only a partial recovery upon addition of IFN-α (Lockwood and Sherman, 1988). This suggested that a possible defect in the *in situ* production of IFN-γ might be present during iron deficiency (Spear and Sherman, 1992). In fact, a later study found that splenocytes from iron-deficient mice secreted less IFN-γ than did cells from iron-sufficient controls (Omara and Blakley, 1994b). The reduced NK cytotoxic activity observed in iron-deficient rats has been correlated with an increase in the incidence of DMBA-induced mammary tumors (Spear and Sherman, 1992), although Hrabinski *et al.* (1995) obtained conflicting results. Similarly, Omara and Blakley (1993) observed increased numbers of urethane-induced lung tumors in iron-deprived mice. Decreased NK cytotoxic activity has also been observed in the peripheral blood of humans who are deficient in iron (Santos and Falcao, 1990). Therefore, it appears that iron deficiency depresses NK cytotoxic activity in both humans and animal models.

In patients with thalassemia with iron overload, although peripheral blood levels of NK cells were normal, their cytotoxic activity against K562 tumor cell targets or virally infected fibroblasts was found to be severely depressed (Kaplan *et al.*, 1984; Akbar *et al.*, 1986, 1987; Sen *et al.*, 1989). It was demonstrated that NK cells from patients with thalassemia produced lower amounts of IFN-α than did their healthy counterparts; incubation of these cells with IFN-α or desferroxamine reversed the inhibition (Akbar *et al.*, 1987). In patients with sickle-cell anemia who have undergone transfusion, peripheral blood NK activity was also depressed (Nualart *et al.*, 1987). Conversely, NK function was not altered in patients suffering from hereditary hemochromatosis (Chapman *et al.*, 1988); this implies that clinical disease conditions might influence NK function. Oddly, although NK function was not altered in hereditary hemochromatosis, this condition is associated with an increased risk of malignancies and susceptibility to transplantable/chemically induced tumors in laboratory animals (Bomford and Williams, 1976; Nelson *et al.*, 1989; Hann *et al.*, 1991).

11.2.5 *Effects on host defense*

Although altered immune parameters in clinical and experimental iron deficiency and overload in animals may explain the association of abnormal iron status with an increased incidence of certain bacterial infections, the effect of iron on host resistance remains controversial. Both an increased incidence of infection (Cantwell, 1972; Feng *et al.*, 1994) and a decreased susceptibility to infection (Masawe *et al.*, 1974; Weinberg, 1984) have been observed in hosts with iron-deficient conditions. Iron-deficient mice were resistant to *Salmonella* infection (Baggs and Miller, 1973), but more susceptible to *Streptococcus pneumoniae* (Chu *et al.*, 1976). Furthermore, mice treated with iron–dextran were more susceptible to *Yersinia* infection (Robins-Browne and Prpic, 1985), while treatment of children with iron salts to correct iron deficiency decreased the incidence of respiratory and enteric infectious diseases (Arbeter *et al.*, 1971; Oppenheimer, 1980). Interestingly, within the same groups of children receiving iron, there was an increased susceptibility to bacterial and malarial infections (Farmer, 1976; Becroft *et al.*, 1977; Murray *et al.*, 1978; Oppenheimer *et al.*, 1986).

In clinical diseases associated with iron overload, a high incidence of bacterial infection is common (Mofenson *et al.*, 1988; Keusch, 1990). Excess iron has been demonstrated to increase susceptibility to transplantable or chemically induced tumors in laboratory animals (Siegers *et al.*, 1988; Nelson *et al.*, 1989;

Hann *et al.*, 1991), but some studies have also found an increased incidence of chemically induced tumors in iron-deficient laboratory rodents (Kuvibidila *et al.*, 1983a; Spear and Sherman, 1992; Omara and Blakley, 1993). Precisely how iron deficiency or excess iron alters resistance to infection is unclear. Iron is generally believed to be a growth nutrient for pathogens and tumour cells (Weinberg, 1984).

In summary, the role of iron in infectious diseases appears to be a complex phenomenon involving interactions between pathogens and the host immune system which, in turn, may depend upon the nature of the invading pathogen(s) and the degree of iron deficiency or overload. Clinical and animal studies appear to indicate that iron deficiency suppresses non-specific host immune defense mechanisms, as well as antibody- and cell-mediated immune responses, thereby giving rise to a generalized immunosuppression. In contrast, excess iron particularly affects non-specific and cell-mediated immune responses, with little impact on humoral responses. As such, the mechanisms by which iron deficiency and/or iron excess influence immune functions appear to be distinct.

11.3 Zinc and the Immune System

11.3.1 *Zinc deficiency and overload*

It is mainly through studies on malnutrition and immunity that it became clear that zinc has a key role in the maintenance of an adequate immune system. Many responses elaborated by the immune system can be impaired by zinc deficiency, which can arise as a result of congenital deficiency (as in acrodermatitis enteropathica: Chandra, 1980; sickle-cell anemia: Ballester *et al.*, 1986; or Down's syndrome: Fabris *et al.*, 1984, 1988; Licastro *et al.*, 1992), in an acquired deficiency subsequent to reduced zinc intake/absorption, or increased loss through excretion. Moreover, in other situations, such as chronic and acute infectious diseases (Sugarman, 1993), malabsorption syndromes (McClain, 1985), cancers (primarily head and neck, lung, hepatocellular, and esophageal: Mathè *et al.*, 1986; Mocchegiani *et al.*, 1994), malnutrition (Castillo-Duran *et al.*, 1987), and AIDS (Fabris *et al.*, 1988; Graham *et al.*, 1991), an associated zinc deficiency could also give rise to immune dysfunction. In cases of severe zinc deficiency, plasma zinc concentrations can be reduced by up to 50 percent; interestingly, in this condition, there is an associated increase in plasma β_2-globulin and decreased α-globulin levels (Sanstead, 1981).

Documented cases of zinc overload are scarce, probably due to the fact that zinc does not accumulate under continued exposure (Underwood, 1977; Prasad, 1983; Bertholf, 1988). Nevertheless, epidemiological data following occupational exposure to zinc suggest that the main effect of this element is metal fume fever; while this syndrome is not specific to zinc alone, it arises

239

subsequent to inhalation of freshly generated zinc oxide fumes. Although increased oral intake of zinc is associated with gastrointestinal disorders, no signs of systemic toxicity could be detected in humans following ingestion of up to 12 g of elemental zinc (Goyer *et al.*, 1979). In animal experiments, results from excess zinc intake are also divergent. Indeed, inhalation of zinc oxide by guinea pigs (at the threshold limit value of 10 mg/m^3 as dust (ATSDR, 1994) for 3 h/day for six consecutive days) was associated with decrements in lung volume as well as a reduced carbon monoxide diffusing capacity. These physiological changes were correlated with interstitial thickening and cellular infiltration of the alveolar ducts and alveoli (Lam *et al.*, 1985).

From a histopathological point of view, it is well documented that children suffering from a zinc deficiency syndrome called acrodermatitis (which leads to a defect in gastrointestinal zinc uptake) present with thymic hypoplasia. The same disease encountered in cattle also leads to thymic atrophy and to an absence of germinal centers in the lymph nodes (Hadden, 1995). On the other hand, Mocchegiani *et al.* (1995a) have recently shown that oral zinc supplementation provided for 30 consecutive days to 22-month-old Balb/c mice with atrophic thymi could induce thymic regrowth and an increase in relative weight. Using bromodeoxyuridine labeling, these same investigators demonstrated that supplementation with zinc could also help restore the numbers of proliferative cells in the thymic cortex.

While these results demonstrate the importance of zinc in the maintenance of thymic integrity, a normal thymic histology is not the only requirement in the ultimate production of functional T-lymphocytes necessary for competent cellular immunity. Indeed, thymic hormones also play a key role in the attraction of precursor T-lymphocytes into the thymus and their subsequent maturation into functional cells. Among them, thymulin (Bach *et al.*, 1972; Dardenne *et al.*, 1982), exclusively produced by the thymus (Savino *et al.*, 1988), is biologically active only in its zinc-bound form (Champion *et al.*, 1986). Thymulin is localized in thymic epithelial cells (Savino *et al.*, 1984), and its secretion is under the regulatory control of estradiol, growth hormone, hydrocortisone, IL-1, progesterone, and prolactin (Savino *et al.*, 1990; Coto *et al.*, 1992). Recently, Mocchegiani *et al.* (1995b) demonstrated that in patients with AIDS, the total amount of plasma thymulin (active and inactive) is not affected, although the amount of biologically active thymulin (zinc-bound form) decreases to near non-detectable levels. When oral zinc supplements were used as an adjunct to azidothymidine (AZT) therapy, zinc helped to stabilize/increase body weight, the absolute number of circulating CD4$^+$ cells, and the amount of plasma active thymulin, as well as to reduce the frequency of opportunistic infections in these patients. At the level of the bone marrow, Murray *et al.* (1983) demonstrated that a single intraperitoneal dose of zinc (at 4 or 12 mg/kg) increased the numbers of bone-marrow precursor cells (i.e. granulocyte/monocyte colony-forming units).

11.3.2 *Humoral-mediated immunity*

It has been demonstrated in animal models of nutritional zinc deficiency that primary and secondary antibody responses to SRBC were significantly suppressed, in conjunction with accelerated thymic hypoplasia and a decreased number of CD4$^+$ cells (Fraker *et al.*, 1978, 1986, 1987; Luecke *et al.*, 1978; Schloen *et al.*, 1979). Moreover, Fraker *et al.* (1984) have documented impaired humoral responses against T-lymphocyte-dependent and -independent antigens in suckling mice born of dams with marginal zinc deficiency that could be overcome by a dietary zinc supplementation. In an *in vitro* study, Lawrence (1981) demonstrated that incubation of mouse spleen cells with 0.1 mM zinc for 5 days provoked a dramatic suppression of SRBC-specific antibody production; under the same conditions, concentrations of 0.1, 1, and 10 µM zinc did not affect the response. Likewise, the lymphoproliferative response of LPS-stimulated B-lymphocytes obtained from Balb/c mice was suppressed following injection of a single intraperitoneal dose of zinc (0.7, 4, or 12 mg/kg; Murray *et al.*, 1983).

Similarly, immunomodulation arising from an abnormal zinc status has also been observed in clinical situations. Children suffering from conditional malnutrition leading to marasmus or kwashiorkor showed a severe impairment of humoral immunity associated with vitamin and metal deficiencies, in particular, a deficiency of zinc (Good and Fernandes, 1979). In addition, zinc treatment of patients over 70 years of age has been shown to increase their anti-tetanus toxoid IgG antibody levels (Duchateau *et al.*, 1981).

11.3.3 *Cell-mediated immunity*

It is well documented that zinc is a key element for the development and maintenance of adequate cellular immunity (Good and Fernandes, 1979; Schloen *et al.*, 1979; Kruse-Jarres, 1989; Keen and Gershwin, 1990). This fact was derived from studies which involved human populations suffering from malnutrition, congenital deficiencies, or AIDS. From a mechanistic point of view, it has been proposed that zinc can prevent DNA fragmentation by endonuclease activity during apoptosis (Wyllie, 1980), leading to an increase in the numbers of CD4$^+$ CD8$^-$ thymocytes. These cells, in turn, give rise to increased numbers of T$_H$ lymphocytes which play a central role in the activation of B and T$_C$ lymphocytes. Zinc was also identified as a key element in the maintenance of adequate membrane fluidity needed to prevent weakening of the cell membrane (Jay *et al.*, 1987; O'Dell *et al.*, 1987). Moreover, zinc was also associated with protection against cellular damage by metal-mediated free radicals (Har-El and Chevion, 1991; Lovering and Dean, 1991) and in the protection of sulfhydryl groups against oxidation (Bray and Bettger, 1990).

However, some deleterious effects of zinc on human immune cells have also been reported. Berger and Skinner (1974) have shown that when human leukocytes are treated *in vitro* for 3 days with 0.1 mM zinc, phytohemagglutinin (PHA)-induced T-lymphocyte proliferation is suppressed. Rao *et al.* (1979), using a similar approach, demonstrated that ConA-induced T-lymphocyte proliferation was stimulated when cells were exposed to 50 or 100 μM zinc for 3 days, but abrogated at 200 μM zinc. Stacey (1986) demonstrated that T_C lymphocyte activity could also be impaired following a 4h incubation of leukocytes with 100 or 200 μM zinc.

Animal studies to examine the role of zinc in the immune system have demonstrated age-related effects. In a series of *in vivo* experiments, Murray *et al.* (1983) reported that, following a single intraperitoneal dose of zinc (0.7, 1.3, 4, or 12 mg/kg), young adult Balb/c mice demonstrated a significant decrease in PHA-induced T-lymphocyte proliferation at all but the lowest zinc concentration; suppression of ConA-induced T-lymphocyte proliferation was also observed. In contrast, Saha *et al.* (1995) have shown that when older mice (9 to 12 months old) were treated first with hydrocortisone (an endocrine inducer of zinc-thymulin; Savino *et al.*, 1990; Coto *et al.*, 1992) and then with zinc (72 μg/day for 5 days), the mice exhibited a significant increase in the proliferative response of their thymocytes to PHA and ConA, as well as to IL-1 and -2. Moreover, Mocchegiani *et al.* (1995a), using 22-month-old Balb/c mice, demonstrated that the proliferative response of splenocytes to PHA and ConA could be significantly increased following oral zinc supplementation. These investigators also showed that incubation of zinc with spleen cells from aged mice increased ConA-induced lymphoproliferation while the PHA-induced response remained unaltered.

11.3.4 *Non-specific immunity*

Although results are conflicting, effects of zinc upon non-specific immunity have also been reported. Lennard *et al.* (1974) reported that zinc deficiency increased the phagocytic capabilities of cells, while studies by Chvapil *et al.* (1977) demonstrated the converse. In addition, Weston *et al.* (1977) failed to demonstrate any association between zinc deficiency and impairment of the phagocytic process. Nevertheless, using human leukocytes, Beswick *et al.* (1986) demonstrated that superoxide production could be significantly suppressed following a brief (10 to 40 min) *in vitro* exposure of the cells to 1 mM zinc. Moreover, Wirth *et al.* (1989) have clearly shown that in A/J mice fed a zinc-deficient diet for 28 days, peritoneal macrophages displayed a reduced *ex vivo* capacity to phagocytose *Trypanosoma cruzi*, as well as a limited ability to kill internalized protozoan parasites. Treatment of the cells with zinc prior to incubation with the parasite restored these two functions. These results suggest that zinc plays an important regulatory role in membrane-associated events, including the binding/uptake of antigens and the subsequent oxidative burst.

Effects on NK activity have also been documented. Oral zinc supplementation for 1 month to aged Balb/c mice (22 months old) was sufficient to increase their cytotoxic activity against YAC-1 target cells (Mocchegiani *et al.*, 1995a). However, in an *in vitro* experiment using cells obtained from unsupplemented aged mice, Mocchegiani *et al.* (1995a) found that pretreatment of the splenocytes with zinc alone at levels of 10^{-5} and 10^{-6} M failed to restore NK function. Moreover, these same researchers have also shown that preincubation of splenocytes (collected from zinc-supplemented aged mice) with IFN-α and IFN-β augmented their NK activity compared with that of cells from age-matched controls. In contrast, using an *in vitro* exposure scenario, Ferry and Donner (1984) demonstrated in three different species of mice that a 4h incubation of spleen cells with 0.5 or 1 mM zinc could dramatically decrease NK activity.

In summary, in contrast to the observed *in vivo* immunopotentiating effect of zinc, data provided from *in vitro* studies demonstrates that zinc is potentially immunosuppressive. However, the mechanism of zinc-induced immunomodulation remains elusive, and further mechanistic studies are required to define clearly any potential adverse immunotoxic effects.

11.4 Copper and the Immune System

11.4.1 Copper deficiency and overload

For normal immune function to occur, adequate intake of the essential trace element copper is required. Nutritional copper deficiency, although rare, can occur in infants (Al-Rashid and Spangler, 1971; Walravens, 1980), in patients on total parenteral nutrition (Karpel and Peden, 1972), and after excessive zinc intake (Botash *et al.*, 1992); domesticated animals can also become copper deficient (Boyne and Arthur, 1981) when maintained on diets containing excessive amounts of molybdenum and sulfur (Arthington *et al.*, 1995; Ladefoged and Sturup, 1995). Copper deficiency is characterized by a decrease in serum ceruloplasmin activity (Lukasewycz and Prohaska, 1990; Kelley *et al.*, 1995). One copper-related heritable disease in humans is Menkes' disease in which a fatal copper deficiency develops in infants (Walravens, 1980). Conversely, elevated blood levels of copper can arise in certain infectious and chronic diseases (Alinger *et al.*, 1978); overall, humans are generally less sensitive to copper toxicity than cattle or sheep. Another copper-related heritable disease in humans is Wilson's disease, which is characterized by accumulation of copper in the body.

Structural and functional alterations in the immune system and lymphoid organs occur during copper deficiency. Splenomegaly and thymic atrophy are consistent findings in copper-deficient mice (Lukasewycz and Prohaska, 1983;

Mulhern and Koller, 1988; Prohaska and Lukasewycz, 1989); however, in rats, splenomegaly occurs only in males (Kramer *et al.*, 1988; Bala *et al.*, 1990). Studies have shown that copper deficiency causes medullary degeneration of the rat thymus (Koller *et al.*, 1987) and severe thymic atrophy (with a loss of cortical thymocytes) in mice (Prohaska *et al.*, 1983). The absolute and relative numbers of splenic mononuclear cells were reported to be higher in copper-deficient mice and rats (Lukasewycz *et al.*, 1985; Bala *et al.*, 1991). In contrast, the total splenic mononuclear cell yield was reduced in young rats nursed by dams fed copper-deficient diets (Bala *et al.*, 1991). Serum levels of thymic hormone (thymulin) were also found to be reduced in copper-deprived rats (Vyas and Chandra, 1983). Overall, the immune system of male rats appears to be more susceptible to copper deficiency than that of females; this same gender effect has not been observed in mice.

11.4.2 *Humoral-mediated immunity*

Alterations in the antibody response and B-lymphocyte function are well documented with experimental copper deficiency. The severity of copper deficiency in mice and rats has been correlated with decreases in the the number of antibody-producing cells and in the total antibody response against SRBC (Sullivan and Ochs, 1978; Prohaska and Lukasewycz, 1981, 1989; Blakley and Hamilton, 1987; Koller *et al.*, 1987; Failla *et al.*, 1988; Lukasewycz and Prohaska, 1990). Although Prohaska and Lukasweycz (1989) reported normal IgM and low IgG plasma levels in copper-deficient mice, Windhauser *et al.* (1991) observed no alterations in the specific antibody response in severely copper-deficient rats. These conflicting observations may be related to species and gender susceptibility to copper deficiency. In contrast, in humans with nutritional copper deficiency (Heresi *et al.*, 1985) and in Menkes' disease (Sullivan and Ochs, 1978), serum antibody levels were reported to be normal.

Several studies have shown that copper deficiency can give rise to alterations in the numbers of B-lymphocytes. The relative percentages of B-lymphocytes in the spleen and the peripheral blood of copper-deficient rats (Bala *et al.*, 1991) or mice (Lukasewycz *et al.*, 1985) were elevated. Although splenocytes in copper-deficient animals expressed normal amounts of class II MHC antigens (Lukasewycz *et al.*, 1985, 1987; Prohaska and Lukasewycz, 1989; Bala *et al.*, 1990), these cells were poor stimulators and responders during a mixed lymphocyte response (MLR) assay (Lukasewycz and Prohaska, 1990). As class II antigens are required for the induction of the MLR, a functional defect in these surface proteins may occur in copper deficiency. In spite of the up-regulation of B-lymphocyte numbers during copper-deficient states, the B-lymphocytes respond poorly to mitogenic stimulation (Lukasewycz and Prohaska, 1983, 1990; Lukasewycz *et al.*, 1985; Kramer *et al.*, 1988; Bala *et al.*, 1991). Bala *et al.* (1991) observed diminished B-lymphocyte blastogenic responses in copper-deficient male rats but not in their female counterparts; splenic B-lymphocyte mitogenic responses in young copper-deficient mice were unaffected (Prohaska and Lukasewycz, 1989).

Copper deficiency in steers did not affect the *in vivo* antibody response to ovalbumin nor the *in vitro* lymphocyte mitogenic response to PWM (Ward *et al.*, 1993). In a study of men on low copper diets, the percentage of circulating B-lymphocytes was found to be increased, but B-lymphocyte mitogenic responses were decreased (Kelley *et al.*, 1995). Human clinical and animal studies therefore demonstrate that copper deficiency is associated with elevated circulating B-lymphocyte numbers and altered B-lymphocyte function.

In the macular mutant mouse (the murine model for Menkes' disease), spleen cells show normal proliferative responses to ConA and LPS, and decreased antibody responses against SRBC (Nakagawa *et al.*, 1993). In these mice, the percentages of splenic Ly-5$^+$ cells (B-lymphocytes) were increased, while the percentages of T-lymphocytes, particularly in the T_H subset, were decreased (Nakagawa *et al.*, 1993). A deficiency in T_H-lymphocyte function could contribute to suppression of humoral immune responses, such as impairment of the IgM to IgG isotype 'switch' mechanism, during copper deficiency. This viewpoint is demonstrated in a study which indicated that serum IgM levels were normal but IgG antibody levels were decreased in copper-deficient mice (Lukasewycz and Prohaska, 1990).

11.4.3 *Cell-mediated immunity*

Although the effect on T-lymphocyte populations during copper deficiency is well characterized, the overall effect on cell-mediated immunity *in vivo* is unclear. Except in the case of female rats, both the relative and absolute numbers of T-lymphocytes and their CD4$^+$ and CD8$^+$ subsets (particularly CD4+) are decreased in the peripheral blood and spleen of copper-deficient laboratory rodents (Lukasewycz *et al.*, 1985; Mulhern and Koller, 1988; Bala *et al.*, 1989, 1990, 1991). The decrease in T-lymphocyte populations could be related to a decreased production of thymulin (Vyas and Chandra, 1983). The quantitative decrease in T-lymphocyte numbers, particularly the CD4$^+$ subset, might be an underlying cause for other observed changes in T-lymphocyte functions which occur during copper deficiency, such as reduced mitogenic responsivity (Prohaska and Lukasewycz, 1981; Lukasewycz *et al.*, 1985; Davis *et al.*, 1987; Kramer *et al.*, 1988; Bala *et al.*, 1989, 1991; Bala and Failla, 1992).

Modulation of cytokine-dependent functionality could also be a contributing factor in immunomodulation during copper deficiency. While the splenocytes of copper-deficient male and female rats responded poorly to stimulation by T-lymphocyte-specific mitogens, the up-regulation of IL-2 and transferrin receptor expression in splenocytes was normal (although somewhat higher in the males) (Bala *et al.*, 1990, 1991). The diminished mitogenic responses were reversible following addition of rat IL-2 (Bala and Failla,

1992), indicating that insufficient IL-2 production may have been responsible. Although one study reported normal IL-2 production in copper-deficient hosts, splenocytes from copper-deficient male rats secreted less IL-2 than those from cells obtained from female animals (Bala and Failla, 1992; Hopkins and Failla, 1995). Splenocytes from copper-deficient mice have also been shown to secrete less IL-2 than cells from their copper-sufficient counterparts (Lukasewycz and Prohaska, 1990).

It is interesting to note that the splenocytes from copper-deficient male (but not female) rats were more unresponsive to mitogenic activation and secreted less IL-2 than did cells from copper-sufficient animals. A similar gender-related effect was also observed in the study by Bala *et al.* (1991), wherein T-lymphocyte PHA-induced mitogenic responses were suppressed in both male and female rats, yet ConA-induced responses were suppressed only in the males. It is not clear whether susceptibility of the male rat immune system is hormonally related. In addition, copper deficiency was also found to decrease PHA-induced *in vivo* cell-mediated immunity in steers but increase *in vitro* lymphocyte mitogenic responses to PHA (Ward *et al.*, 1993).

Clinical studies involving healthy men on low copper diets (i.e. 0.38 mg Cu/day) indicated a decrease in peripheral blood lymphocyte mitogenic responses; no effects on circulating IL-2 receptors nor on the numbers of total T-lymphocytes or their $CD4^+$ and $CD8^+$ subsets were observed, although the amount of IL-2 secreted into the media by peripheral mononuclear cells stimulated with ConA was reduced (Kelley *et al.*, 1995). The disparity between this human clinical study and animal studies with respect to circulating T-lymphocytes and the $CD4^+$ and $CD8^+$ subsets may be due to differences in the degree and duration of copper deficiency. Despite considerable published information regarding the alterations of T-lymphocyte functions during copper deficiency, the effects on *in vivo* cell-mediated immune responses remain controversial.

Modulation of the numbers and functionality of T-lymphocytes during copper deficiency can also impact upon inflammatory and hypersensitivity responses. Cellular immunity determined by DTH reaction to bovine serum albumin was unaffected (Koller *et al.*, 1987), while the DCH reaction to oxazolone was decreased (Kishore *et al.*, 1984) in copper-deficient rats. In contrast, both the DCH reaction to oxozalone and the DTH reaction to SRBC were enhanced in copper-deficient mice (Jones, 1984). The mechanism(s) responsible for, or associated with, this copper-deficiency-related enhancement of DTH and DCH remain(s) unclear. It has been shown that histamine-induced acute and delayed inflammatory responses were increased in copper-deficient mice (Jones, 1984); these responses are thought to be the result of increases in the numbers of inflammatory cells (monocytes) and IL-1 production in the copper-deficient host (Bala *et al.*, 1991; Lukasewycz and Prohaska, 1990). In addition, the activity of ceruloplasmin, a plasma copper transport protein which acts as a free radical scavenger to protect cells against damage by free radicals released from macrophages and neutrophils (Saenko *et al.*,

1994), is decreased during copper deficiency (Lukasewycz and Prohaska, 1990; Kelley *et al.*, 1995).

Lastly, copper deficiency impairs the ability of mice to be immunized against syngeneic malignant lymphocytes, once again indicating that cellular immunity (e.g. T_C function) may be compromised during this condition (Lukasewycz and Prohaska, 1982, 1990). A study by Greene *et al.* (1987) also reported that copper deficiency increased the incidence of 1,2-dimethyhydrazine-induced colonic tumors in rats. In contrast, the growth of gliosarcoma tumors was suppressed in copper-deficient rats (Yoshida *et al.*, 1995).

11.4.4 *Non-specific immunity*

Macrophages are phagocytic cells that are critical for natural (innate) and acquired immunity as they function as both APC and effector cells. In copper-deficient rats, Bala *et al.* (1991) reported relative increases in the percentages of monocytes and/or macrophages in the peripheral blood and spleen, but a normal expression of class II MHC antigens. In addition, splenic and peritoneal macrophages from copper-deficient mice produced higher amounts of IL-1 than did cells recovered from copper-sufficient hosts (Lukasewycz and Prohaska, 1990). Whether these quantitative changes in the number of macrophages/monocytes and IL-1 production contribute to an enhanced cell-mediated immunity in copper deficiency is unknown. Indeed, even though the cellular copper content and Cu/Zn-superoxide dismutase activities were lower in peritoneal macrophages from rats with copper deficiency (Babu and Failla, 1990a), phagocytosis of opsonized erythrocytes was normal; however, candidacidal and respiratory burst activities were depressed.

Although not observed in laboratory animals, neutrophils appear to be very sensitive to copper deficiency, with neutropenia being a characteristic clinical presentation during copper deficiency (Al-Rashid and Spangler, 1971; Zidar *et al.*, 1977; Kelley *et al.*, 1995). Neutropenia is postulated to be linked to a copper requirement in neutrophil production (Brewer, 1987) or to the production of anti-neutrophil antibodies during copper deficiency (Higuchi *et al.*, 1991). The latter mechanism is supported by the fact that these antibodies disappeared from the blood after copper supplementation (Higuchi *et al.*, 1991).

As with macrophages, neutrophilic phagocytosis of erythrocytes was unaffected, but copper content, the activities of Cu/Zn-superoxide dismutase and the oxidative burst, and fungicidal/microbicidal activities were decreased in neutrophils from copper-deprived rats (Babu and Failla, 1990b), sheep, and cattle (Boyne and Arthur, 1981, 1986; Jones and Suttle, 1981). Although neutropenia was present, the bacterial phagocytic activity of neutrophils from healthy men fed low copper diets was not impaired (Kelley *et al.*, 1995). In the animal models, neutrophil function is reparable since abrogation of the copper deficiency restored both neutrophil copper status and neutrophil function to normal (Babu and Failla, 1990b). Because ceruloplasmin has been shown to up-regulate phagocytosis and oxidative burst activities in macrophages and neutrophils (Saenko *et al.*, 1994), the decreased concentrations of ceruloplasmin during copper deficiency may account for the diminished oxidative burst activities in these cells. That copper deficiency appears to suppress neutrophil antimicrobial activity without altering phagocytosis, and resupplementation induced increases in the numbers of circulating neutrophils and a decreased incidence of respiratory tract infections, suggests a critical role for copper in neutrophil function and/or production (Castillo-Duran *et al.*, 1983).

Studies on the effect of copper deficiency on NK activity are limited. In rats fed copper-depleted (0 ppm Cu) diets, NK cytotoxic activity was markedly suppressed (Koller *et al.*, 1987). Conversely, consumption of low copper diets by healthy men had no effect on the number of circulating NK cells (Kelley *et al.*, 1995).

11.4.5 *Effects on host resistance*

Epidemiological, clinical, and animal studies all indicate that copper deficiency impairs host resistance, primarily through major suppressive effects upon antibody-mediated responses and the anti-microbicidal activities of macrophages and neutrophils. The alterations in these immune parameters may account for the increased susceptibility to microbial infections observed in conditions of copper deficiency: infants with Menkes' disease succumb to pneumonia at an early age (Menkes *et al.*, 1962; Danks *et al.*, 1972); nutritional copper deficiency in humans is characterized by recurrent bacterial infections (Al-Rashid and Spangler, 1971; Karpel and Peden, 1972; Sann *et al.*, 1978); laboratory and domestic animals with copper deficiency are more susceptible to infection (Newberne *et al.*, 1968; Jones and Suttle, 1983; Pletcher and Banting, 1983); and copper deficiency enhances the susceptibility of rodents to certain tumors (Lukasewycz and Prohaska, 1982; Greene *et al.*, 1987).

Limited data describing the effect of high levels of copper on the immune system are available. Elevated serum copper concentrations were associated with poor mitogenic responses by splenocytes in aged mice, yet increased responses in cells from young mice (Massie *et al.*, 1993). Although the basis for the high serum copper levels in these reports was not known (i.e. elevation due to infection or other chronic diseases; Markowitz *et al.*, 1955; Alinger *et al.*, 1978), levels of the plasma copper-transport protein ceruloplasmin (which has immunosuppressive effects upon lymphocytes: Kelley *et al.*, 1995; and increases in relation to copper status: Nève *et al.*, 1988) were increased with age (Massie *et al.*, 1979). Whether the decreased lymphocyte mitogenic responses observed in the aged mice in the study by Massie *et al.* (1993) were, therefore, attributable to ceruloplasmin alone or other factors is unclear.

The effect of excess dietary copper on immune function was evaluated in rats and mice fed 50 to 400 ppm copper in the diet for up to 10 weeks (Pocino *et al.*, 1991). In rats, mitogen-induced T-lymphocyte blastogenesis was inhibited depending on the dose and duration of copper exposure. In mice fed 50 and 100 ppm copper for 3 weeks, B-lymphocyte mitogenic responses were enhanced, although prolonged exposure also eventually suppressed the mitogenic response. In addition, autoantibodies against bromelain-treated mouse erythrocytes were enhanced by the presence of the excess copper. This phenomenon could have been a result of a stimulatory effect of copper upon autoreactive B-lymphocytes or an alteration in the function(s) of regulatory T-lymphocytes. Prolonged exposure to excess copper also diminished cellular immunity as assessed by the DTH response to SRBC.

An excess of copper has been reported to decrease resistance against *Candida albicans* infection in mice (Vaughn and Weinberg, 1978) and *Salmonella* infection in chickens (Hill, 1974, 1979, 1980). However, excess copper has also been shown to result in a delay in the development of chemically induced hepatic tumors in rats (Sharpless, 1946). In catfish (*Saccobranchus fossilis*) exposed to copper in the water, hematocrit and antibody responses to SRBC were suppressed (Khangarot *et al.*, 1988). Antibody titers, numbers of antibody-producing spleen cells, phagocytosis, and cellular immune response were also suppressed in catfish exposed to high levels of copper in the water for 28 days (Khangarot and Tripathi, 1991). However, feeding weanling pigs excess copper in the diet (200 to 400 ppm Cu) for 5 weeks did not suppress the humoral immune response against SRBC (Kornegay *et al.*, 1989), suggesting that the porcine immune system is relatively more resistant to excess copper than that of fish or mice.

Overall, it appears that the mammalian immune system is much more sensitive to copper deficiency than excess.

11.5 Summary

One common finding for each of the three metals described in this chapter is the impairment of humoral and cell-mediated responses in patients or in laboratory animal species that experienced a nutritional deficiency. In contrast, the consequences from an excess of any of the three metals are quite different. Indeed, they range from the immunoenhancement due to an excess of zinc in the elderly and in patients with AIDS, to no immunomodulation whatsoever arising from the presence of an excess of copper, to immunosuppression of non-specific responses during iron overload. The mechanism(s) that mediate these immunomodulatory effects are not yet fully elucidated and, consequently, further studies are required to define clearly the immunomodulatory potential of these microelements.

References

AKBAR, A.N., FITZGERALD-BOCARSLY, P.A., DE SOUSA, M., GIARDINA, P.J., HILGARTNER, M.W., and GRADY, R.W. (1986). Decreased natural killer activity in thalassemia major: a possible consequence of iron overload. *J. Immunol.* **136**, 1635–1640.

AKBAR, A.N., FITZGERALD-BOCARSLY, P.A., GIARDINA, P.J., HILGARTNER, M.W., and GRADY, R.W. (1987). Modulation of the defective natural killer activity seen in thlassemia major with desferrioxamine and α-interferon. *Clin. Exp. Immunol.* **70**, 345–353.

ALFORD, C.E., KING, T.E., JR, and CAMPBELL, P.A. (1991). Role of transferrin, transferrin receptors, and iron in macrophage listericidal activity. *J. Exp. Med.* **174**, 459–466.

ALINGER, P., KOLARZ, G., and WILLVONSEDER, R. (1978). Copper in ankylosing spondylitis and rheumatoid arthritis. *Scand. J. Rheumatol.* **7**, 75–78.

AL-RASHID, R.A. and SPANGLER, J. (1971). Neonatal copper deficiency. *N. Engl. J. Med.* **285**, 841–843.

ARBETER, A., ECHEVERRI, L., FRANCO, D., MUNSON, D., VELEZ, H., and VITALE, J. (1971). Nutrition and infection. *Fed. Proc.* **30**, 1421–1428.

AROSA, F.A. and DE SOUSA, M. (1995). Iron differentially modulates the CD4–lck and CD8-lck complexes in resting peripheral blood T-lymphocytes. *Cell. Immunol.* **161**, 138–142.

ARTHINGTON, J.D., CORAH, L.R., BLECHA, F., and HILL, D.A. (1995). Effect of copper depletion and repletion on lymphocyte blastogenesis and neutrophil bactericidal functions in beef heifers. *J. Animal Sci.* **73**, 2079–2085.

ATSDR (Agency for Toxic Substances and Disease Registry) (1994). *Toxicological Profile for Zinc (Update)*, Atlanta, GA: United States Department of Health and Human Services, Public Health Service.

BABU, U. and FAILLA, M.L. (1990a). Respiratory burst and candidacidal activity of peritoneal macrophages are impaired in copper-deficient rats. *J. Nutr.* **120**, 1692–1699.

(1990b). Copper status and function of neutrophils are reversibly depressed in marginally and severely copper-deficient rats. *J. Nutr.* **120**, 1700–1709.

BACH, J.F., PAPIERNIK, M., LEVASSEUR, P., DARDENNE, M., BAROIS, A., and LE BRIGAND, H. (1972). Evidence for a serum thymic factor produced by the human thymus. *Lancet* **ii**, 1056–1058.

BAGGS, R.B. and MILLER, S.A. (1973). Nutritional iron deficiency as a determinant of host resistance in the rat. *J. Nutr.* **103**, 1554–1560.

BALA, S. and FAILLA, M.L. (1992). Copper deficiency reversibly impairs DNA synthesis in activated T-lymphocytes by limiting interleukin 2 activity. *Proc. Natl Acad. Sci. USA* **89**, 6794–6797.

BALA, S., FAILLA, M.L., and LUNNEY, J. (1989). T-cell numbers and mitogenic responsiveness of peripheral blood mononuclear cells are decreased in copper deficient rats. *Nutr. Res.* **10**, 749–760.

(1990). Phenotypic and functional alterations in peripheral blood mononuclear cells of copper-deficient rats. *Ann. NY. Acad. Sci.* **587**, 283–285.

(1991). Alterations in splenic lymphoid cell subsets and activation antigens in copper-deficient rats. *J. Nutr.* **121**, 745–753.

BALLART, I.J., ESTEVEZ, M.E., SEN, L., DIEZ, R.A., GIUNTOLI, R.A., DE MIANI, S.A., and PENALVER, J. (1986). Progressive dysfunction of monocytes associated with iron overload and age in patients with thalassemia major. *Blood* **67**, 105–109.

BALLESTER, O.F., ABDALLAH, M., and PRASAD, A.S. (1986). Lymphocyte subpopulation abnormalities in sickle cell anemia: a distinctive pattern from that of AIDS. *Am. J. Haematol.* **21**, 23–27.

BECROFT, D.M., DIX, M.R., and FARMER, K. (1977). Intramuscular iron-dextran and susceptibility of neonates to bacterial infections. *In vitro* studies. *Arch. Dis. Childhood* **52**, 778–781.

BERGER, N.A. and SKINNER, A.M. (1974). Characterization of lymphocyte transformation induced by zinc ions. *J. Cell. Biol.* **61**, 45–55.

BERTHOLF, R.L. (1988). Zinc, in Seiler, H.G. and Sigel, H. (Eds) *Handbook on Toxicity of Inorganic Compounds*, pp. 787–800, New York: Marcel Dekker.

BESWICK, P.H., BRANNEN, P.C., and HURLES, S.S. (1986). The effects of smoking and zinc on the oxidative reactions of human neutrophils. *J. Clin. Lab. Immunol.* **21**, 71–75.

BHASKARAM, P., SHARADA, K., SIVAKUMA, B., RAO, K.V., and NAIR, M. (1989). Effect of iron and vitamin A deficiency on macrophage function in children. *Nutr. Res.* **9**, 35–45.

BLAKLEY, B.R. and HAMILTON, D.L. (1987). The effect of copper deficiency on the immune response in mice. *Drug. Nutr. Interact.* **5**, 103–111.

(1988). The effect of iron deficiency on the immune response in mice. *Drug. Nutr. Interact.* **5**, 249–255.

BOMFORD, A. and WILLIAMS, R. (1976). Long-term results of venesection therapy in idiopathic haemochromatosis. *Q. J. Med.* **45**, 611–623.

BOTASH, A.S., NASCA, J., DUBOWY, R., WEINBERGER, H.L., and OLIPHANT, M. (1992). Zinc-induced copper deficiency in an infant. *Am. J. Dis. Children* **146**, 709–711.

BOYNE, R. and ARTHUR, J.R. (1981). Effects of selenium and copper deficiency on neutrophil function in cattle. *J. Comp. Pathol.* **91**, 271–276.

(1986). Effects of molybdenum- or iron-induced copper deficiency on the viability and function of neutrophils from cattle. *Res. Vet. Sci.* **41**, 417–419.

BRAY, T. and BETTGER, W.J. (1990). The physiological role of zinc as antioxidant. *Free Rad. Biol. Med.* **8**, 281–291.

BREWER, N.R. (1987). Comparative metabolism of copper. *J. Am. Vet. Med. Assoc.* **190**, 654–658.

BRYAN, C.F. (1991). The immunogenetics of hereditary hemochromatosis. *Am. J. Med. Sci.* **301**, 47–49.

BRYAN, C.F. and LEECH, F. (1983). The immunoregulatory nature of iron. I. Lymphocyte proliferation. *Cell. Immunol.* **75**, 71–79.

BRYAN, C.F., NISHIYA, K., POLLACK, M.S., DUPONT, B., and DE SOUSA, M. (1981). Differential inhibition of the MLR by iron: association with HLA phenotype. *Immunogenetics* **12**, 129–140.

BRYAN, C.F., LEECH, S.H., DUCOS, R., EDWARDS, C.Q., KUSHNER, J.P., SKOLNICK, M.H., BOZELKA, B., LINN, J.C., and GAUMER, R. (1984). Thermostable erythrocyte rosette-forming lymphocytes in hereditary hemochromatosis. I. Identification in peripheral blood. *J. Clin. Immunol.* **4**, 134–142.

BRYAN, C.F. LEECH, S.H., KUMAR, P., GAUMER, R., BOZELKA, B., and MORGAN, J. (1991). The immune system in hereditary hemochromatosis: a quantitative and functional assessment of the cellular arm. *Am. J. Med. Sci.* **301**, 55–61.

CANTINIEAUX, B., BOELAERT, J., HARIGA, C., and FONDU, P. (1988). Impaired neutrophil defense against *Yersinia enterocolitica* in patients with iron overload who are undergoing dialysis. *J. Lab. Clin. Med.* **111**, 524–528.

CANTWELL, R.J. (1972). Iron deficiency anemia of infancy: some clinical principles illustrated by the response of Maori infants to neonatal parenteral iron administration. *Clin. Paediat.* **11**, 443–449.

CARDIER, J.E., ROMANO, E., and SOYANO, A. (1995). Lipid peroxidation and changes in T-lymphocyte subsets and proliferative response in experimental iron overload. *Immunopharmacol. Immunotoxicol.* **17**, 705–717.

CARVALHO, G.S. and DE SOUSA, M. (1988). Iron exerts a specific inhibitory effect on CD2 expression on human PBL. *Immunol. Lett.* **19**, 163–167.

CASTILLO-DURAN, C., FISBERG, M., VALENZUELA, A., EGANA, J.I., and UAUY, R. (1983). Controlled trial of copper supplementation during recovery from marasmus. *Am. J. Clin. Nutr.* **37**, 898–903.

CASTILLO-DURAN, C., HERESI, G., FISBERG, M., and UAUY, R. (1987). Controlled trial of zinc supplementation during recovery for malnutrition: effects on growth and immune function. *Am. J. Clin. Nutr.* **45**, 602–608.

CHAMPION, S., IMHOF, B.A., SAVAGNER, P., and THIERY, J.P. (1986). The embryonic thymus produces chemotactic peptides involved in the homing of hemopoietic precursors. *Cell* **44**, 781–790.

CHANDRA, R.K. (1980). Acrodermatitis enteropathica: zinc levels and cell-mediated immunity. *Paediatrics* **66**, 789–791.

CHANDRA, R.K. and SARAYA, A.K. (1975). Impaired immunocompetence associated with iron deficiency. *J. Paediat.* **86**, 899–902.

CHAPMAN, D.E., GOOD, M.F., POWELL, L.W., and HALLIDAY, J.W. (1988). The effect of iron, iron-binding proteins and iron overload on human natural killer cell activity. *J. Gastroenterol. Hepatol.* **3**, 9–16.

CHU, S.H.W., WELCH, K.J., MURRAY, E.S., and HEGSTED, D.M. (1976). Effect of iron deficiency on susceptibility to *Streptococcus pneumoniae* infection in the rat. *Nutr. Rep. Intl.* **14**, 605–609.

CHVAPIL, M., STANKOVA, L., ZUKOSKI, C., 4th, and ZUKOSKI, C., 3rd (1977). Inhibition of some functions of polymorphonuclear leukocytes by *in vitro* zinc. *J. Lab. Clin. Med.* **89**, 135–146.

COTO, J.A., HADDEN, E.M., SAURO, M., ZORN, N., and HADDEN, J.W. (1992). Interleukin-1 regulates secretion of zinc-thymulin by thymic epithelial cells and its action on T-lymphocyte proliferation and nuclear protein kinase C. *Proc. Natl Acad. Sci. USA* **89**, 7752–7756.

COTRAN, R.S., KUMAR, V., and ROBBINS, S.L. (1989). *Pathologic Basis of Diseases*, 4th Edn, Toronto, Canada: WB Saunders Co.

DANKS, D.M., CAMPBELL, P.E., STEVENS, B.J., MAYNE, V., and CARTWRIGHT, E. (1972). Menkes' kinky-hair syndrome. An inherited defect in copper absorption with widespread effects. *Pediatrics* **50**, 188–201.

DARDENNE, M., PLEAU, J.M., NABARRA, B., LEFRANCIER, P., DERRIEN, M., CHOAY, J., and BACH, J.F. (1982). Contribution of zinc and other metals to the biological activity of the serum thymic factor. *Proc. Natl Acad. Sci. USA* **79**, 5370–5373.

DAVIS, M.A., JOHNSON, W.T., BRISKEAN, M., and KRAMER, T.R. (1987). Lymphoid cell functions during copper deficiency. *Nutr. Res.* **7**, 211–222.

DEISS, A. (1983). Iron metabolism in reticuloendothelial cells. *Sem. Hematol.* **20**, 81–90.

DHUR, A., GALAN, P., PREZIOSI, P., and HERCBERG, S. (1991). Lymphocyte subpopulations in the thymus, lymph nodes and spleen of iron-deficient and rehabilitated mice. *J. Nutr.* **121**, 1418–1424.

DUCHATEAU, J., DELESPESSE, G., VRIJENS, R., and COLLET, H. (1981). Beneficial effects or oral zinc supplementation on the immune response of old people. *Am. J. Med.* **70**, 1001–1004.

DWYER, J., WOOD, C., McNAMARA, J., WILLIAMS, A., ANDIMAN, W., RINK, L., O'CONNOR, T., and PEARSON, H. (1987). Abnormalities in immune system of children with β-thalassemia major. *Clin. Exp. Immunol.* **68**, 621–629.

FABRIS, N., MOCCHEGIANI, E., GALLI, M., IRATO, L., LAZZARIN, A., and MORONI, M. (1988). AIDS, zinc deficiency, and thymic hormone failure. *J. Am. Med. Assoc.* **259**, 839–840.

FABRIS, N., MOCCHEGIANI, E., ZANNOTI, M., LICASTRO, F., and FRANCESCHI, C. (1984). Thymic hormone deficiency in normal aging and Down's syndrome: is there a primary failure of the thymus? *Lancet* **i**, 983–986.

FAILLA, M.L., BABU, U., and SEIDEL, K.E. (1988). Use of immunoresponsiveness to demonstrate that the dietary requirements for copper in young rats is greater with dietary fructose than dietary starch. *J. Nutr.* **118**, 487–496.

FARMER, K. (1976). The disadvantages of routine administration of intramuscular iron to neonates. *N.Z. Med. J.* **84**, 286–287.

FENG, X.B., YANG, X.Q., and SHEN, J. (1994). Influence of iron deficiency on serum IgG subclass and pneumococcal polysaccharides specific IgG subclass antibodies. *Chinese Med. J. (Engl.)* **107**, 813–816.

FERRY, F. and DONNER, M. (1984). *In vitro* modulation of murine natural killer cytotoxicity by zinc. *Scand. J. Immunol.* **19**, 435–445.

FRAKER, P.J., DEPASQUALE-JARDIEU, P.M., ZWICKL, C.M., and LUECKE, R.W. (1978). Regeneration of T-cell helper function in zinc deficient mice. *Proc. Natl Acad. Sci. USA* **75**, 5660–5664.

FRAKER, P.J., HILDEBRANDT, K., and LUECKE, R.W. (1984). Alteration of antibody-mediated responses of suckling mice to T-cell dependent and independent antigens by maternal marginal zinc deficiency. *J. Nutr.* **114**, 170–179.

FRAKER, P.J., GERSHWIN, M.E., GOOD, R.A., and PRASAD, A.S. (1986). Interrelationship between zinc and immune function. *Fed. Proc.* **45**, 1474–1479.

FRAKER, P.J., JARDIEU, P., and COOK, J. (1987). Zinc deficiency and immune function. *Arch. Dermatol.* **123**, 1699–1701.

GALÀN, P., DAVILA, M., MEKKI, N., and HERCBERG, S. (1988). Iron deficiency, inflammatory processes and humoral immunity in children. *Int. J. Vit. Nutr.* **58**, 225–230.

GALÀN, P., THIBAULT, H., PREZIOSI, P., and HERCBERG, S. (1992). Interleukin-2 production in iron deficient children. *Biol. Trace Elem. Res.* **32**, 421–426.

GEBRAN, S.J., ROMANO, E.L., and SOYANO, A. (1993). Iron polymers impair the function and maturation of macrophages. *Immunopharmacol. Immunotoxicol.* **15**, 397–414.

GOOD, M.F., CHAPMAN, D.E., POWELL, L.W., and HALLIDAY, J.W. (1986). The effect of iron (Fe^{3+}) on the cloning efficiency of human memory T4+ lymphocytes. *Clin. Exp. Immunol.* **66**, 340–347.

(1987a). The effect of experimental iron overload on splenic T-cell function: analysis using cloning techniques. *Clin. Exp. Immunol.* **68**, 375–383.

GOOD, M.F., POWELL, L.W., and HALLIDAY, J.W. (1987b). The effect of non-transferrin-bound iron on murine T-lymphocyte subsets: analysis by clonal techniques. *Clin. Exp. Immunol.* **70**, 164–172.

(1988). Iron status and cellular immune competence. *Blood. Rev.* **2**, 43–49.

GOOD, R.A. and FERNANDES, G. (1979). Nutrition, immunity and cancer – a review. Part I: Influence of protein or protein calorie malnutrition and zinc deficiency on immunity. *Clin. Bull.* **9**, 3–12.

GOYER, R.A., APGAR, J., and PISCATOR, M. (1979). Toxicity of zinc, in Henkin, R.I. (Ed.) *Zinc*, pp. 249–268, Baltimore: University Park Press.

GRADY, R.W., AKBAR, A.N., GIARDINA, P.J., HILGARTNER, M.W., and DE SOUSA, M. (1985). Disproportionate lymphoid cell subsets in thalassemia major: the relative contribution of transfusion and splenectomy. *Br. J. Haematol.* **59**, 713–724.

GRAHAM, N.M., SORESON, D., ODAKA, N., BROOKMEYER, R., CHAN, D., WILLET, W.C., MORRIS, J.S., and SAAH, A.J. (1991). Relationship of serum copper and zinc levels to HIV-1 seropositivity and progression in AIDS. *J. AIDS* **4**, 976–980.

GREENE, F.L., LAMB, L.S., BARWICK, M., and PAPPAS, N.J. (1987). Effect of dietary copper on colonic tumor production and aortic integrity in the rat. *J. Surg. Res.* **42**, 503–512.

GUGLIELMO, P., CUNSOLO, F., LOMBARDO, T., SORTINO, G., GIUSTOLISI, R., CACCIOLA, E., and CACCIOLA, E. (1984). T-subset abnormalities in thalassemia intermedia: possible evidence for a thymus functional deficiency. *Acta Haematol.* **72**, 361–367.

HADDEN, J.W. (1995). The treatment of zinc deficiency is an immunotherapy. *Int. J. Immunopharmacol.* **17**, 697–701.

HALLQUIST, N.A., MCNEIL, L.K., LOCKWOOD, J.F., and SHERMAN, A.R. (1992). Maternal-iron-deficiency effects on peritoneal macrophage and peritoneal natural-killer-cell cytotoxicity in rat pups. *Am. J. Clin. Nutr.* **55**, 741–746.

HANN, H.W., STAHLHUT, N.W., and MENDUKE, H. (1991). Iron enhances tumor growth: observations on spontaneous mammary tumors in mice. *Cancer* **68**, 2407–2410.

HARADA, T., BABA, M., TORII, I., and MORIKAWA, S. (1987). Ferritin selectively suppresses delayed-type hypersensitivity responses at induction or effector phase. *Cell. Immunol.* **109**, 75–88.

HAR-EL, R. and CHEVION, M. (1991). Zinc protects against metal-mediated free radical induced damage: studies on single and double-strand DNA breakage. *Free Rad. Res. Commun.* **12–13**, 509–515.

HASAN, S.M., AZIZ, M., AHMAD, P., and AGGARWAL, M. (1989). Phagocyte metabolic functions in iron deficiency anaemia of Indian children. *J. Trop. Pediat.* **35**, 6–9.

HELYAR, L. and SHERMAN, A.R. (1987). Iron deficiency and interleukin-1 production by rat leukocytes. *Am. J. Clin. Nutr.* **46**, 346–352.

HERESI, G., CASTILLO-DURAN, C., MUNOZ, C., ARAVELO, M., and SCHLESINGER, L. (1985). Phagocytosis and immunoglobulin levels in hypocupremic infants. *Nutr. Res.* **5**, 1327–1334.

HIGUCHI, S., HIGASHI, A., NAKAMURA, T., YANABE, Y., and MATSUDA, T. (1991). Anti-neutrophil antibodies in patients with nutritional copper deficiency. *Eur. J. Pediat.* **150**, 327–330.

HILL, C.H. (1974). Influence of high levels of minerals on the susceptibility of chicks to *Salmonella gallinarum. J. Nutr.* **104**, 1221–1226.

——— (1979). Dietary influences on resistance to *Salmonella* infection in chicks. *Fed. Proc.* **38**, 2129–2133.

——— (1980). Influence of time of exposure to high levels of minerals on the susceptibility of chicks to *Salmonella gallinarum. J. Nutr.* **110**, 433–436.

HOPKINS, R.G. and FAILLA, M.L. (1995). Chronic intake of marginally low copper diet impairs *in vitro* activities of lymphocytes and neutrophils from male rats despite minimal impact on conventional indicators of copper status. *J. Nutr.* **125**, 2658–2668.

HRABINSKI, D., HERTZ, J.L., TANTILLO, C., BERGER, V., and SHERMAN, A.R. (1995). Iron repletion attenuates the protective effect of iron deficiency in DMBA-induced mammary tumors in rats. *Nutr. Cancer* **24**, 133–142.

JAY, M., STUART, S.M., McCLAIN, C.J., PALMIERI, D.A., and BUTTERFIELD, D.A. (1987). Alterations in lipid membrane fluidity and the physical state of cell surface sialic acid in zinc-deficient rat erythrocyte ghosts. *Biochim. Biophys. Acta* **897**, 507–511.

JIANG, X. and BALDWIN, C.L. (1993). Iron augments macrophage-mediated killing of *Brucella abortus* alone and in conjunction with interferon-γ. *Cell. Immunol.* **148**, 397–407.

JONES, D.G. (1984). Effects of dietary copper depletion on acute and delayed inflammatory responses in mice. *Res. Vet. Sci.* **37**, 205–210.

JONES, D.G. and SUTTLE, N.F. (1981). Some effects of copper deficiency on leukocyte function in sheep and cattle. *Res. Vet. Sci.* **31**, 151–156.

JONES, D.G. and SUTTLE, N.F. (1983). The effect of copper deficiency on the resistance of mice to infection with *Pasteurella haemolytica. J. Comp. Pathol.* **93**, 143–149.

JOYNSON, D.H., WALKER, D.M., JACOBS, A., and DOLBY, A.E. (1972). Defect of cell-mediated immunity of patients with iron deficiency anemia. *Lancet* **ii**, 1058–1059.

KAPLAN, J., SARNAIK, S., GITLIN, J., and LUSHER, J. (1984). Diminished helper/suppressor lymphocyte ratios and natural killer activity in recipients of repeated blood transfusions. *Blood* **64**, 308–310.

KAPLAN, S.S., QUIE, P.G., and BASFORD, R.E. (1975). Effect of iron on leukocyte function: inactivation of H_2O_2 by iron. *Infect. Immun.* **12**, 303–308.

KARPEL, J.T. and PEDEN, V.H. (1972). Copper deficiency in long-term parenteral nutrition. *J. Pediat.* **80**, 32–36.

KEEN, C.L. and GERSHWIN, M.E. (1990). Zinc deficiency and immune function. *Annu. Rev. Nutr.* **10**, 415–431.

KELLEY, D.S., DAUDU, P.A., TAYLOR, P.C., MACKEY, B.E., and TURNLUND, J.R. (1995). Effects of low-copper diets on human immune response. *Am. J. Clin. Nutr.* **62**, 412–416.

KEUSCH, G.T. (1990). Micronutrients and susceptibility to infection. *Ann. N.Y. Acad. Sci.* **587**, 181–188.

KHANGAROT, B.S. and TRIPATHI, D.M. (1991). Changes in humoral and cell-mediated immune response and in skin and respiratory surfaces of catfish, *Saccobranchus fossilis*, following copper exposure. *Ecotoxicol. Environ. Safety* **22**, 291–308.

KHANGAROT, B.S., RAY, P.K., and SINGH, K.P. (1988). Influence of copper treatment on the immune response in an air-breathing teleost, *Saccobranchus fossilis. Bull. Environ. Contam. Toxicol.* **41**, 222–226.

KISHORE, V., LATMAN, N, ROBERTS, D.W., BARNETT, J.B., and SORENSON, J.R. (1984). Effect of nutritional copper deficiency on adjuvant arthritis and immunocompetence in the rat. *Agents Action.* **14**, 274–282.

KOCHANOWSKI, B.A. and SHERMAN, A.R. (1984). Phagocytosis and lysozyme activity in granulocytes from iron-deficient rat dams and pups. *Nutr. Res.* **4**, 511–520.

(1985). Decreased antibody formation in iron-deficient rat pups – effect of iron repletion. *Am. J. Clin. Nutr.* **41**, 278–284.

KOLLER, L.D., MULHERN, S.A., FRANKEL, N.C., STEVEN, M.G., and WILLIAMS, J.R. (1987). Immune dysfunction in rats fed a diet deficient in copper. *Am. J. Clin. Nutr.* **45**, 997–1006.

KORNEGAY, E.T., VAN HEUGTEN, P.H.G., LINDEMANN, M.D., and BLODGETT. D.J. (1989). Effects of biotin and high copper levels on performance and immune response of weanling pigs. *J. Animal Sci.* **67**, 1471–1477.

KRAMER, T.R., JOHNSON, W.T., and BRISKE-ANDERSON, M. (1988). Influence of iron and sex of rats on hematological, biochemical, and immunological changes during copper deficiency. *J. Nutr.* **118**, 214–221.

KRANTMAN, H.J., YOUNG, S.R., ANK, B.J., O'DONNELL, C.M., RACHELEFSKY, G.S., and STIEHM, E.R. (1982). Immune function in pure iron deficiency. *Am. J. Dis. Children* **136**, 840–844.

KRUSE-JARRES, J.D. (1989). The significance of zinc for humoral and cellular immunity. *J. Trace Elem. Electrolyte Health Dis.* **3**, 1–8.

KUVIBIDILA, S.R. and SARPONG, D. (1990). Mitogenic response of lymph nodes and spleen lymphocytes from mice with moderate and severe iron deficiency anemia. *Nutr. Res.* **10**, 195–210.

KUVIBIDILA, S.R. and WADE, S. (1987). Macrophage function as studied by the clearance of ^{125}I-labelled polyvinylpyrrolidone in iron-deficient and iron replete mice. *J. Nutr.* **117**, 170–176.

KUVIBIDILA, S.R., BALIGA, B.S., and SUSKIND, R.M. (1981). Effects of iron deficiency anemia on delayed cutaneous hypersensitivity in mice. *Am. J. Clin. Nutr.* **34**, 2635–2640.

(1982). Generation of plaque forming cells in iron-deficient anemic mice. *Nutr. Rep. Intl* **26**, 861–871.

(1983a). The effect of iron-deficiency anemia on cytolytic activity of mice spleen and peritoneal cells against allogeneic tumor cells. *Am. J. Clin. Nutr.* **38**, 238–244.

KUVIBIDILA, S.R., DARDENNE, M., SAVINO, W., and LEPAULT, F. (1990). Influence of iron-deficiency anemia on selected thymus functions in mice: thymulin biological activity, T-cell subsets, and thymocyte proliferation. *Am. J. Clin. Nutr.* **51**, 228–232.

KUVIBIDILA, S.R., NAUSS, K.M., BALIGA, B.S., and SUSKIND, R.M. (1983b). Impairment of blastogenic response of splenic lymphocytes from iron-deficient mice: *in vivo* repletion by hemin, transferrin, and ferric chloride. *Am. J. Clin. Nutr.* **37**, 557–565.

LADEFOGED, O. and STURUP, S. (1995). Copper deficiency in cattle, sheep and horses caused by excess molybdenum from fly ash: a case report. *Vet. Human Toxicol.* **37**, 63–65.

LAM, H.F., CONNER, M.W., ROGERS, A.E., FITZGERALD, S., and AMDUR, M.O. (1985). Functional and morphological changes in the lungs of guinea pigs exposed to freshly generated ultrafine zinc oxide. *Toxicol. Appl. Pharmacol.* **78**, 29–38.

LATUNDE-DADA, G.O. and YOUNG, S.P. (1992). Iron deficiency and immune responses. *Scand. J. Immunol.* **11**, 207–209.

LAWRENCE, D.A. (1981). Heavy metal modulation of lymphocyte activities. I. *In vitro* effects of heavy metals on primary humoral immune responses. *Toxicol. Appl. Pharmacol.* **57**, 439–451.

LENNARD, E.S., BJORNSON, A.B., PETERING, H.G., and ALEXANDER, J.W. (1974). An immunologic and nutritional evaluation of burn neutrophil function. *J. Surg. Res.* **16**, 286–298.

LICASTRO, F., MOCCHEGIANI, E., ZANNOTI, M., ARENA, G., MASI, M., and FABRIS, N. (1992). Zinc affects the metabolism of thyroid hormones in children with Down's syndrome: normalization of thyroid stimulating hormone and reversal triiodothyronine plasmic levels by dietary zinc supplementation. *Int. J. Neurosci.* **65**, 259–268.

LIMA, M.F. and KIERSZENBAUM, F. (1987). Lactoferrin effects on phagocytic cell function. II. The presence of iron is required for the lactoferrin molecule to stimulate intracellular killing by macrophages but not to enhance the uptake of particles and microorganisms. *J. Immunol.* **139**, 1647–1651.

LOCKWOOD, J.F. and SHERMAN, A.R. (1988). Spleen natural killer cells from iron-deficient rat pups manifest an altered ability to be stimulated by interferon. *J. Nutr.* **118**, 1558–1563.

LOVERING, K.E. and DEAN, R.T. (1991). Restriciton of the participation of copper in radical-generating systems by zinc. *Free Rad. Res. Commun.* **14**, 217–225.

LU, J. and HAYASHI, K. (1995). Iron overload and iron deficiency regulates ED1 and ED2 expression in splenic macrophages of rats. *Chinese J. Pathol.* **24**, 21–24.

LUECKE, R.W., SIMONEL, C.E., and FRAKER, P.J. (1978). The effect of restricted dietary intake on the antibody mediated response of the zinc-deficient A/J mouse. *J. Nutr.* **108**, 881–887.

LUKASEWYCZ, O.A. and PROHASKA, J.R. (1982). Immunization against transplantable leukemia impaired in copper-deficient mice. *J. Natl Cancer Inst.* **69**, 489–493.

(1983). Lymphocytes from copper-deficient mice exhibit decreased mitogen activity. *Nutr. Res.* **3**, 335–341.

(1990). The immune response in copper deficiency. *Ann. N.Y. Acad. Sci.* **587**, 147–159.

LUKASEWYCZ, O.A., PROHASKA, J.R., MEYER, S.G., SCHMIDTKE, J.R., HATFIELD, S.M., and MARDER, P. (1985). Alterations in lymphocyte subpopulations in copper-deficient mice. *Infect. Immun.* **48**, 644–647.

LUKASEWYCZ, O.A., KOLQUIST, K.L., and PROHASKA, J.R. (1987). Splenocytes from copper-deficient mice are low responders and weak stimulators in mixed lymphocyte reactions. *Nutr. Res.* **7**, 43–52.

LUO, C., CHEN, J., LIAO, Q., LI, Q., YANG, X., and LI, Y. (1990). Influence of iron deficiency anemia on development of thymus and spleen adenosine deaminase activity in rats. *J. West China Univ. Med. Sci.* **21**, 63–66.

MACDOUGALL, L.G. and JACOBS, M.R. (1978). The immune response in iron-deficient children. Isohemagglutinin titres and antibody response to immunization. *S. Afr. Med. J.* **53**, 405–407.

MACDOUGALL, L.G., ANDERSON, R., McNAB, G.M., and KATZ, J. (1975). The immune response in iron-deficient children: impaired cellular defense mechanisms with altered humoral components. *J. Pediat.* **86**, 833–843.

MACKLER, B., PERSON, R., OCHS, H., and FINCH, C.A. (1984). Iron deficiency in the rat: effects on neutrophil activation and metabolism. *Pediat. Res.* **18**, 549–551.

MAINOU-FOWLER, T. and BROCK, J.H. (1985). Effect of iron deficiency on the response of mouse lymphocytes to concanavalin A: the importance of transferrin-
bound iron. *Immunology* **54**, 325–332.

MARKOWITZ, H., GUBLER, C.J., MAHONEY, J.P., CARTWRIGHT, G.E., and WINTROBE,
M.M. (1955). Studies on copper metabolism. XIV. Copper, ceruloplasmin, and
oxidase activity in sera of normal human subjects, pregnant women, and patients
with infection, hepatolenticular degeneration, and the nephrotic syndrome. *J. Clin.
Invest.* **34**, 1498–1508.

MASAWE, A.E., MUINDI, J.M., and SWAI, G.B. (1974). Infections in iron deficiency
and other types of anemia in the tropics. *Lancet* **ii**, 314–317.

MASSIE, H.R., COLACICCO, J., and AIELLO, V.R. (1979). Changes with age in copper
and ceruloplasmin in serum from humans and C57BL/6 mice. *Age* **2**, 97–106.

MASSIE, H.R., OFOSU-APPIAH, W.O., and AIELLO, V.R. (1993). Elevated serum
copper is associated with reduced immune response in aging mice. *Gerontology*
39, 136–145.

MATHÈ, G., MISSET, J.L., GIL-DELGADO, M., MUSSET, M., REIZENSTEIN, P. and
CANON, C. (1986). A phase II trial of immunorestoration with zinc gluconate in
immunodepressed cancer patients. *Biomed. Pharmacother.* **40**, 383–385.

MATZNER, Y., HERSHKO, C., POLLIACK, A., KONIJN, A., and IZAK, G. (1979). Suppressive effect of ferritin on *in vitro* lymphocyte function. *Br. J. Haematol.* **42**, 345–
353.

MCCLAIN, C.S. (1985). Zinc metabolism in malabsorption syndromes. *J. Am. Coll.
Nutr.* **4**, 49–64.

MENKES, J.H., ALTER, M., STEIGLEDER, G.H., WEAKLEY, D.R., and SUNG, J.H.
(1962). A sex-linked recessive disorder with retardation of growth, peculiar hair,
and focal cerebral and cerebellar degeneration. *Pediatrics* **29**, 764–779.

MOCCHEGIANI, E., PAOLUCCI, P., GRANCHI, D., CAVALLAZZI, L., SANTARELLI, L.,
and FABRIS, N. (1994). Plasma zinc level and thymic hormone activity in young
cancer patients. *Blood* **83**, 749–757.

MOCCHEGIANI, E., SANTARELLI, L., MUZZIOLI, M., and FABRIS, N. (1995a). Revers-
ibility of the thymic involution and of age-related peripheral immune dysfunctions
by zinc supplementation in old mice. *Int. J. Immunopharmacol.* **17**, 703–718.

MOCCHEGIANI, E., VECCIA, S., ANCARANI, F., SCALISE, G., and FABRIS, N. (1995b).
Benefit of oral zinc supplementation as an adjunct to zidovudine (AZT) therapy
against opportunistic infections in AIDS. *Int. J. Immunopharmacol.* **17**, 719–727.

MOFENSON, H.C., CARACCIO, T.R., and SHARIEFF, N. (1988). Iron sepsis: *Yersinia
enterocolitica* septicemia possibly caused by an overdose of iron. *New Engl. J. Med.*
316, 1092–1093.

MOORE, L.L. and HUMBERT, J.R. (1984). Neutrophil bactericidal dysfunction towards
oxidant radical-sensitive microorganisms during experimental iron deficiency.
Pediat. Res. **18**, 684–689.

MULHERN, S.A. and KOLLER, L.D. (1988). Severe or marginal copper deficiency
results in a graded reduction in immune status in mice. *J. Nutr.* **118**, 1041–1047.

MUNN, C.G., MARKENSON, A.L, KAPADIA, A., and DE SOUSA, M. (1981). Impaired T-
cell mitogenic responses in some patients with thalassemia intermedia. *Thymus* **3**,
119–128.

MUNOZ, C., OLIVARES, M., SCHLESINGER, L., LOPEZ, M., and LETELIER, A. (1994).
Increased *in vitro* tumor necrosis-alpha production in iron deficiency anemia. *Eur.
Cytokine Network* **5**, 401–404.

MURAKAWA, H., BLAND, C.E., WILLIS, W.T., and DALLMAN, P.R. (1987). Iron deficiency and neutrophil function: different rates of correction of the depressions in oxidative burst and myeloperoxidase activity after iron treatment. *Blood* **69**, 1464–1468.

MURRAY, M.J., MURRAY, A.B., MURRAY, M.B., and MURRAY, C.J. (1978). The adverse effect of iron repletion on the course of certain infections. *Br. Med. J.* **2**, 1113–1115.

MURRAY, M.J., WILSON, F.D., FISHER, G.L., and ERIKSON, K.L. (1983). Modulation of murine lymphocyte and macrophage proliferation by parenteral zinc. *Clin. Exp. Immunol.* **53**, 744–749.

NAKAGAWA, S., FUKATA, Y., NAGATA, H., MIYAKE, M., and HAMA, T. (1993). The decreased immune responses in macular mouse, a model of Menkes' kinky hair disease. *Res. Commun. Chem. Pathol. Pharmacol.* **79**, 61–73.

NELSON, R.L., YOO, S.J., TANURE, J.C., ANDRIANOPOULOS, G., and MISUMI, A. (1989). The effect of iron on experimental colorectal carcinogenesis. *Anticancer Res.* **9**, 1477–1482.

NÈVE, J., FONTAINE, J., PERETZ, A., and FAMAEY, F. (1988). Changes in zinc, copper and selenium status during adjuvant induced arthritis in rats. *Agents Action.* **25**, 146–155.

NEWBERNE, P.M., HUNT, C.E., and YOUNG, V.R. (1968). The role of diet and the reticuloendothelial system in the response of rats to *Salmonella typhimurium* infection. *Br. J. Exp. Pathol.* **49**, 448–457.

NUALART, P., ESTEVEZ, M.E., BALLART, I.J., DE MIANI, S.A., PENALVER, J., and SEN, L. (1987). Effect of alpha interferon on the altered T/B-cell immunoregulation in patients with thalassemia major. *Am. J. Hematol.* **24**, 151–159.

O'DELL, B.L., BROWNING, J.D., and REEVES, P.G. (1987). Zinc deficiency increases the osmotic fragility of rat erythrocytes. *J. Nutr.* **117**, 1883–1889.

OMARA, F.O. and BLAKLEY, B.R. (1993). Influence of low dietary iron and iron overload on urethan-induced lung tumors in mice. *Can. J. Vet. Res.* **57**, 209–211.

(1994a). The effects of iron deficiency and iron overload on cell-mediated immunity in the mouse. *Br. J. Nutr.* **72**, 899–909.

(1994b). The IgM and IgG antibody responses in iron-deficient and iron-loaded mice. *Biol. Trace Elem. Res.* **46**, 155–161.

OMARA, F.O., BLAKLEY, B.R., and HUANG, H.S. (1994). Effects of iron status on endotoxin-induced mortality, phagocytosis, and interleukin-1α and tumor necrosis factor-α production. *Vet. Human Toxicol.* **36**, 423–428.

OPPENHEIMER, S.J. (1980). Anaemia of infancy and bacterial infection in Papua-New Guinea. *Ann. Trop. Med. Parasitol.* **74**, 69–72.

OPPENHEIMER, S.J., GIBSON, F.D., MACFARLANE, S.B., MOODY, J.B., HARRISON, C., SPENCER, A., and BUNARI, O. (1986). Iron supplementation increases prevalence and effects of malaria: report on clinical studies in Papua-New Guinea. *Trans. Royal Soc. Trop. Med. Hyg.* **80**, 603–612.

PARDALOS, G., KANAKOUDI-TSAKALIDIS, F., MALAKA-ZAFIRIU, M., TSANTALI, H., ATHANASIOU-METAXA, M., KALLINIKOS, G., and PAPAEVANGELOU, G. (1987). Iron-related disturbances of cell-mediated immunity in multitransfused children with thalassemia major. *Clin. Exp. Immunol.* **68**, 138–145.

PATTANAPANYASAT, K. (1990). Expression of cell-surface transferrin receptor following *in vitro* stimulation of peripheral blood lymphocytes in patients with β-thalassemia and iron-deficiency anaemia. *Eur. J. Haematol.* **44**, 190–195.

PERKKIO, M.V., JANSSON, L.T., DALLMAN, P.R., SIIMES, M.A., and SAVILAHTI, E. (1987). sIgA- and IgM-containing cells in the intestinal mucosa of iron-deficient rats. *Am. J. Clin. Nutr.* **46**, 341–345.

PLETCHER, J.M. and BANTING, L.F. (1983). Copper deficiency in piglets characterized by spongy myelopathy and degenerative lesions in the great blood vessels. *J.S. Afr. Vet. Assoc.* **54**, 43–46.

POCINO, M., BAUTE, L., and MALAVÉ, I. (1991). Influence of the oral administration of excess copper on the immune response. *Fundam. Appl. Toxicol.* **16**, 249–256.

PONS, H.A., SOYANO, A., and ROMANO, E. (1992). Interaction of iron polymers with blood mononuclear cells and its detection with the Prussian Blue reaction. *Immunopharmacology* **23**, 29–35.

PORTO, G., REIMAO, R., GONCALVES, C., VICENTE, C., JUSTICA, B., and DE SOUSA, M. (1994). Haemachromatosis as a window into the study of immunological system: a novel correlation between CD8+ lymphocytes and iron overload. *Eur. J. Haematol.* **52**, 283–290.

POWELL, L.W., BASSETT, M.L., and HALLIDAY, J.W. (1980). Hemochromatosis: 1980 update. *Gastroenterology* **78**, 374–381.

PRASAD, A.S. (1983). Human zinc deficiency, in Sarkar, B. (Ed.) *Biological Aspects of Metals and Metal-Related Diseases*, pp. 107–119, New York: Raven Press.

PRASAD, J.S. (1979) Leukocyte function in iron-deficiency anemia. *Am. J. Clin. Nutr.* **32**, 550–552.

PROHASKA, J.R. and LUKASEWYCZ, O.A. (1981). Copper deficiency suppresses the immune response in mice. *Science* **21**, 559–561.

(1989). Copper deficiency during perinatal development: effects on the immune response of mice. *J. Nutr.* **119**, 922–931.

PROHASKA, J.R., DOWNING, S.W., and LUKASEWYCZ, O.A. (1983). Chronic dietary copper deficiency alters biochemical and morphological properties of mouse lymphoid tissues. *J. Nutr.* **113**, 1583–1590.

RAO, K.M., SCHWARTZ, S.A., and GOOD, R.A. (1979). Age-dependent effects of zinc on the transformation of human lymphocytes to mitogens. *Cell. Immunol.* **42**, 270–278.

ROBINS-BROWNE, R.M. and PRPIC, J.K. (1985). Effect of iron and desferroxamine on infections with *Yersinia enterocolitica*. *Infect. Immun.* **47**, 774–779.

SAENKO, E.L., SKOROBOGAT'KO, O.V., TARASENKO, P., ROMASHKO, V., ZHURAVETZ, L., ZADOROZHNAYA, L., SENJUK, O.F., and YAROPOLOV, A.I. (1994). Modulatory effects of ceruloplasmin on lymphocyte, neutrophils, and monocytes of patients with altered immune status. *Immunol. Invest.* **23**, 99–114.

SAHA, A.R., HADDEN, E.M., and HADDEN, J.W. (1995). Zinc induces thymulin secretion from human thymic epithelial cells *in vitro* and augments splenocyte and thymocyte responses *in vivo*. *Int. J. Immunopharmacol.* **17**, 729–733.

SANN, L., DAVID, L., GALY, G., and ROMAND-MONIER, R. (1978). Copper deficiency and hypocalcemic rickets in small-for-date infant. *Acta Paediat. Scand.* **67**, 303–307.

SANSTEAD, H.H. (1981). Zinc in human nutrition, in Bronner, F. and Coburn, J.W. (Eds) *Disorders of Mineral Metabolism*, pp. 94–159, New York: Academic Press.

SANTOS, P.C. and FALCAO, R.P. (1990). Decreased lymphocyte subsets and K-cell activity in iron deficiency anemia. *Acta. Haematol.* **84**, 118–121.

SAVINO, W., HUANG, P.C., CORRIGAN, A., BERRIH, S., and DARDENNE, M. (1984). Thymic hormone containing cells. V. Immunohistological detection of metallothioneine within the cells bearing thymulin (a zinc-containing hormone) in human and mouse thymuses. *J. Histochem. Cytochem.* **32**, 942–946.

SAVINO, W., BARTOCCIONI, E., HOMO-DELARCHE, F., GAGNERAULT, M.C., ITOH, T., and DARDENNE, M. (1988). Thymic hormone containing cells. IX. Steroids *in vitro* modulate thymulin secretion by human and murine thymic epithelial cells. *J. Steroid Biochem.* **30**, 479–484.

SAVINO, W., BAN, E., VILLA VERDE, D.M., and DARDENNE, M. (1990). Modulation of thymic endocrine function, cytokeratin expression and cell proliferation, by hormones and neuropeptides. *Int. J. Neurosci.* **51**, 201–204.

SCHLOEN, L.H., FERNANDEZ, G., GARAFALO, J.A., and GOOD, R.A. (1979). Nutrition, immunity, and cancer-A review. Part II: Zinc, immune function, and cancer. *Clin. Bull.* **9**, 63–75.

SEN, L., GOICOA, M.A., NUALART, P.J., BALLART, I.J., PALACIOS, F., DIEZ, R.A., and ESTEVEZ, M.E. (1989). Immunologic studies in thalassemia major: the immune alterations in patients with thalassemia major may be secondary to transfusion-related antigenic stimulation together with iron overload. *Medicina (Buenos Aires).* **49**, 131–134.

SHARPLESS, R.G. (1946). The effects of copper on liver tumor induction by *p*-dimethylaminoazobenzene. *Fed. Proc.* **5**, 239–240.

SHERMAN, A.R. (1990). Influence of iron on immunity and disease resistance. *Ann. N.Y. Acad. Sci.* **587**, 140–146.

SHERMAN, A.R. and LOCKWOOD, J.F. (1987). Impaired natural killer cell activity in iron-deficient rat pups. *J. Nutr.* **117**, 567–571.

SIEGERS, C.P., BUMANN, D., BARETTON, G., and YOUNES, M. (1988). Dietary iron enhances the tumor rate in dimethylhydrazine-induced colon carcinogenesis in mice. *Cancer Lett.* **41**, 251–256.

SOYANO, A., CANDELLET, D., and LAYRISSE, M. (1982). Effect of iron-deficiency on the mitogen-induced proliferative response of rat lymphocytes. *Int. Arch. Allergy Appl. Immunol.* **69**, 353–357.

SPEAR, A.T. and SHERMAN, A.R. (1992). Iron deficiency alters DMBA-induced tumor burden and natural killer cell cytotoxicity in rats. *J. Nutr.* **122**, 46–55.

STACEY, N.H. (1986). Effects of cadmium and zinc on spontaneous and antibody-dependent cell-mediated cytotoxicity. *J. Toxicol. Environ. Health* **18**, 293–300.

SUGARMAN, B. (1983). Zinc and infections. *Rev. Infect. Dis.* **5**, 137–147.

SULLIVAN, J.L. and OCHS, H.D. (1978). Copper deficiency and the immune system. *Lancet.* **ii**, 686.

THORSON, J.A., SMITH, K.M., GOMEZ, F., NAUMANN, P.W., and KEMP, J.D. (1991). Role of iron in T-cell activation: T_H1 clones differ from T_H2 clones in their sensitivity to inhibition of DNA synthesis caused by IgG monoclonal antibodies against the transferrin receptor and the iron chelator desferroxamine. *Cell. Immunol.* **134**, 126–137.

TURGEON-O'BRIEN, H., AMIOT, J. LEMIEUX, L., and DILLON, J.C. (1985). Myeloperoxidase activity of polymorphonuclear leukocytes in iron deficiency anemia and anemia of chronic disorders. *Acta Haematol.* **74**, 151.

UNDERWOOD, E.J. (Ed.) (1977). *Trace Elements in Human and Animal Nutrition*, 4th Edn, New York: Academic Press.

VAN ASBECK, B.S., VERBRUGH, H.A., VAN OOST, B.A., MARX, J.J., IMHOF, H., and VERHOEF, J. (1982). *Listeria monocytogenes* meningitis and decreased phagocytosis associated with iron overload. *Br. Med. J.* **284**, 542–544.

VAN ASBECK, B.S., MARX, J.J., STRUYVENBERG, J., and VERHOEF, J. (1984). Functional defects in phagocytic cells from patients with iron overload. *J. Infect.* **8**, 232–240.

VAUGHN, V.J. and WEINBERG, E.D. (1978). *Candida albicans* dimorphism and virulence: role of copper. *Mycopathologica* **64**, 39–42.

VYAS, D. and CHANDRA, R.K. (1983). Thymic factor activity, lymphocyte stimulation response and antibody-producing cells in copper deficiency. *Nutr. Res.* **3**, 343–349.

WALRAVENS, P.A. (1980). Nutritional importance of copper and zinc in neonates and infants. *Clin. Chem.* **26**, 185–189.

WALTER, T., ARREDONDO, S., ARÉVALO, M., and STEKEL, A. (1986). Effect of iron therapy on phagocytosis and bactericidal activity in neutrophils of iron-deficient infants. *Am. J. Clin. Nutr.* **44**, 877–882.

WARD, J.D., SPEARS, J.W., and KEGLEY, E.B. (1993). Effect of copper level and source (copper lysine vs copper sulfate) on copper status, performance, and immune response in growing steers fed diets with or without supplemented molybdenum and sulfur. *J. Animal Sci.* **71**, 2748–2755.

WEINBERG, E.D. (1984). Iron withholding: a defense against infection and neoplasia. *Physiol. Rev.* **64**, 65–102.

WEISS, G., WACHTER, H., and FUCHS, D. (1995). Linkage of cell-mediated immunity to iron metabolism. *Immunol. Today* **16**, 495–500.

WESTON, W.L., HUFF, J.C., HUMBERT, J.R., HAMBIDGE, K.M., NELDNER, K.H., and WALRAVENS, P.A. (1977). Zinc correction of defective chemotaxis in acrodermatitis enteropathica. *Arch. Dermatol.* **113**, 422–425.

WINDHAUSER, M.M., KAPPEL, L.C., McCLURE, J., and HEGSTED, M. (1991). Suboptimal levels of dietary copper vary immunoresponsiveness in rats. *Biol. Trace Elem. Res.* **30**, 205–217.

WIRTH, J.J., FRAKER, P.J., and KIERSZENBAUM, F. (1989). Zinc requirement for macrophage function: effect of zinc deficiency on uptake and killing of a protozoan parasite. *Immunology* **68**, 114–119.

WYLLIE, A.H. (1980). Glucocorticoid-induced thymocyte apoptosis is associated with endogenous endonuclease activation. *Nature* **284**, 555–556.

YETGIN, S., ALTAY, C., CILIV, G., and LALELI, Y. (1979). Myeloperoxidase activity and bactericidal function of PMN in iron deficiency. *Acta Haematol.* **61**, 10–14.

YOSHIDA, D., IKEDA, Y., and NAKAZAWA, S. (1995). Suppression of tumor growth in experimental 9L gliosarcoma model by copper depletion. *Neurol. Med.-Chir.* **35**, 133–135.

ZIDAR, B.L., SHADDUCK, R.K., ZEIGLER, Z., and WINKELSTEIN, A. (1977). Observations on the anemia and neutropenia of copper deficiency. *Am. J. Hematol.* **3**, 177–185.

APPENDIX

Immunotoxicology Summary Tables

MITCHELL D. COHEN

Institute of Environmental Medicine, New York University Medical Center, Long Meadow Road, Tuxedo, NY 10987, USA

Introduction

The purpose of the tables in this chapter is to complement the material in several of the individual chapters herein, and to provide the reader with a general overview of the literature which has examined the immunotoxicity associated with the major environmentally/occupationally encountered metals discussed earlier. Studies were selected that demonstrated the immunosuppressive and/or immunoenhancing effects of each of the individual metals on a variety of immunological responses in both humans and animal models, following either *in vivo* or *in vitro* exposure scenarios.

Table 1 Immunomodulating effects of arsenic compounds

In vivo exposures

Compound	Means of exposure	Species	Immune parameter analyzed	Effects	Citation
Arsenic	Per os	Human	PBL proliferation	↓	Ostrosky-Wegman et al., 1991
			PHA-induced mononuclear cell mitogenic responses (cells from cancer patients)	↑	Yu et al., 1992
Arsenic trioxide	Inhalation	Mouse	Mortality due to *Streptococcus pneumoniae*	↑	Aranyi et al., 1985
			Bactericidal activity against inhaled *Klebsiella pneumoniae*	↓	
			Mortality due to *Streptococcus pneumoniae*	↑	
			Bactericidal activity against inhaled *Klebsiella pneumoniae*	↓	
	SC		Mortality due to Pseudorabies virus infection	↑	Gainer and Pry, 1972
Arsine	Inhalation	Mouse	Splenic lymphocytes (total) percentage	↓	Rosenthal et al., 1989
			Splenic T-lymphocyte percentage	↓	
			Splenic B-lymphocyte percentage	↓	
			PEM levels and cytotoxic activity	N.E.	
			NK cell activity (YAC-1 target)	↓	
			T$_C$ cell activity	↓	
			Splenic T- and B-lymphocyte proliferation	N.E.	
			Host resistance to PYB6 tumor cells, B16F10 melanoma, or influenza virus	N.E.	
			Host resistance to *Listeria monocytogenes* and *Plasmodium yoelii*	↓	
			RBC levels, hematocrit, hemoglobin levels	↓	Hong et al., 1989
			White blood cell counts	↑	

Compound	Route	Species	Endpoint	Effect	Reference
Gallium arsenide	IT	Mouse	Marrow-associated erythroid precursor level	→	Sikorski et al., 1991a
			Splenic erythropoesis	←	
			Bone marrow cellularity	N.E.	
			Granulocyte/macrophage precursor levels	→	
			Primary IgM antibody response to SRBC	→	
			Splenic cell SRBC antigen presentation and processing	→	
			Macrophage Ia$^+$ percentage/expression	N.E./↓	
			Splenic cell phagocytosis	N.E.	
			Splenic cell cytochrome c/KLH antigen presentation	N.E./N.E.	
			Splenic cell IL-1 production	N.E.	
			Splenocyte proliferation	→	Burns and Munson, 1993b
			CD8$^+$/CD4$^+$ cell levels	N.E./↓	
			LPS-induced B-lymphocyte proliferation	N.E.	
			Anti-CD3ε-induced T-lymphocyte proliferation	→	
			CD25 expression	→	
			Leukocyte function antigen-1 expression	→	
			Intercellular adhesion molecule-1 expression	→	
			ConA-/PHA-induced lymphocyte proliferation	→	
			Spleen cellularity	←	Sikorski et al., 1989
			PFC formation	→	
			Serum complement levels	←	
			ConA-/PHA-induced splenic T-lymphocyte proliferation	N.E.	
			LPS-induced splenic B-lymphocyte proliferation	←	
			Splenic MLR activity	←	
			DTH reaction (vs KLH)	←	
			NK cell activity (YAC-1 target)	←	
			Percentage monocytes in PEC	←	

Table 1 (cont.)

In vivo exposures

Compound	Means of exposure	Species	Immune parameter analyzed	Effects	Citation
Sodium arsenate	Per os	Mouse	Percentage lymphocytes in PEC	↓	Kerkvliet et al., 1980
			PEC phagocytosis of covaspheres	↓	
			PEC phagocytosis of chicken erythrocytes	↑	
			Host resistance to *Plasmodium yoelii* and *Streptococcus pneumoniae*	N.E.	
			Host resistance to viable *Listeria monocytogenes*	↑	
			Host resistance to viable B16F10 tumor cells	↓	
	IT	Rat	Tumor latency/tumor incidence: from injected sarcoma cells	↑/↓	Lantz et al., 1994
			from injected virus	N.E.	
			Cell-mediated cytotoxicity *in vivo*	↑	
			Cell-mediated cytotoxicity *in vitro*	↓/↑	
			BAL PGE$_2$/TNF-α levels	N.E.	
			PAM induced TNF-α, O$_2^-$, and PGE$_2$ production	↓/↑/N.E.	
Sodium arsenite	IP	Mouse	Splenic T-lymphocyte proliferation	↓	Zelikoff et al., 1989
			Splenic MLR activity	↑	
			PEM phagocytic activity	↑	
	Per os		Mortality due to Pseudorabies virus infection	↑	
			Tumor incidence/tumor growth	↓/↑	Schrauzer and Ishmael, 1974
	SC		Mortality due to Pseudorabies virus infection	↑	Gainer, 1972
	IT	Rat	BAL PGE$_2$/TNF-α levels	N.E./N.E.	Lantz et al., 1994
			PAM induced TNF-α, O$_2^-$, and PGE$_2$ production	↓/N.E./N.E.	

Compound		Host cell	Immune parameter analyzed	Effects	Citation
Arsenic-associated fly ash	Inhalation	Human	Serum IgA, IgM, and IgG levels	N.E.	Bencko et al., 1988
			Blood α_1-antitrypsin and α_2-macroglobulin levels	N.E.	
			Blood transferrin, orosomucoid, and ceruloplasmin levels	↑	

In vitro exposures

Compound	Host cell	Immune parameter analyzed	Effects	Citation
Arsenic trioxide	Human	PHA-stimulated lymphocyte DNA synthesis	↑/↓	Meng, 1993
Gallium arsenide	Mouse	PFC formation	↓	Burns et al., 1991
Sodium arsenate	Cow	PHA-induced PBL proliferation	↑/↓	McCabe et al., 1983
	Mouse	Vesicular stomatitis virus-induced IFN production	↓	Gainer, 1972
	Rat	PAM O_2^{-} production	↓	Lantz et al., 1994
		PAM TNF-α production	↓	
		PAM PGE_2 production	↓	
	Human	PBL spontaneous proliferation	↓	Gonsebatt et al., 1992
		PHA-induced PBL proliferation	↓	
		PHA-induced PBL proliferation	↑/↓	McCabe et al., 1983
		PHA-stimulated lymphocyte DNA synthesis	↑/↓	Meng, 1993
Sodium arsenite	Cow	PAM phagocytic activity	N.E./↓	Fisher et al., 1986
		PAM viability	N.E.	
		PAM adherence	N.E.	
		PHA-induced PBL proliferation	↑/↓	McCabe et al., 1983
	Mouse	PFC formation	↓	Burns et al., 1991
		Vesicular stomatitis virus-induced IFN production	↓	Gainer, 1972
		PFC formation	↑/↓	Yoshida et al., 1987
		Viable B- and T-lymphocyte levels	↓	

Table 1 (cont.)

In vitro exposures

Compound	Host cell	Immune parameter analyzed	Effects	Citation
	Rabbit	Vesicular stomatitis virus-induced IFN production	↓	Gainer, 1972
		Zymosan-induced PAM $O_2^{\cdot-}$ production	N.E./↓	Labedzka et al., 1989
		Zymosan-induced PAM H_2O_2 production	N.E./↓	
		Zymosan-induced PAM chemiluminescence	↓	
		Lactate dehydrogenase release	N.E./↑	
	Rat	PAM $O_2^{\cdot-}$ production (induced)	→	Lantz et al., 1994
		PAM TNF-α production (induced)	→	
		PAM PGE_2 production (induced)	N.E.	
	Human	PBL spontaneous proliferation	→	Gonsebatt et al., 1992
		PHA-induced PBL proliferation	→	
		PHA-induced PBL proliferation	↑/↓	McCabe et al., 1983
		PHA-stimulated lymphocyte DNA synthesis	↑/↓	Meng, 1993

Table 2 Immunomodulating effects of beryllium compounds

In vivo exposures

Compound	Means of exposure	Species	Immune parameter analyzed	Effects	Citation
Beryllium sulfate	Inhalation	Mouse	PAM levels in BAL	↑	Sendelbach *et al.*, 1986
			PAM membrane damage	↑	
	IP		PEM Ia⁺ expression	↑	Behbehani *et al.*, 1985
			PEM ROI production	↑	Pfeifer *et al.*, 1994
			Splenic index	↑	
			PEM PMA-induced ROI production	N.E.	
			Splenocyte mitogen responsiveness	↑	
			Splenocyte BeSO₄-induced proliferation	↓/N.E.	
	IT		Lung immune cell levels (total)	↑	Huang *et al.*, 1992
			Lung monocyte/PAM/lymphocyte levels	↑	
			PAM MAC-1, -2, -3 marker expression	↑	
			Splenic BeSO₄-induced lymphoproliferation	↑	
	IV		Spleen red pulp size/RBC cell levels	→	Moatamed *et al.*, 1975
			Splenic white pulp T-lymphocyte zone size	→	
	Ear painting	Guinea pig	DTH reaction (vs BeSO₄)	↑	Chiappino *et al.*, 1969
			Lymph node neutrophil levels	↑	
	Inhalation (for 60 d after):				Krivanek and Reeves, 1972
	Per os		DTH reaction (vs BeSO₄)	N.E.	
			DTH reaction (vs Be-citrate)	N.E.	
			DTH reaction (vs Be-albuminate)	N.E.	
			DTH reaction (vs Be-aurintricarboxylate)	N.E.	

269

Table 2 (cont.)

In vivo exposures

Compound	Means of exposure	Species	Immune parameter analyzed	Effects	Citation
	ID		DTH reaction (vs BeSO$_4$)	↑	
			DTH reaction (vs Be-albuminate)	↑	
			DTH reaction (vs Be-aurintricarboxylate)	N.E.	
			DTH reaction (vs Be-citrate)	N.E./↓	
	ID toe-pad		DTH reaction (vs Be salts)	↑	Marx and Burrell, 1973
			PEM MIF production (vs BeSO$_4$ or BeF$_2$)	↑	
			PEM lymphocytotoxin production	↑	
	IP		DTH reaction (vs Be salts)	↑	Marx and Burrell, 1973
			PEM MIF production (vs BeSO$_4$ or BeF$_2$)	↑	
			PEM lymphocytotoxin production	↑	
	ID toe-pad	Rabbit	DTH reaction (vs BeSO$_4$)	↑	Kang et al., 1977
			PAM MIF production	↑	
			Popliteal lymph node lymphocyte mitogen responsiveness	↓	
	Inhalation	Rat	PAM levels in BAL	↑	Sendelbach et al., 1986
			PAM membrane damage	↑	
Beryllium hydroxide	SC	Sheep	Regional lymph node lymphoblast proliferation	↑	Haley et al., 1990
Beryllium oxide	IT	Guinea pig	Lung PAM and neutrophil levels in BAL	↑	Chiappino et al., 1969
			DTH reaction (vs BeSO$_4$)	↑	
	IT	Mouse	Lymphocyte levels in BAL	↑	Huang et al., 1992
			Splenic lymphocyte proliferation (vs BeSO$_4$)	↑	

Material	Species	Exposure	DTH reaction (vs BeF$_2$) / Endpoint	Result	Reference
Beryllium sulfosalicylate	Guinea pig	Treated lymphocytes (injected IP)	DTH reaction (vs BeF$_2$)	↑	Jones and Amos, 1975
Beryllium metal	Rat	Inhalation	PAM and neutrophil levels in BAL	↑	Hall, 1984
			PAM morphological changes	↑	
			BAL protein and lactate dehydrogenase levels	↑	
Beryllium (undefined)	Human	Occupational and/or environmental	Patients with chronic beryllium disease:		Bost *et al.*, 1994
			PAM TNF-α and IL-6 mRNA	↑	
			PAM IL-1β mRNA	N.E.	
			BAL TNF-α levels	↑	
			BAL IL-1β and IL-6 levels	N.E.	
			Patients with chronic beryllium disease:		Cullen *et al.*, 1987
			Peripheral blood percentage CD4$^+$ cells	→	
			Percentage CD4$^+$ cells in BAL	↑	
			Percentage CD8$^+$ cells in BAL	→	
			Ex vivo BeF$_2$-/BeSO$_4$-induced lymphoproliferation	↑	
			Patients with chronic beryllium disease:		Newman *et al.*, 1989
			Total cell recovery in BAL	↑	
			PAM levels in BAL	↑	
			Lymphocyte levels in BAL	↑	
			Ex vivo BeSO$_4$-induced lymphoproliferation	↑	
			Patients with chronic beryllium disease:		Price *et al.*, 1976
			PBM BeSO$_4$-induced migration inhibition	→	
			PBL viability	N.E.	
			Patients with chronic beryllium disease:		Tinkle *et al.*, 1996
			Ex vivo lung PAM/T-lymphocyte BeSO$_4$-induced production of:		
			TNF-α, IL-2, IL-6, and IFN-γ	↑	
			IL-4	N.E.	

Table 2 (cont.)

In vivo exposures

Compound	Means of exposure	Species	Immune parameter analyzed	Effects	Citation
			Patients with chronic beryllium disease: *Ex vivo* lung T-lymphocyte BeSO$_4$-induced production of:		Tinkle *et al.*, 1997
			IL-2 and IFN-γ	↑	
			IL-4 and IL-7	N.E.	
			Serum soluble IL-2 receptor levels	↑	
			Serum IFN-γ levels	N.E.	
Coal fly ash (with Be)	Inhalation	Human	Anti-organ autoantibody levels	↑	Bencko *et al.*, 1980
			Serum IgA, IgM, and IgG levels	↑	

In vitro exposures

Compound	Host cell	Immune parameter analyzed	Effects	Citation
Beryllium sulfate	Dog	PAM viability	→	Finch *et al.*, 1988
	Guinea pig	PEM MIF production	N.E.	Marx and Burrell, 1973
	Mouse	PEM viability	→	Behbehani *et al.*, 1985
		PEM IL-1α release	N.E.	
		PEM Ia expression	N.E.	
		BeSO$_4$-induced splenic lymphocyte proliferation	N.E.	Huang *et al.*, 1992
		B-lymphocyte Ig cap formation	→	Morita *et al.*, 1982
		BeSO$_4$-induced splenic lymphocyte proliferation	↑	Newman and Campbell, 1987
		BeSO$_4$-induced thymocyte proliferation	N.E.	

Compound	Species	Parameter	Effect	Reference
		Splenic lymphocyte viability	↓	Price and Skilleter, 1985
		BeSO$_4$-induced splenic lymphocyte proliferation	↑/↓	Unanue et al., 1976
		PEM Ia expression	↑	
		PEM IL-1α release	↑	Kang et al., 1979
	Rabbit	PAM enzyme release (β-Glu, β-NAG, and acid phosphatase)	↑	
		PAM viability	↓	Kang et al., 1977
		PAM membrane integrity/viability	↓	Deodhar et al., 1973
		PAM protein synthesis	↓	Saltini et al., 1989
	Human	BeSO$_4$-induced PBL maturation	N.E.	
		BeSO$_4$-induced PBL or lung lymphocyte proliferation	N.E.	
Beryllium chloride	Rat	PAM uptake of Be	↑	Hart and Pittman, 1980
		Splenic B-lymphocyte-dependent immune hemolysis	↓	Seko et al., 1982
Beryllium oxide	Dog	PAM viability	↓	Finch et al., 1988
	Rabbit	PAM enzyme release (β-Glu, β-NAG, and acid phosphatase)	↑	Kang et al., 1979
		PAM viability	↓	
Beryllium sulfosalicylate	Mouse	Lymphocyte viability	↓	Price and Skilleter, 1985
		Lymphocyte mitogen responsiveness	↑/↓	
Coal fly ash (with Be)	Rat	PAM viability	↓	Finch et al., 1991
		PAM phagocytic index	↓	

273

Table 3 Immunomodulating effects of cadmium compounds

In vivo exposures

Compound	Means of exposure	Species	Immune parameter analyzed	Effects	Citation
Cadmium acetate	Per os	Mouse	Resistance to encephalomyocarditis virus	↑	Exon et al., 1979
			Resistance to encephalomyocarditis virus	↑	Exon et al., 1986
			Splenic PFC (IgG, IgM) formation (vs SRBC)	N.E.	Muller et al., 1979
	IV	Rat	Resistance to Escherichia coli	→	Cook et al., 1975
			Resistance to Salmonella enteritidis endotoxin	→	Cook et al., 1974
Cadmium chloride	Gavage	Goat	Cutaneous hypersensitivity reaction	→	Haneef et al., 1995
	Gavage	Mouse	DTH reaction	→	Barnes and Munson, 1978
			Kupffer cell phagocytosis of viable Listeria	→	
	Inhalation		Serum antibody titer (vs influenza virus)	N.E.	Chaumard et al., 1991
			Resistance to murine cytomegalovirus	N.E.	Daniels et al., 1987
			Resistance to Streptococcus pneumoniae	→	Gardner et al., 1977
			Splenic PFC (IgM) formation	→	Graham et al., 1978
			PHA-/allogeneic antigen-induced splenic lymphocyte proliferation	→	Krzystyniak et al., 1987
			Spleen cell viability	→	
			Splenic B-lymphocyte LPS-induced proliferation	↑	
			Splenic PFC (IgM) formation	→	
	IM		Splenic PFC (IgM) formation	N.E.	Graham et al., 1978

Route	Parameter	Effect	Reference
IP	Resistance to *Listeria monocytogenes*	↓	Berche et al., 1980
	Resistance to *Klebsiella pneumoniae*	↓	
	Resistance to *Pseudomonas aeruginosa*	↓	Bozelka and Burkholder, 1979
	Resistance to *Mycobacterium bovis*	↑	Bozelka et al., 1978
	Splenic PFC (IgG) formation	↑	
	Splenic PFC (IgM) formation	↑	Daniels et al., 1987
	Murine cytomegalovirus-induced IFN-γ:		
	Serum titers	N.E.	
	Splenic titers	N.E.	
	Splenic PFC (IgG, IgM) formation	↑	Koller et al., 1976
	Resistance to *Listeria monocytogenes*	↓	Simonet et al., 1984
	Resistance to Japanese encephalitis virus	↓/N.E.	Suzuki et al., 1981
	Thymic atrophy	↑	Yamada et al., 1981
	Splenomegaly	↑	
IV	DTH reaction	↓	Kojima and Tamuar, 1981
Per os	Time to rejection of allograft	↓	Balter et al., 1982
	Time to isograft acceptance	↑	
	LPS-induced splenic B-lymphocyte proliferation	↑	Blakley, 1985
	Splenic PFC formation (vs SRBC)	↓	Borgman et al., 1987
	Splenic PFC formation (vs SRBC)	↓	Cay, 1981
	Splenic PFC (IgM) formation (vs SRBC)	↑	
	Resistance to encephalomyocarditis virus	N.E./↑	Exon et al., 1986
	Resistance to *Hexamita muris* protozoa	↓	Exon et al., 1975
	MOPC-104E tumor-induced mortality	↓/N.E.	Gray et al., 1982
	LPS-induced splenic B-lymphocyte proliferation	↑	Gaworski and Sharma, 1978

Table 3 (cont.)

In vivo exposures

Compound	Means of exposure	Species	Immune parameter analyzed	Effects	Citation
			PHA-induced splenic lymphocyte proliferation/blastogenesis	→	
			ConA-induced splenic lymphocyte proliferation/blastogenesis	N.E.	Ilback et al., 1994
			Resistance to Coxsackie B3 virus	N.E.	
			NK cell activity (YAC-1 target)	N.E.	Kerkvliet et al., 1979
			Maloney sarcoma virus-induced tumors	→	Knutson et al., 1980
			Reticuloendothelial cell binding/catabolism of immune complexes	→	
			Splenic PFC (IgG, IgM) formation	→	Koller et al., 1975
			Splenic PFC (IgG, IgM) formation	→	Koller et al., 1976
			Splenic B-lymphocyte EAC rosetting	→	Koller and Brauner, 1977
			PPD-induced splenic lymphocyte proliferation	↓/↑	Koller et al., 1979
			LPS-induced splenic B-lymphocyte proliferation	↑	Koller and Roan, 1980a
			ConA-induced splenic lymphocyte proliferation/blastogenesis	N.E.	Koller and Roan, 1980b
			Splenocyte MLR activity	N.E.	
			Splenic PFC (IgM) formation (vs SRBC)	↑/↓	Malave and DeRuffino, 1984
			ConA-/PHA-induced splenic lymphocyte proliferation/blastogenesis	↑	

Compound	Route	Species	Endpoint	Effect	Reference
			ConA-/PHA-induced splenic lymphocyte proliferation/blastogenesis	↑	Muller et al., 1979
			DTH reaction	→	Schulte et al., 1994
			Antibody formation (vs DNP-BSA)	N.E.	
			Splenic/thymic atrophy	↑	Thomas et al., 1985
			Resistance to *Listeria monocytogenes*	N.E.	
			Resistance to influenza A2/Taiwan, HIV-1, or HIV-2 virus	N.E.	
	SC		ConA-/PHA-induced splenic lymphocyte proliferation/blastogenesis	↓/N.E.	Ohsawa et al., 1986
			ConA-/PHA-induced splenic lymphocyte proliferation/blastogenesis	↑	
			LPS-induced splenic B-lymphocyte lymphoproliferation (species-dependent)	↓/N.E.	Suzuki et al., 1981
			Serum antibody titers vs Japanese encephalitis virus	N.E.	
	Per os	Rabbit	Serum antibody titer (vs pseudorabies virus)	→	Koller, 1973
	Gavage	Rat	NK cell activity (YAC-1 target)	N.E.	Stacey et al., 1988
	IP		NK cell activity (YAC-1 target)	→	Stacey et al., 1988
	Per os		NK cell activity (YAC-1 target)	↓/N.E.	Cifone et al., 1989a
Cadmium oxide	Inhalation	Mouse	Resistance to *Pasteurella multocida*	→	Chaumard et al., 1983
			Resistance to Orthomyxovirus influenza A	↑	
			Thymic atrophy	←	Ohsawa and Kawai, 1981
			Splenomegaly	←	
			Anemia, lymphopenia, and neutropenia	←	

Table 3 (cont.)

In vivo exposures

Compound	Means of exposure	Species	Immune parameter analyzed	Effects	Citation
	SC		Numbers of large lymphocytes	↑	
			Numbers of small lymphocytes	→	
			Thymic atrophy	↑	
			Splenomegaly	↑	
			Anemia, lymphopenia, and neutropenia	↑	
			Numbers of large lymphocytes	↑	
			Numbers of small lymphocytes	→	Ohsawa and Kawai, 1981
	Inhalation	Rat	Thymic atrophy	↑	
			Splenomegaly	↑	
			Anemia, lymphopenia, and neutropenia	↑	
			Numbers of large lymphocytes	↑	
			Numbers of small lymphocytes	→	
	SC		Thymic atrophy	↑	
			Splenomegaly	↑	
			Anemia, lymphopenia, and neutropenia	↑	
			Numbers of large lymphocytes	↑	
			Numbers of small lymphocytes	→	
Cadmium sulfate	Diet	Chicken	Resistance to *Salmonella gallinarium*	↑	Hill, 1979
			Resistance to *Escherichia coli*	N.E.	
	Per os	Mouse	Resistance to encephalomyocarditis virus	↑	Exon *et al.*, 1986
			Resistance to encephalomyocarditis virus	→	Gainer, 1977a,b
	SC		Serum antibody titer (vs human γ-globulin)	↓/↑	Jones *et al.*, 1971

Compound		Host cell	Immune parameter analyzed	Effects	Citation
Cadmium (undefined)	Occupational	Human	Serum IgA, IgG, IgM levels	N.E.	Karakaya et al., 1994
			Antibody light chain catabolism	→	Vigliani, 1969

In vitro exposures

Compound	Host cell	Immune parameter analyzed	Effects	Citation
Cadmium acetate	Mouse	PEM phagocytic activity	→	Loose et al., 1978a
		PEM microbicidal activity	→	
		PAM phagocytic and microbicidal activity	→	
		PEM viability	→	Loose et al., 1978b
		PAM viability	→	
	Rat	PAM O$_2$ consumption	→	Castranova et al., 1980
		PAM glucose metabolism	→	
Cadmium chloride	Guinea pig	PEM mobility/reaction to MIF	→	Kiremidjian-Schumacher et al., 1981a,b
		Splenocyte mobility/reaction to MIF	→	
		PAM phagocytic activity	N.E.	Kramer et al., 1990
		PAM chemiluminescence	N.E.	
	Mouse	Splenic B-lymphocyte DNA/RNA synthesis	→	Daum et al., 1993
		Splenocyte PFC (IgM) formation	↑/↓	Fujimaki et al., 1982
		Splenocyte PFC (IgM) formation (vs DNP-Ficoll)	↑/↓	Fujimaki, 1985
		Lymphocyte RNA synthesis	↑	Gallagher and Gray, 1982
		Spontaneous splenocyte blastogenesis	↑	Gallagher et al., 1979
		LPS-induced splenocyte blastogenesis	N.E.	
		PEM phagocytic activity	→	Hilbertz et al., 1986
		PEM viability	→	
		PEM chemiluminescence	↑	

Table 3 *(cont.)*

In vitro exposures

Compound	Host cell	Immune parameter analyzed	Effects	Citation
		Splenocyte PFC (IgM) formation (vs SRBC)	↓/N.E.	Lawrence, 1981
		PEM phagocytic activity	→	Loose et al., 1978a
		PEM microbicidal activity	→	
		PAM phagocytic and microbicidal activity	→	
		PEM viability	→	Loose et al., 1978b
		PAM viability	→	
		ConA-induced splenic lymphocyte proliferation/blastogenesis	↑	Otsuka and Ohsawa, 1991
		ConA-induced IL-2 production/release	→	Payette et al., 1995
		ConA-induced IL-2 receptor expression	→	
		ConA-induced lymphocyte cell cycling	→	
		LPS-induced splenocyte blastogenesis	↑/↓	Shenker et al., 1977
	Rabbit	PAM EAC rosette formation	→	Hadley et al., 1977
		PAM phagocytic activity	→	Graham et al., 1975
	Rat	PAM O$_2$ consumption	→	Castranova et al., 1980
		PAM glucose metabolism	→	
		PAM ROI production	→	
	Human	PHA-induced lymphocyte blastogenesis	→	Borella et al., 1990
		Lymphocyte MLR activity	→	
		PHA-/PMA-induced IL-2 production/release	→	Cifone et al., 1989b

PHA-/PMA-induced IL-2 receptor expression	→	Cifone et al., 1990
Peripheral blood NK cell activity (YAC target)	→	
PBL ADCC activity	→	Funkhouser et al., 1994
Monocytic cell line IL-6 production/release	→	
Monocytic cell line IL-6 mRNA levels	→	Herberman et al., 1987
IL-2-enhanced NK cell activity (YAC-1 target)	→	
PBM spontaneous IL-8 production/release	←	Horiguchi et al., 1993
PBM IL-8 mRNA levels	←	
PHA-induced lymphocyte blastogenesis	→	Kastelan et al., 1981
Lymphocyte MLR activity	→	
PHA-induced PBM production of IL-1β and TNF-α	→	Theocharis et al., 1994
PHA-induced PBM IL-1β and TNF-α mRNA levels	→	

Table 4 Immunomodulating effects of chromium compounds

In vivo exposures

Compound	Means of exposure	Species	Immune parameter analyzed	Effects	Citation
Calcium chromate	IT	Rat	Zymosan-induced PAM chemiluminescence	N.E.	Galvin and Oberg, 1984
			PAM O$_2$ consumption	N.E.	
Chromic chloride	IM	Mouse	Splenic PFC formation	N.E.	Graham et al., 1978
Chromic nitrate	Inhalation	Rabbit	PAM levels in BAL/PAM size	←	Johansson et al., 1986a,b
			PAM morphological changes (lamellar inclusions, enlarged Golgi)	←	
			PAM ROI production	→	
			PAM phagocytic activity	N.E.	
Chromium phosphate	IV	Rat	Serum neutrophil levels	←	Kinnaert et al., 1970
			Serum mononuclear levels	→	
			Antibody synthesis/titer (vs SRBC)	→	
			Whole-body carbon clearance	→	
			Skin graft survival	N.E.	
Chromium trioxide	IT	Rat	Zymosan-induced PAM chemiluminescence	N.E.	Galvin and Oberg, 1984
			PAM O$_2$ consumption	N.E.	
Sodium chromate	Inhalation	Rabbit	PAM levels in BAL/PAM size	N.E.	Johansson et al., 1986a,b
			PAM ROI production	N.E.	
			PAM phagocytic activity	N.E.	
			PAM morphological changes (lamellar inclusions, enlarged Golgi)	←	
	SC	Rat	Antibody formation (vs *Escherichia coli* T-1 phage)	→	Figoni and Treagan, 1975
Sodium dichromate	Inhalation	Rat	Serum Ig levels	N.E./↑/↓	Glaser et al., 1985
			Splenic PFC formation	N.E./↑/↓	

Compound		Host cell	Immune parameter analyzed	Effects	Citation
	Per os	Guinea pig	PAM levels (percentage in BAL)	N.E./↓	van Hoogstraten et al., 1992
			Lymphocyte and neutrophil levels in BAL	↑/↓	
			PAM phagocytic activity	↑/↓	
			Chromium skin tolerance	↑	

In vitro exposures

Compound	Host cell	Immune parameter analyzed	Effects	Citation
Calcium chromate	Rat	PAM zymosan-induced chemiluminescence	↓	Galvin and Oberg, 1984
		PAM O_2 consumption	↓	
Chromic chloride	Rat	Thymocyte ATP/GTP production	N.E.	Lazzarini et al., 1985
		Thymocyte O_2 consumption	N.E.	
		PAM O_2 consumption (stimulated)	↓	Castranova et al., 1980
		PAM ROI production	↓	
		PAM viability	↓	
	Rabbit	PAM phagocytosis	↓	Graham et al., 1975
	Human	PBL mitogen responsiveness	N.E.	Borella et al., 1990
Chromium (metallic)	Monkey	LLC-MK2 kidney cell IFN-α/-β production	↓	Hahon and Booth, 1984
Chromium trioxide	Rat	Zymosan-induced PAM chemiluminescence	↓	Galvin and Oberg, 1984
		PAM O_2 consumption	↓	
Potassium chromate	Rat	PAM viability	↓	Pasanen et al., 1986
		PAM LDH release	↑	
Potassium dichromate	Rat	Thymocyte ATP/GTP production	↓	Lazzarini et al., 1985
		Thymocyte O_2 consumption	↓	
	Human	PBL mitogen responsiveness	↑/↓	Borella et al., 1990

Table 4 (cont.)

In vitro exposures

Compound	Host cell	Immune parameter analyzed	Effects	Citation
Sodium chromate	Cow	PAM viability/phagocytosis	↓	Hooftman *et al.*, 1988
	Mouse	PEM random migration	↓	Christensen *et al.*, 1992
		PEM phagocytic activity	↓	
		PEM IFN-α/-β production	↓	
		L-929 fibroblast IFN-α/-β production	↓	Pribyl and Treagan, 1977
Sodium dichromate	Human	T-lymphocyte mitogen responsiveness	N.E./↓	Kucharz and Sierakowski, 1987
		T-lymphocyte IL-2 production	↓	

Table 5 Immunomodulating effects of lead compounds

In vivo exposures

Compound	Means of exposure	Species	Immune parameter analyzed	Effects	Citation
Lead acetate	Diet	Mouse	In Pb-exposed offspring of Pb-exposed dams:		Talcott and Koller, 1983
			Splenic index	→	
			Generation of anti-albumin antibody	→	
			DTH responses	N.E.	
			PEM phagocytic activity	N.E.	
	IP		Sensitivity to LPS endotoxin	←	Dentener *et al.*, 1989
			Sensitivity to TNF-α	←	
			LPS-induced TNF-α production	→	
			DTH reaction (vs SRBC)	←	Descotes *et al.*, 1984
	Per os		Resistance to viable *Listeria monocytogenes*	→	Bincoletto and Queiroz, 1996
			Serum colony-stimulating factor levels	N.E.	Blakley and Archer, 1981
			Antibody formation:		
			Macrophage-dependent antigen	→	
			Macrophage-independent antigen	N.E.	
			Resistance to encephalomyocarditis virus	→	Gainer, 1974
			Incidence of splenomegaly during infection	←	
			In situ induction of IFN	N.E.	
			In situ binding/functionality of IFN	N.E.	
			Resistance to viable *Listeria monocytogenes*	→	Kowolenko *et al.*, 1991
			Responsiveness of bone marrow cells and splenocytes to macrophage growth factor	→	

Table 5 (cont.)

In vivo exposures

Compound	Means of exposure	Species	Immune parameter analyzed	Effects	Citation
			Serum colony-stimulating activity	↑	
			PFC response	N.E./↑	Mudzinski et al., 1986
			Serum antibody levels (vs SRBC)	N.E.	
			ADCC response against SRBC	→	Neilan et al., 1983
			NK cell activity (YAC-1 target)	N.E.	
			Bone marrow nucleated and pluripotent stem cell levels/recovery:		Schlick and Friedberg, 1982
			10-day exposure	N.E.	
			1-month exposure	↓/N.E.	
			3-month exposure	↓/↓	
	Per os	Rat	In Pb-exposed offspring of Pb-exposed dams:		Faith et al., 1979
			Thymic weight	→	
			Splenic lymphocyte mitogen responsiveness	→	
			DTH responses	→	
			In Pb-exposed offspring of Pb-exposed dams:		Luster et al., 1978
			Serum IgG levels	→	
			Serum IgA and IgM levels	N.E.	
			Serum antibody levels (vs SRBC)	→	
			Serum antibody levels (vs LPS)	N.E.	
	IV	Rat	Resistance to viable *Escherichia coli* infection	→	Cook et al., 1975
			Resistance to heat-killed *Escherichia coli*	→	

Compound	Route	Species	Endpoint	Effect	Reference
			Resistance to viable *Staphylococcus epidermidis* infection	N.E./↓	Descotes *et al.*, 1984
Lead carbonate	IP	Mouse	DTH reaction (vs SRBC)	→	Descotes *et al.*, 1984
Lead chloride	IP	Mouse	DTH reaction (vs SRBC)	↑	Descotes *et al.*, 1984
	Per os		Resistance to encephalomyocarditis virus	→	Gainer, 1977
			PEM phagocytosis of *Listeria*	N.E.	Kowolenko *et al.*, 1988
			Splenic macrophage phagocytosis of *Listeria*	N.E.	
			PEM antigen presentation	→	
			PEM IL-1 production	N.E.	
			PEM MLR activity	↑	
	SC		*Ex vivo* splenic lymphocyte IL-4 production	↑	Heo *et al.*, 1996
			Ex vivo splenic lymphocyte IFN-γ production	→	
			Plasma IFN-γ levels	→	
			Plasma IL-4 and IgE levels	↑	
	Inhalation	Rat	PAM levels in BAL	N.E.	Bingham *et al.*, 1972
			PAM size	↑	
Lead nitrate	IP	Mouse	DTH reaction (vs SRBC)	→	Descotes *et al.*, 1984
			Resistance to viable *Salmonella typhimurium*	→	Hemphill *et al.*, 1972
Lead oxide	IP	Mouse	DTH reaction (vs SRBC)	→	Descotes *et al.*, 1984
	IT	Rat	PAM levels in BAL/PAM size	↑	Kaminski *et al.*, 1977
			PAM viability	N.E.	

Table 5 (*cont.*)

In vivo exposures

Compound	Means of exposure	Species	Immune parameter analyzed	Effects	Citation
Lead trioxide	Inhalation	Rat	PAM morphological changes (*ex vivo*): without added lymphocytes	N.E.	Bingham *et al.*, 1972
			with added lymphocytes (pseudopodia and vacuolization)	↑	
			PAM levels in BAL	→	
			PAM size	↑	
Lead (undefined)	Occupational and/or environmental	Human	PBM E-rosetting activity	N.E.	Cohen *et al.*, 1989
			ConA-/PHA-induced PBL blast transformation	N.E.	
			Numbers of PBL CD4$^+$, CD8$^+$ cells	N.E.	Bergeret *et al.*, 1990
			ConA-induced PBL suppressor activity	↑	
			Neutrophil chemotaxis	→	
			Ex vivo neutrophil phagocytic activity	N.E.	
			Serum C3/total Ig levels	→	Ewers *et al.*, 1982
			Salivary IgA levels	→	
			Neutrophil chemotaxis	→	Governa *et al.*, 1988
			Neutrophil/monocyte levels (whole blood)	↑	Guillard and Lauwerys, 1989
			PMA-induced chemiluminescence in blood	N.E.	
			PHA-induced PBL blast transformation	→	Jaremin, 1983
			PBM E-rosetting/migration activity	N.E.	
			Circulating B-lymphocyte levels	→	Jaremin, 1990

	Effect	Reference
Serum IgG and IgM levels	→	Queiroz et al., 1993
Serum IgG production (vs typhoid antigen)	→	
Neutrophil chemotaxis	→	Queiroz et al., 1994a
Neutrophil NBT reduction	→	
Ex vivo neutrophil phagocytosis of Candida albicans and Candida pseudotropicalis	N.E.	
Splenic phagocytosis of C. albicans and C. pseudotropicalis	N.E.	
Ex vivo neutrophil lysis of C. albicans	→	
Ex vivo neutrophil lysis of C. pseudotropicalis	N.E.	
Splenic lysis of C. albicans	→	
Splenic lysis of C. pseudotropicalis	N.E.	
PHA-induced PBL proliferation	N.E.	Queiroz et al., 1994b
Serum IgM, IgG, and IgA levels	N.E.	
Serum Ig/complement levels	N.E.	Reigart and Graber, 1976
Anamnestic response to tetanus toxoid	N.E.	
Neutrophil chemotactic activity	→	Valentino et al., 1991
Neutrophil random migration	N.E.	
Neutrophil spontaneous ROI formation	N.E.	
Neutrophil leukotriene B4 (LTB$_4$) release	↑	
Neutrophil arachidonic acid content	↑	
Serum Ig/complement levels	N.E.	Kimber et al., 1986
PBL mitogen responsiveness	N.E.	
Natural killer cell activity (vs K562 target)	N.E.	

Tetraethyl lead

Table 5 *(cont.)*

In vitro exposures

Compound	Host cell	Immune parameter analyzed	Effects	Citation
Lead acetate	Mouse	PEM O$_2^-$ production	↓	Buchmuller-Rouiller *et al.*, 1989
		PEM H$_2$O$_2$ production	↓	
		PEM HMP shunt activation by PMA	↑	
		PEM HMP shunt activation by LPS or MAF	↓	
		PEM HMP shunt activity	↓	
		PEM G6PDH enzyme activity	N.E.	
		PEM 6PGDH enzyme activity	N.E.	
		PEM glucose uptake	↓	
		PEM intracellular killing of *Leishmania*	↓	Mauel *et al.*, 1989
		PEM HMP shunt activity	↓	
		PEM activation by LPS or MAF	↓	
		PEM binding of IFN-γ	N.E.	
		PEM post-binding processing of IFN-γ	↓	
	Rabbit	Zymosan-induced PAM O$_2^-$ production	↓	Labedzka *et al.*, 1989
		Zymosan-induced PAM H$_2$O$_2$ production	↓	
		Zymosan-induced PAM chemiluminescence	↓	
		Lactate dehydrogenase release	N.E.	
	Rat	Resting PAM O$_2$ consumption	↓	Castranova *et al.*, 1980
		Resting PAM glucose metabolism	↓	
		Phagocytosing PAM O$_2$ consumption	↓	
		Phagocytosing PAM ROI formation/ release	↓	
		PAM membrane integrity	N.E.	

	Human	Neutrophil chemotactic activity	→	Governa et al., 1987
		Neutrophil phagocytic activity	→	
		Neutrophil zymosan-induced $O_2^{\cdot-}$ formation	N.E.	
		Neutrophil fluorescence polarization:		
		without stimulation	N.E./↑	
		with n-fMLP stimulation	N.E./↑	
Lead chloride	Mouse	T_H2 lymphocyte IL-4 production	←	Heo et al., 1996
		T_H2 lymphocyte proliferation	←	
		T_H1 lymphocyte IFN-γ production	→	
		T_H1 lymphocyte proliferation	→	
		PEM PMA-induced oxidative metabolism:		Hilbertz et al., 1986
		1-h exposure	←	
		20-h exposure	→	
		PEM zymosan-induced oxidative metabolism:		
		1- or 20-h exposure	→	
		PEM phagocytic activity	↓/N.E.	
		PEM viability		
		PEM phagocytosis of Listeria	N.E.	Kowolenko et al., 1988
		Splenic macrophage phagocytosis of Listeria	N.E.	
		PEM antigen presentation	→	
		PEM IL-1 production	N.E.	
		PEM MLR activity	↑	
		Splenic lymphocyte mitogen responsiveness:		Lawrence, 1981c
		Spontaneous proliferation	←	
		to LPS	←	
		to 2-ME	←	

Table 5 (cont.)

In vitro exposures

Compound	Host cell	Immune parameter analyzed	Effects	Citation
		to ConA	↑/N.E.	
		to PHA	↑/N.E.	
		Splenic lymphocyte PFC formation	↑/N.E./↓	
		Splenic lymphocyte MLR reactivity:		
		Pb-treated stimulator cells	N.E.	
		Pb-treated responder cells	↑	
		Untreated stimulator and responder cells:		
		Allogeneic stimulators	↑	
		Syngeneic stimulators	N.E.	
		Splenic lymphocyte mitogen responsiveness	↑	Lawrence, 1981b
		Splenic PFC formation	↑	
		B-lymphocyte LPS-induced IgM production	↑	McCabe and Lawrence, 1990
		B-lymphocyte Class II antigen expression	↑	
		B-lymphocyte Class I antigen expression	N.E.	
		B-lymphocyte sIgD, F$_{c\varepsilon}$ receptor expression	↑	
		Splenic B-lymphocyte maturation to AFC	↑	McCabe and Lawrence, 1991
		Splenic B-lymphocyte IgM/IgG formation	↑	
		APC-induced T$_H$1 lymphocyte activation	↓	
		APC-induced T$_H$2 lymphocyte activation	↑	

B-lymphocyte Class II antigen expression	↑	McCabe et al., 1991
B-lymphocyte Class II mRNA	N.E.	
B-lymphocyte Class II mRNA translation and post-translational product stability	↑	
APC-induced DO-11.10 T-lymphocyte hybridoma IL-2 production:		Smith and Lawrence, 1988
Pb present in reaction	N.E./↓	
Pb pretreatment of APC	→	
ConA-induced splenocyte IL-2 production	N.E.	
IL-2-induced HT-2 lymphocyte cell line proliferation	N.E.	
DO-11.10 T-lymphocyte hybridoma viability	N.E.	
ConA-induced DO-11.10 T-lymphocyte hybridoma proliferation	N.E.	Smith and Lawrence, 1988
Pb-induced DO-11.10 T-lymphocyte hybridoma proliferation	N.E.	
Splenic macrophage IFN-γ-/TNF-α-/ ConA-induced NO production/release	→	Tian and Lawrence, 1995
Splenic lymphocyte PFC formation (vs SRBC)	↑	Warner and Lawrence, 1986a
Splenic lymphocyte spontaneous proliferation	↑	
Splenic lymphocyte spontaneous proliferation	↑	Warner and Lawrence, 1986b
Numbers of Thy[+], L3T4[+], Lyt2[+], and sIgM[+] cells	↑	

Table 5 *(cont.)*

In vitro exposures

Compound	Host cell	Immune parameter analyzed	Effects	Citation
		Numbers of splenic lymphocytes cell cycling	↑	Warner and Lawrence, 1988
		IL-2-induced HT-2 lymphocyte cell line proliferation	N.E.	
		ConA-induced splenocyte IL-2 production	N.E.	
		Pb-induced splenocyte IL-2 production	N.E./↑	
		Pb-induced splenocyte IL-2 receptor expression	↑	
		Pb-induced splenocyte proliferation:		
		Pb only	↑	
		Pb/anti-IL-2 antibody (vs Ni only)	N.E.	
		Pb/anti-IL-2 receptor antibody (vs Ni only)	→	
		Pb/anti-IFN-γ antibody (vs Ni only)	→	
		Pb/anti-Thy1.2 antibody (vs Ni only)	N.E.	
	Rabbit	PAM phagocytic activity	→	Jian et al., 1985
		PAM F_C rosetting capacity	→	
	Rat	Resting PAM O_2 consumption	→	Castranova et al., 1980
		Resting PAM glucose metabolism	→	
		Phagocytosing PAM O_2 consumption	→	
		Phagocytosing PAM ROI formation/release	→	
		PAM membrane integrity	N.E.	
	Human	PBL Ig production		Borella and Giardino, 1991
		Spontaneous	←	
		PWM-induced	←	

Substance	Species	Endpoint	Effect	Reference
Lead oxide		PMA-induced chemiluminescence (using whole blood analysis)	→	Guillard and Lauwerys, 1989
	Rabbit	PAM LPS-induced TNF-α production	→	Cohen et al., 1994
		PAM viability	→	
		PAM viability	→	DeVries et al., 1983
		PAM morphological changes: swelling of mitochondria, nuclear membrane, and endoplasmic reticuli	←	
		Zymosan-induced PAM $O_2^{\cdot-}$ production	N.E./↓	Labedzka et al., 1989
		Zymosan-induced PAM H_2O_2 production	N.E./↓	
		Zymosan-induced PAM chemiluminescence	N.E./↓	
		Lactate dehydrogenase release	N.E./↑	
Lead tetroxide	Rabbit	PAM viability	→	DeVries et al., 1983
		PAM morphological changes: swelling of mitochondria, nuclear membrane, and endoplasmic reticuli	←	
Lead oxide-coated fly ash	Rabbit	PAM viability	→	DeVries et al., 1983
		PAM morphological changes: swelling of mitochondria, nuclear membrane, and endoplasmic reticuli	←	
Lead (atmospheric)	Human	PHA-induced PBL blast transformation	→	Jaremin, 1983

Table 5 (cont.)

In vitro exposures

Compound	Host cell	Immune parameter analyzed	Effects	Citation
Triethyl lead	Human	PBL viability	N.E.	Stournaras *et al.*, 1990
		PBL CD3$^+$, CD4$^+$, or CD8$^+$ percentages	N.E.	
		PBL ATP content	N.E./↓	
		PBL MLR reactivity:		
		without IL-2 present	→	
		with IL-2 present	→	
		PHA-induced PBL T-lymphocyte proliferation:		
		without IL-2 present	→	
		with IL-2 present	N.E./↓	
		PHA-induced PBL IL-2 production	N.E./↓	
		PHA-induced PBL surface expression:		
		Tac/IL-2 receptor p55	→	
		Tf-R/transferrin receptor	N.E.	
		Ia/Class II MHC molecule	N.E.	
		PHA-induced PBL IL-2 Tac mRNA levels	→	

Table 6 Immunomodulating effects of mercury compounds

In vivo exposures

Compound	Means of exposure	Species	Immune parameter analyzed	Effects	Citation
Mercuric chloride	IM	Chicken	Antibody formation (vs SRBC):		Bridger and Thaxton, 1983
			Primary and secondary response	N.E.	
			Serum IgM and IgG levels	N.E.	
	Per os	Chicken	Antibody formation (vs SRBC):		Bridger and Thaxton, 1983
			Primary and secondary response	→	
			Antibody formation (vs *Brucella abortus*):		
			Primary response	→	
			Secondary response	←	
			Serum IgM, IgG levels	→	
	IP	Mouse	HSV-2 clearance (hepatic)	→	Ellermann-Eriksen *et al.*, 1994
	Per os		PEM phagocytic activity	↑	Brunet *et al.*, 1993
			Bone marrow lymphoid/total cell ratios	N.E.	
			Bone marrow B-lymphocyte maturation	N.E.	
			Bone marrow cell surface marker expression	N.E.	
			Splenic B-lymphocyte sIg expression	N.E.	
			Splenic T-lymphocyte marker (Thy 1.2, Lyt 2, L3T4) expression	↓/N.E.	
	Per os		Thymic and splenic index	→	Dieter *et al.*, 1983
			Splenic PFC formation	→	
			Splenic PFC formation (vs LPS)	N.E.	
			Splenic T-lymphocyte mitogen responsiveness	→	
			Splenic T-lymphocyte MLR activity	→	
			Splenic B-lymphocyte mitogen responsiveness	↑	

297

Table 6 (cont.)

In vivo exposures

Compound	Means of exposure	Species	Immune parameter analyzed	Effects	Citation
			Total blood white blood cell levels	↑/↓	Hultman and Johansson, 1991
			Total blood RBC levels	N.E.	
			Serum Ig levels	N.E.	
			Antinucleolar antibody formation	↑	Hultman and Enestrom, 1992
			Splenic T-lymphocyte mitogen responsiveness	↑/N.E.	
			Splenic B-lymphocyte mitogen responsiveness	↑/N.E.	
	SC		Antinucleolar antibody formation	↑	Hultman et al., 1993
			Antinuclear antibody formation	↑	Robinson et al., 1986
			Antinucleolar antibody formation	↑/N.E.	Saegusa et al., 1990, 1991
	IM	Rabbit	Serum anti-basement membrane autoantibody levels	↑	Roman-Franco et al., 1978
	IP	Rat	Splenic IFN-γ-producing cell levels	↓/N.E.	van der Meide et al., 1993
	SC		Serum anti-lamanin antibody levels	N.E./↑	Kosuda et al., 1993
			Spleen/lymph node cell levels (total)	↑	
			Spleen/lymph node T-lymphocyte ratio $(RT6^-:RT6^+)$	N.E./↑	
		Rat	Serum IgE levels	↓/N.E.	Prouvost-Danon et al., 1981
Methyl mercury chloride	Diet	Mouse	Thymic index/total cell levels	↓	Ilback, 1991
			Splenic index	N.E.	
			NK cell activity (blood/splenic) (YAC target)	↓	

	Parameter	Result	Reference
	Lymphocyte (splenic and thymic) mitogen responsiveness	↑	
	Total blood RBC levels	↑	Koller et al., 1977a
	Total blood white blood cell levels	N.E.	
	Antibody formation (vs SRBC):		Koller et al., 1979
	Primary response	→	
	Secondary response	↑	
	Antibody formation (vs SRBC):		
	Primary response	→	
	Secondary response	N.E./↓	
IP	Splenic lymphocyte microtubule integrity/assembly	→	Brown et al., 1988
	Splenic lymphocyte mitogen responsiveness	→	
SC	Thymic index	→	Hirokawa and Hayashi, 1980
	Thymic lymphocyte mitogen responsiveness	↑	
	Splenic index	↑	
	Splenic follicle size	→	
	Splenic PFC formation	→	
	Splenic lymphocyte mitogen responsiveness	→	
Rabbit / Diet	Antibody formation (vs A/PR8 virus):		Koller et al., 1977b
	Primary response	N.E./↓	
	Secondary response	N.E./↓	
Rat pups / Diet	Placentally exposed only:		Ilback et al., 1991
	Thymic and splenic total cell levels	N.E.	
	Thymic and splenic index	N.E.	
	NK cell activity (splenic) (YAC-1 target)	N.E.	
	Lymphocyte (splenic and thymic) mitogen responsiveness	N.E.	
	Total circulating white blood cell levels	↑	

Table 6 (cont.)

In vivo exposures

Compound	Means of exposure	Species	Immune parameter analyzed	Effects	Citation
			Milk-exposed only:		Ilback et al., 1991
			Total circulating white blood cell levels	N.E.	
			Thymic total cell levels	N.E.	
			Thymic index	N.E.	
			Thymic lymphocyte mitogen responsiveness	↑	
			Splenic total cell levels	N.E.	
			Splenic index	→	
			Splenic lymphocyte mitogen responsiveness	→	
			NK cell activity (splenic) (YAC-1 target)	N.E.	
			Milk- and placentally exposed:		
			Total blood white cell levels	N.E.	
			Thymic and splenic total cell levels	N.E.	
			Thymic and splenic index	N.E.	
			NK cell activity (splenic) (YAC-1 target)	→	
			Splenic lymphocyte mitogen responsiveness	→	
			Thymic lymphocyte mitogen responsiveness	↑	

In vitro exposures

Compound	Host cell	Immune parameter analyzed	Effects	Citation
Mercuric chloride	Guinea pig	Lymphocyte (splenic, thymic, and lymph node) proliferation	↑	Nordlind, 1982
		Lymphocyte (splenic, thymic, and lymph node) proliferation	↑/↓	Nordlind, 1983
	Mouse	PEM viability	→	Christensen et al., 1988
		PEM random migration	→	
		PEM phagocytosis	N.E./↓	

Measure	Result	Reference
PEM viability	→	Christensen et al., 1993a
PEM random migration	→	
PEM phagocytosis	→	
PEM IFN-α production	→	
PEM viability	→	Christensen et al., 1993b
PEM random migration	→	
PEM phagocytosis	→	
PEM morphological changes	←	
Splenic B-lymphocyte viability	N.E.	Daum et al., 1993
B-lymphocyte surface marker expression (ICAM, LFA-1, MHC Class I, Class II)	N.E.	
B-lymphocyte RNA and DNA synthesis	→	
LPS-stimulated B-lymphocyte cell cycling	→	
B-lymphocyte Ig isotype expression (IgM, IgG$_1$, IgG$_{2a}$, IgG$_{2b}$ and IgG$_3$)	→	
PEM viability	→	Ellermann-Eriksen et al., 1994
PMA-induced PEM ROI production	→	
HSV-2-induced PEM ROI production	→	
HSV-2-induced PEM TNF-α/IFN-α production	→	
PMA-induced PEM ROI production	→	Lison et al., 1988
PEM lactate dehydrogenase release	←	
PEM plasminogen activator levels	↑/↓	
Splenic lymphocyte mitogen responsiveness	→	Nakatsuru et al., 1985
Splenic PFC formation (vs TNP-SRBC)	→	
Splenic MLR activity	→	

Table 6 (cont.)

In vitro exposures

Compound	Host cell	Immune parameter analyzed	Effects	Citation
		LPS toxicity toward PEM	N.E./↓	Patierno et al., 1983
		PEM phagocytic index	N.E.	Tam and Hindsill, 1984
		PEM phagocytic capacity	N.E.	
		PEM intracellular killing	N.E.	
		PEM glass adherence capacity	N.E.	
	Rabbit	PAM ROI production	→	Geertz et al., 1994
		PAM leukotriene B4 (LTB$_4$) production	↑	Kudo and Waku, 1994
		PAM free arachidonic acid levels	↑	
		Zymosan-induced PAM O_2^- production	N.E./↓	Labedzka et al., 1989
		Zymosan-induced PAM H_2O_2 production	→	
		Zymosan-induced PAM chemiluminescence	→	
		Lactate dehydrogenase release	N.E.	
	Rat	PAM O_2 consumption (stimulated)	→	Castranova et al., 1980
		PAM ROI production	→	
		PAM viability	→	
		Strain-dependent effects:		
		PEM ROI production	↓/↑	Contrino et al., 1992
		PEM phagocytic activity	N.E./↓	
		Peritoneal neutrophil ROI production	N.E./↓/↑	
		Peritoneal cell (non-elicited) ROI production:		
		Preincubated before Hg added	↑	
		Without preincubation before Hg added	→	
		T-lymphocyte (splenic) Ca^{2+} levels:		Tan et al., 1993
		extracellular Ca^{2+} present	↑	
		extracellular Ca^{2+} absent	→	

Measurement	Effect	Reference
Splenic IFN-γ-producing cells	→	van der Meide et al., 1993
Splenocyte viability	→	
Human Neutrophil chemiluminescence	→	Baginski, 1988
Neutrophil phagocytic activity	→	
Neutrophil lysozyme release	←	
Neutrophil β-glucuronidase release	N.E.	
Neutrophil viability	N.E.	
Neutrophil ROI production	←	Jansson and Harms-Ringdahl, 1993
fMLP-induced neutrophil ROI production	←	
U937 macrophage cell line tissue factor activity	←	Kaneko et al., 1994
Cytosolic Ca²⁺ levels	←	Kimata et al., 1983
PBL PFC (IgE) formation:		
−PWM	N.E.	
+PWM	←	
PBL PFC (IgA, IgG, IgM) formation:		
±PWM	N.E.	
Neutrophil O_2 consumption (stimulated)	→	Malamud et al., 1985
Neutrophil chemiluminescence	→	
fMLP-/zymosan-induced neutrophil ROI production	→	
Neutrophil viability	N.E./↓	
Neutrophil viability	→	Obel et al., 1993
Neutrophil ROI production	→	
Neutrophil chemotaxis	→	
PBL (B-lymphocyte) mitogen responsiveness	→	Shenker et al., 1992a

Table 6 (cont.)

In vitro exposures

Compound	Host cell	Immune parameter analyzed	Effects	Citation
		PBL (B-/T-lymphocyte) GSH, GSSG content	→	
		PBL (T-lymphocyte) GSH/GSSG ratio	↑	
		PBL (B-lymphocyte) GSH/GSSG ratio	N.E.	
		PBL (B-/T-lymphocyte) GSHPX and GSHRX enzyme activities	N.E.	
		PBM GSH content	→	
		PBM GSSG content	N.E.	
		PBM GSH/GSSG ratio	→	
		PBM GSHPX and GSHRX eznyme activities	N.E.	Shenker *et al.*, 1992b
		PBL (T-lymphocyte) viability	→	
		PBL (T-lymphocyte) membrane integrity	→	
		PBL (T-lymphocyte) energy metabolism	N.E.	
		PBL (T-lymphocyte) Ca^{2+} levels	↑	
		PBL (T-lymphocyte) nuclear changes	↑	
		PBM viability/membrane integrity	→	
		PBM Ca^{2+} levels/nuclear changes	↑	
		PBM energy metabolism	→	
		PBL (B-lymphocyte) viability	→	
		PBL (B-lymphocyte) membrane integrity	→	
		PBL (B-lymphocyte) mitogen-induced CD23 and transferrin receptor expression	→	
		PBL (B-lymphocyte) mitogen-induced CD69 expression	N.E.	
		PBL (B-lymphocyte) energy metabolism	N.E.	
		PBL (B-lymphocyte) Ca^{2+} levels	↑	
		PBL (B-lymphocyte) nuclear changes	↑	Shenker *et al.*, 1993a

Compound	Species	Parameter	Effect	Reference
Methyl mercury chloride		PBL (B-lymphocyte) mitogen responsiveness	→	Shenker et al., 1993b
		PBL (B-lymphocyte) IgG and IgM production	→	
		PBL (T-lymphocyte) IL-2 production	→	Brown et al., 1988
		PBL (T-lymphocyte) mitogen-induced CD69, and IL-2 and transferrin receptors expression	→	
	Mouse	Splenic lymphocyte microtubule integrity/assembly	→	
		Splenic lymphocyte mitogen responsiveness	→	Christensen et al., 1993a
		PEM viability	→	
		PEM random migration	→	
		PEM phagocytosis	→	
		PEM IFN-α production	→	
		Splenic T-lymphocyte mitogen responsiveness	↑/N.E.	Hultman and Enestrom, 1992
		Splenic B-lymphocyte mitogen responsiveness	↑/↓	
		Splenic lymphocyte mitogen responsiveness	→	Nakatsuru et al., 1985
		Splenic MLR activity	→	
		Splenic PFC formation (vs TNP-SRBC)	N.E./↓	
	Rat	Splenic T-lymphocyte Ca^{2+} levels:		Tan et al., 1993
		extracellular Ca^{2+} present	←	
		extracellular Ca^{2+} absent	←	
	Human	Neutrophil viability	→	Obel et al., 1993
		Neutrophil ROI production	→	
		Neutrophil chemotaxis	→	

Table 6 (*cont.*)

In vitro exposures

Compound	Host cell	Immune parameter analyzed	Effects	Citation
		PBL (B-lymphocyte) mitogen responsiveness	→	Shenker et al., 1993a
		PBL (B-lymphocyte) IgG, IgM production	→	
		PBL (B-lymphocyte) viability	→	
		PBL (B-lymphocyte) energy metabolism	N.E.	
		PBL (B-lymphocyte) mitogen-induced CD69 expression	→	
		PBL (B-lymphocyte) mitogen-induced CD23 and transferrin receptor expression	→	
		PBL (B-lymphocyte) Ca^{2+} levels	←	
		PBL (B-lymphocyte) nuclear changes	←	
		PBL (T-lymphocyte) IL-2 production	→	Shenker et al., 1993b
		PBL (T-lymphocyte) mitogen-induced CD69, and IL-2 and transferrin receptors expression	→	
		PBL (T-lymphocyte) mitogen responsiveness	→	Shenker et al., 1992b
		PBL (T-lymphocyte) viability	→	
		PBL (T-lymphocyte) membrane integrity	→	
		PBL (T-lymphocyte) energy metabolism	→	
		PBL (T-lymphocyte) Ca^{2+} levels	←	
		PBL (T-lymphocyte) nuclear changes	←	
		PBM viability	→	
		PBM membrane integrity	→	
		PBM energy metabolism	→	
		PBM Ca^{2+} levels	←	
		PBM nuclear changes	←	
Phenyl mercuric acetate	Human	U937 macrophage cell line tissue factor activity	↑	Kaneko et al., 1994

Table 7 Immunomodulating effects of nickel compounds

In vivo exposures

Compound	Means of exposure	Species	Immune parameter analyzed	Effects	Citation
Nickel acetate	SC	Rat	Antibody formation (vs *Escherichia coli* T-1 phage)	↓	Figoni and Treagan, 1975
Nickel chloride	Inhalation	Hamster	*Ex situ* ciliary beating in tracheal rings	↓	Adalis *et al.*, 1978
	Inhalation	Mouse	Mortality due to *Streptococcus pyogenes*	N.E./↑	Adkins *et al.*, 1979
			Clearance of viable *Streptococcus pyogenes*	↓	
			Mortality due to murine cytomegalovirus	N.E.	Daniels *et al.*, 1987
			Splenic NK cell activity (YAC-1 target)		
			Uninfected host	N.E.	
			Virally infected host	N.E.	
			Splenic PFC formation	↓	Graham *et al.*, 1978
	IM		Mortality due to murine cytomegalovirus	N.E./↑	Daniels *et al.*, 1987
			Splenic NK cell activity (YAC-1 target)		
			Uninfected host	N.E.	
			Virally infected host	↓	
			Serum IFN levels in virally infected host	N.E.	
			Splenic IFN levels in virally infected host	N.E.	
			Splenic PFC formation	N.E./↓	Graham *et al.*, 1975b
			Spleen weight	N.E./↓	
			Splenic index	N.E./↓	

Table 7 (*cont.*)

In vivo exposures

Compound	Means of exposure	Species	Immune parameter analyzed	Effects	Citation
			Splenic PFC formation	↓	Graham *et al.*, 1978
			Splenic NK cell activity (YAC-1 target):		Rogers *et al.*, 1983
			Simultaneous injection:		
			Ni alone	→	
			Ni + Mn²⁺	↑	
			Time-spaced injections:		
			Ni–24 h → Mn²⁺	N.E.	
			Mn²⁺–24 h → Ni	↑	
			Spleen cellularity	N.E./↓	Smialowicz *et al.*, 1984
			Spleen cell viability	N.E.	
			Ex vivo splenic NK cell activity (YAC-1 target)	→	
			In vivo splenic NK cell activity (YAC-1 target)	↑	
			In vivo pulmonary NK cell activity (YAC-1 target)	→	
			Suppressor cell levels (splenic)	N.E.	
			Resistance against viable B16F10 tumor cells	→	
			Using single injection of NiCl₂:		
			Splenic index	N.E./↓	
			Thymic index	N.E./↓	
			ConA-induced splenic lymphoproliferation:		
			1 day post-Ni	→	
			8 days post-Ni	N.E.	

PHA-induced splenic lymphoproliferation:	
1 day post-Ni	→
8 days post-Ni	N.E.
LPS-induced splenic lymphoproliferation:	
1 or 8 days post-Ni	N.E.
Splenic NK cell activity (YAC-1 target):	
1 day post-Ni	N.E./↓
3 days post-Ni	N.E.
Mortality from *Salmonella typhimurium* endotoxin	N.E.
Percentage splenic T-lymphocytes	→
Percentage splenic B-lymphocytes	N.E.
Splenic PFC formation	N.E./↓
Splenic PFC formation (vs polyvinyl pyrrolidine)	N.E.
Ex vivo PEM phagocytic activity (using chicken RBC)	
With mouse anti-CRBC	N.E.
Without mouse anti-CRBC	N.E.
Using multiple injections of NiCl₂:	
ConA-induced splenic lymphocyte proliferation (3 days post-Ni)	N.E./↑
PHA-induced splenic lymphocyte proliferation (3 days post-Ni)	N.E.
LPS-induced splenic lymphocyte proliferation (3 days post-Ni)	↑

Table 7 (cont.)

In vivo exposures

Compound	Means of exposure	Species	Immune parameter analyzed	Effects	Citation
			PWM-induced splenic lymphocyte proliferation (3 days post-Ni)	N.E./↑	Smialowicz et al., 1985
			Spleen cellularity	N.E./↓	
			Spleen cell viability	N.E.	
			Ex vivo splenic NK cell activity (YAC-1 target)	→	
			In vivo splenic NK cell activity (YAC-1 target)	↑	
			In vivo pulmonary NK cell activity (YAC-1 target)	→	
			Suppressor cell levels (splenic)	N.E.	
			Resistance against viable B16F10 tumor cells	→	
			Spleen weight	N.E./↓	Smialowicz et al., 1987b
			Splenic index	N.E./↓	
			Thymus weight	N.E./↓	
			Thymic index	N.E./↓	
			Splenic NK cell activity (YAC-1 target):		
			Simultaneous injection/single site:		
			Ni alone	→	
			Ni + Mn^{2+}	←	
			Ni + Mg^{2+}	N.E.	
			Ni + Zn^{2+}	N.E.	
			Ni + Ca^{2+}	N.E.	
			Simultaneous injection/separate sites:		
			Ni alone	→	

		Parameter	Effect	Reference
		Ni + Mn^{2+}	→	
		Ni + Mg^{2+}	→	
		Ni + Zn^{2+}	→	
		Ni + Ca^{2+}	→	
		Time-spaced injections/separate sites:		
		Ni–0.5h → Mn^{2+}	→	
		Mn^{2+}–0.5h → Ni	→	
		Mn^{2+}–1.0h → Ni	←	
		Mn^{2+}–2.0h → Ni	←	
		Mn^{2+}–4.0h → Ni	←	
		Mn^{2+}–6.0h → Ni	←	
IP		Splenic NK cell activity (YAC-1 target)		Smialowicz *et al.*, 1984
		1 day post-Ni	→	
		2 days post-Ni	→	
		3 days post-Ni	N.E./↓	
In utero	Mouse pup	Spleen weight	N.E.	Smialowicz *et al.*, 1986
		Splenic index	N.E.	
		Thymus weight	N.E.	
		Thymic index	N.E./↓	
		ConA-induced splenic lymphocyte proliferation	N.E.	
		PHA-induced splenic lymphocyte proliferation	N.E.	
		LPS-induced splenic lymphocyte proliferation	N.E./↓	
		Splenic MLR activity	N.E.	
		Splenic NK cell activity (YAC-1 target)	N.E./↓	
		Splenic PFC formation	N.E./↑	
		Resistance against viable B16F10 tumor cells	N.E.	

Table 7 (cont.)

In vivo exposures

Compound	Means of exposure	Species	Immune parameter analyzed	Effects	Citation
	Per os	Mouse	Resistance to encephalomyocarditis virus	↓	Gainer, 1977
			In situ induction of IFN	↓	
	Inhalation	Rabbit	PAM phagocytic activity	↑	Wiernik *et al.*, 1983
			PAM size (in BAL)	↑	
			PAM ultrastructural changes (presence of microvilli, vacuoles, and laminated structures)	↑	
			Ex vivo PAM NBT reduction (basal)	↑	
			Ex vivo PAM NBT reduction (particle-induced)	↑	
			Ex vivo PAM bactericidal activity	N.E./↓	
			PAM containing laminated structures	↑	Johansson *et al.*, 1987
			Recovered PAM lysozyme level	↓	
			BAL fluid lysozyme level	↓	Lundborg and Camner, 1982
			BAL fluid lysozyme level	↓	Lundborg and Camner, 1984
			PAM levels in BAL	↑	
			Recovered PAM lysozyme level	↓	
			PAM lysozyme release (cultured post-harvest)	↓	
	Inhalation	Rat	PAM levels in BAL	N.E./↓	Adkins *et al.*, 1979
			PAM viability in BAL	N.E.	

Route	Parameter	Effect	Reference
	Ex vivo PAM phagocytic activity:		
	Adherent cells	↑/↓	
	Suspended cells	N.E./↓	
	PAM levels in BAL	N.E.	
	PAM size	N.E.	Bingham *et al.*, 1972
IM	Spleen weight	N.E./↓	
	Thymus weight	N.E.	
	Thymic and splenic index	N.E.	
	Ni-induced splenic lymphocyte proliferation	N.E.	
	ConA-induced splenic lymphocyte proliferation	N.E.	
	PHA-induced splenic lymphocyte proliferation	N.E.	
	PWM-induced splenic lymphocyte proliferation	N.E.	
	Splenic NK cell activity (YAC-1 target)	↓	
	Splenic PFC formation	N.E.	
	Mortality due to viable MADB106 tumor cells	N.E.	Smialowicz *et al.*, 1987a
IT	In BAL fluid:		
	Total protein levels	N.E./↑	
	Sialic acid levels	N.E./↑	
	Lactate dehydrogenase activity	N.E./↑	
	β-glucuronidase activity	N.E./↑	
	GSHPX activity	N.E.	
	GSHRX activity	N.E./↑	
	Total nucleated cells	N.E./↑	
	Neutrophil levels	N.E./↑	
	PAM levels	N.E./↑	Benson *et al.*, 1986b

Table 7 (cont.)

In vivo exposures

Compound	Means of exposure	Species	Immune parameter analyzed	Effects	Citation
	SC		Lymphocyte levels	N.E.	Knight et al., 1987
			Eosinophil levels	N.E.	
			Thymus weight	↓	
			Thymic index	↓	
			Thymus gland lymphoid depletion	N.E./↑	
			Thymic cortex lymphocyte levels	N.E./↓	
			Thymus gland lipoperoxide levels	N.E./↑	
			Thymus cortex lymphocyte pyknosis	N.E./↑	
			Thymic cortex lymphocyte levels	↓	Milicivec and Milicevic, 1989
			Thymic medulla lymphocyte levels	←	
			PAM cAMP levels (1–4h post-injection)	↑/↓	
			Thymus weight	↓	Sunderman et al., 1985
			Liver and lung tissue peroxide levels	↑	
			PAM levels in BAL (1–4h post-injection)	N.E.	Sunderman et al., 1989
			PAM viability (1–4h post-injection)	N.E.	
			PAM cAMP levels (1–4h post-injection)	↑/↓	
			PAM 5'-nucleotidase activity levels (1–4h post-injection)	↓	
			Serum 5'-nucleotidase activity levels (1–4h post-injection)	N.E.	

Substance	Route	Species	Endpoint	Result	Reference
			PAM levels in BAL (≥24h post-injection)	N.E./↓	
			PAM viability (≥24h post-injection)	N.E./↓	
			PAM cAMP levels (≥24h post-injection)	N.E.	
			PAM 5′-nucleotidase activity levels (≥24h post-injection)	↓/N.E.	
			Ex vivo PAM phagocytic activity (≥24h post-injection)	N.E./↓	
			PAM malondialdehyde levels (≥24h post-injection)	N.E./↑	
			Serum 5′-nucleotidase activity levels (≥24h post-injection)	↓	
			Plasma malondialdehyde levels (>24h post-injection)	↑	Camner et al., 1978
Nickel (metallic)	Inhalation	Rabbit	Ex vivo PAM phagocytic activity	↑	
			PAM size (in BAL)	↑	
			PAM ultrastructural changes (presence of microvilli, vacuoles, and laminated structures)	↑	Jarstrand et al., 1978
			Ex vivo PAM phagocytic activity	↑	
			PAM size (in BAL)	↑	
			Ex vivo PAM basal NBT reduction	↑	
			Ex vivo PAM induced NBT reduction	↑	Johansson et al., 1980
			PAM size (in BAL)	↑	
			PAM surface smoothing	↑	
			PAM ultrastructural changes (presence of microvilli and lamellar bodies)	↑	
			Ex vivo PAM basal NBT reduction	↑	
			Ex vivo PAM induced NBT reduction	↑	

Table 7 (cont.)

In vivo exposures

Compound	Means of exposure	Species	Immune parameter analyzed	Effects	Citation
			Ex vivo PAM NBT reduction (increase above resting levels)	↓	
			Presence of foamy PAM in alveoli	↑	Johansson *et al.*, 1981
			Ex vivo PAM phagocytic activity	N.E.	Johansson *et al.*, 1983
			Ex vivo PAM basal NBT reduction	↑	
			Ex vivo PAM induced NBT reduction	N.E.	
			Ex vivo PAM NBT reduction (increase above resting levels)	↓	
			BAL fluid lysozyme activity	↓	Lundborg and Camner, 1982
	IM	Rat	Tumor incidence after single Ni injection *and* IFN-γ treatment (10 weeks post-Ni)	N.E.	Judde *et al.*, 1987
			PBM NK cell activity (YAC-1 target):		
			Without IFN-γ treatment (10 weeks post-Ni)	↓/N.E.	
			With IFN-γ treatment (10 weeks post-Ni)	↓/N.E.	
Nickel oxide	IT	Hamster	Mortality due to influenza A/PR/8 virus		Port *et al.*, 1975
			Ni exposure 1 h post-infection	N.E.	
			Ni exposure 24 h post-infection	↑	
			Ni exposure 48 h post-infection	N.E./↑	
			Ni exposure prior to infection	↓	

Route	Species	Parameter	Effect	Reference
Inhalation	Mouse	Neutrophil and PAM infiltration into virally infected lungs	↑	Benson et al., 1989
		Total protein levels, and lactate dehydrogenase and β-glucuronidase activity in BAL fluid	↑	
		Total nucleated cells in BAL	↓/↑	Haley et al., 1990
		Percentage of neutrophils in BAL	↑	
		Percentage of PAM in BAL	↓	
		Nucleated cells in BAL	N.E./↑	
		PAM and neutrophil levels in BAL	N.E./↑	
		Nucleated cells in lung-associated lymph nodes	N.E./↑	
		Nucleated cells in spleen	N.E.	
		Lymph node PFC formation	N.E./↑	
		Splenic PFC formation	↓/N.E.	
		Splenic MLR activity	N.E.	
		Ex vivo PAM phagocytic activity	↓	
		Ex vivo PEM phagocytic activity	N.E.	
		Resistance against viable B16F10 tumor cells	N.E.	
		Splenic NK cell activity (YAC-1 target)	N.E.	
IM		Splenic PFC formation	N.E.	Graham et al., 1978
Inhalation	Rat	Total protein levels, and lactate dehydrogenase and β-glucuronidase activity in BAL fluid	↑	Benson et al., 1989
		Total nucleated cells in BAL	↑	
		Percentage of neutrophils in BAL	↑	
		Percentage of PAM in BAL	↓	
		PAM levels/size in BAL	↑	Bingham et al., 1972

Table 7 (cont.)

In vivo exposures

Compound	Means of exposure	Species	Immune parameter analyzed	Effects	Citation
			Ultrastructural changes in recovered PAM	↑	Murthy and Niklowitz, 1983
			Ultrastructural changes in recovered PAM	↑	Migally et al., 1982
			PAM levels in BAL (4 week exposure)	N.E./↓	Spiegelberg et al., 1984
			Granulocyte and lymphocyte levels in BAL (4 month exposure)	↑	
			PAM levels in BAL (4 month exposure)	↑/↓	
			PAM size in BAL (4 week exposure)	N.E./↑	
			PAM size in BAL (4 month exposure)	↑	
			Polynucleated PAM (4 week or 4 month exposure)	↑	
			Ex vivo PAM phagocytic activity (4 week or 4 month exposure)	↑	
			Serum antibody levels (vs SRBC) (4 week or 4 month exposure)	→	
			Splenic PFC formation (4 week or 4 month exposure)	→	
	IT		In BAL fluid:		Benson et al., 1986b
			Total nucleated cells	N.E.	
			Neutrophil levels	N.E.	
			PAM levels	N.E.	
			Lymphocyte levels	N.E.	
			Eosinophil levels	N.E.	

Substance	Route	Species	Endpoint	Effect	Reference
Nickel oxide (Ni$_2$O$_3$)	IM	Rat	Total protein levels	N.E./↑	Judde et al., 1987
			Sialic acid levels	N.E.	
			Lactate dehydrogenase activity	N.E.	
			β-glucuronidase activity	N.E.	
			GSHPX activity	N.E.	
			GSHRX activity	N.E.	
Nickel subsulfide	IT	Monkey	PBM NK cell activity (YAC-1 target)	↑	Haley et al., 1987
			Lung PFC formation	N.E.	
			Lung NK cell conjugate formation	N.E.	
			Lung NK cell activity (K562 target)	↑	
			Ex vivo PAM phagocytic activity	↓	
	Inhalation	Mouse	Bronchial lymph node lymphoid depletion	N.E./↑	Benson et al., 1987
			Thymus gland lymphoid depletion	N.E./↑	
			Thymic atrophy	N.E./↑	
			Splenic lymphoid depletion	N.E./↑	
			Splenic atrophy	N.E./↑	
			Splenic NK cell activity (YAC-1 target)	N.E.	
			In BAL fluid:		Benson et al., 1989
			Total protein levels, and lactate dehydrogenase and β-glucuronidase activity	↑	
			Total nucleated cells	↑	
			Percentage of neutrophils	↑	
			Percentage of PAM	↓/N.E.	
			Nucleated cells in BAL	N.E./↑	Haley et al., 1990
			PAM and neutrophil levels in BAL	N.E./↑	
			Nucleated cells in lung-associated lymph nodes	N.E./↑	
			Nucleated cells in spleen	N.E.	

Table 7 (cont.)

In vivo exposures

Compound	Means of exposure	Species	Immune parameter analyzed	Effects	Citation
	Inhalation	Rat	Lymph node PFC formation	N.E./↑	
			Splenic PFC formation	N.E./↓	
			Splenic MLR activity	N.E./↓	
			Ex vivo PAM phagocytic activity	→	
			Ex vivo PEM phagocytic activity	N.E.	
			Resistance against viable B16F10 tumor cells	N.E.	
			Splenic NK cell activity (YAC-1 target)	N.E./↓	Benson *et al.*, 1987
			Bronchial lymph node lymphoid depletion	N.E./↑	Benson *et al.*, 1989
			Thymus gland lymphoid depletion	N.E./↑	
			Thymic atrophy	N.E./↑	
			Splenic lymphoid depletion	N.E./↑	
			Splenic atrophy	N.E./↑	
			In BAL fluid:		
			Total protein levels, and lactate dehydrogenase and β-glucuronidase activity	↑	
			Total nucleated cells	↑	
			Percentage of neutrophils	↑	
			Percentage of PAM	→	
	IM		PBM NK cell activity (YAC-1 target):		
			Without IFN-γ treatment (10 weeks post-Ni)	↓	Judde *et al.*, 1987
			With IFN-γ treatment (10 weeks post-Ni)	N.E.	

		PBM NK cell activity (YAC-1 target)	N.E.	Kasprzak et al., 1987
		Splenic NK cell activity (YAC-1 target)	N.E.	
		Injection site NK cell and cytotoxic- and/or suppressor-T-lymphocyte levels	N.E.	
	IT	In BAL fluid:		Benson et al., 1986b
		Total protein levels	N.E./↑	
		Sialic acid levels	N.E./↑	
		Lactate dehydrogenase activity	N.E./↑	
		β-glucuronidase activity	N.E./↑	
		GSHPX activity	N.E./↑	
		GSHRX activity	N.E./↑	
		Total nucleated cells	N.E./↑	
		Neutrophil levels	N.E./↑	
		PAM levels	N.E./↑	
		Lymphocyte levels	N.E.	
		Eosinophil levels	N.E.	
Nickel sulfate	Diet	Mouse	Serum antibody (IgG and IgM) levels (vs KLH)	→	Schiffer et al., 1991
		KLH-induced splenic lymphocyte proliferation	→	
		LPS-induced splenic lymphocyte proliferation	→	
		ConA-induced splenic lymphocyte proliferation	N.E.	
	Epicutaneous	Host sensitization (adjuvant present)	↑	Ishii et al., 1995
	Inhalation	Mortality due to Streptococcus pyogenes	↑	Adkins et al., 1979
		Bronchial lymph node lymphoid depletion	N.E./↑	Benson et al., 1988a

Table 7 *(cont.)*

In vivo exposures

Compound	Means of exposure	Species	Immune parameter analyzed	Effects	Citation
			Bronchial lymph node lymphoid hyperplasia	N.E.	
			Thymus gland lymphoid depletion	N.E./↑	
			Thymic and splenic atrophy	N.E./↑	
			Splenic lymphoid depletion	N.E./↑	
			Splenic NK cell activity (YAC-1 target)	N.E.	Benson *et al.*, 1989
			In BAL fluid:		
			Total protein levels, and lactate dehydrogenase and β-glucuronidase activity	↑	
			Percentage of neutrophils	↑/N.E.	
			Percentage of PAM	↓/N.E.	
			Nucleated cells in BAL	N.E./↑	
			PAM and neutrophil levels in BAL	N.E./↑	
			Nucleated cells in lung-associated lymph nodes	N.E./↑	Haley *et al.*, 1990
			Nucleated cells in spleen	N.E.	
			Lymph node PFC formation	N.E.	
			Splenic PFC formation	N.E.	
			Splenic MLR activity	N.E.	
			Ex vivo PAM phagocytic activity	↑/↓	
			Ex vivo PEM phagocytic activity	N.E.	
			Resistance against viable B16F10 tumor cells	N.E.	
			Splenic NK cell activity (YAC-1 target)	N.E.	

Route	Parameter	Effect	Reference
ID	Host sensitization (adjuvant present)	↑	van Hoogstraten et al., 1993
IM	Splenic PFC formation	↓	Graham et al., 1978
Per os	Liver weight	↓	Dieter et al., 1988
	Liver index	N.E.	
	Spleen weight	N.E./↓	
	Splenic index	N.E.	
	Thymus weight	↓	
	Thymic index	↓	
	Splenic and thymic atrophy	↑	
	Splenic PFC formation	N.E./↑	
	ConA-induced splenic lymphocyte proliferation	N.E.	
	LPS-induced splenic lymphocyte proliferation	↓	
	Splenic NK cell activity (YAC-1 target)	N.E.	
	Mortality from viable *Listeria monocytogenes*	N.E.	
	Bone marrow cellularity	↓	
	Bone marrow cell G6PDH enzyme activity	↓	
	Bone marrow cell lactate and malate dehydrogenase enzyme activity	N.E.	
	Bone marrow stem cell proliferative response	↓	
	Bone marrow granulocyte–macrophage proliferative response	↓	
	Bone marrow granulocyte–macrophage G6PDH enzyme activity	↓	

Table 7 (cont.)

In vivo exposures

Compound	Means of exposure	Species	Immune parameter analyzed	Effects	Citation
	Inhalation	Rat	Bronchial lymph node lymphoid depletion	N.E./↑	Benson *et al.*, 1988b
			Bronchial lymph node lymphoid hyperplasia	↑	
			Mediastinal lymph node lymphoid depletion	N.E./↑	
			Mediastinal lymph node lymphoid hyperplasia	↑	
			Thymus gland lymphoid depletion	↑	
			Splenic lymphoid depletion	↑	
			In BAL fluid:		Benson *et al.*, 1989
			Total protein levels, and lactate dehydrogenase and β-glucuronidase activity	↑	
			Total nucleated cells	↑	
			Percentage of neutrophils	↑	
			Percentage of PAM	↓	
	IT		In BAL fluid:		Benson *et al.*, 1986b
			Total protein levels	N.E./↑	
			Sialic acid levels	N.E./↑	
			Lactate dehydrogenase activity	N.E./↑	
			β-glucuronidase activity	N.E./↑	
			GSHPX activity	N.E.	
			GSHRX activity	N.E./↑	
			Total nucleated cells	N.E./↑	

Nickel (unspecified)	Occupational and/or environmental	Human	Neutrophil levels	N.E./↑	Bencko et al., 1983
			PAM levels	N.E./↑	
			Lymphocyte levels	N.E.	
			Eosinophil levels	N.E.	
			Serum IgG, IgA, and IgM levels	↑	Waksvik et al., 1981
			Serum α_1-antitrypsin, transferrin, α_2-macroglobulin, and lysozyme levels	↑	
			PBL sister chromatid exchange incidence	N.E.	
			Using nickel-sensitive patients:		
			HLA-A, -B, -C, or -DR antigen expression	N.E.	Braathen et al., 1983
			Asthma events (NiSO$_4$-induced)	↑	Davies, 1986
			HLA-A, -B, -C, or -DR antigen expression	N.E.	Dumont-Fruytier et al., 1980
			NiSO$_4$-induced PBL proliferation	↑	Everness et al., 1990
			ConA-induced PBL proliferation	N.E.	
			Keratinocyte ICAM-1 expression	↑	Garioch et al., 1991
			Dermis lymphocyte LFA-1 expression	↑	
			Epidermis lymphocyte LFA-1 expression	↑	
			HLA-A or HLA-B antigen expression	N.E.	Hansen et al., 1982
			Nickel-specific T-lymphocytes: Requirement for HLA-DR/DQ molecules	↑	Kapsenberg et al., 1988

Table 7 (cont.)

In vivo exposures

Compound	Means of exposure	Species	Immune parameter analyzed	Effects	Citation
			TNF-α, GM-CSF, IL-2, and IFN-γ production	↑	Kapsenberg et al., 1991
			IL-4 and IL-5 production	N.E.	
			TNF-α, GM-CSF, IL-2, and IFN-γ production	↑	Kapsenberg et al., 1992
			IL-4 and IL-5 production	N.E.	
			HLA-A, -B, -C, or -DR antigen expression	N.E.	Karvonen et al., 1984
			Asthma events ($NiSO_4$-induced)	↑	Malo et al., 1982
			Asthma events ($NiSO_4$-induced)	↑	Malo et al., 1985
			Serum nickel-specific IgE levels	N.E.	
			HLA-DRw6 antigen expression	↑	Mozzanica et al., 1990
			Asthma events ($NiSO_4$-induced)	↑	Novey et al., 1983
			Serum nickel-specific IgE levels	↑	
			Nickel-specific T-lymphocytes: IFN-γ production	N.E.	Probst et al., 1995
			IL-4 and IL-5 production	↑	
			Asthma events (work dust-induced)	↑	Shirakawa et al., 1990
			Serum nickel-specific IgE levels	↑	
			Nickel-specific T-lymphocytes: CD3 and CD4 antigen expression	↑	Sinigaglia et al., 1985
			IL-2 and IFN-γ production	↑	
			HLA-B_{35} antigen expression	↑	Walton et al., 1986

Route	Host	Immune parameter analyzed	Effects	Citation
		In hard metal asthmatic patients:		
		ConA-induced PBL proliferation	N.E.	Kusaka et al., 1991
		NiSO$_4$-induced PBL proliferation	N.E./↑	
Inhalation	Rabbit	PAM levels in BAL	↑	Reichrtova et al., 1986
		PAM lysosomal enzyme (acid phosphatase and β-glucuronidase) activity	↑	
		Vacuolization in PAM cytoplasm	↑	
		PAM rosetting capacity	↓	
		Levels of tissue immune complexes	↑	Reichrtova et al., 1989
		Serum circulating immune complex levels	↑	
		Non-specific serum tumoricidal activity	↑	
Inhalation	Rat	Chromosomal aberration incidence in PAM	↑	Chorvatovicova and Kovacikova, 1992
		Chromosomal aberration incidence in bone marrow cells	↑	
IV		PBL lactate dehydrogenase activity	↓	Reichrtova et al., 1989
		Liver weight	↓	
		Spleen weight	↓	
		Allograft survival	↑	

In vitro exposures

Compound	Host cell	Immune parameter analyzed	Effects	Citation
Ferro-nickel alloy	Rat	PEM viability	↓	Kuehn et al., 1982
Nickel antimonide	Rat	PEM viability	N.E.	Kuehn et al., 1982
Nickel chloride	Dog	PAM viability	↓	Benson et al., 1986a

Table 7 (cont.)

In vitro exposures

Compound	Host cell	Immune parameter analyzed	Effects	Citation
	Goat	PAM viability	N.E./↓	Waseem et al., 1993
		PAM lactate dehydrogenase release	N.E.	
		PAM phagocytic activity	N.E./↓	
	Hamster	Ciliary beating in tracheal rings	↓	Adalis et al., 1978
	Mouse	PEM (non-elicited) adherence to glass:		Hernandez et al., 1991
		Male mice – early (10 min)	N.E./↑	
		Male mice – late (45 min)	↑	
		Female mice – early (10 min)	N.E./↑	
		Female mice – late (45 min)	N.E./↑	
		PEL (non-elicited) adherence to glass:		
		Male mice – early (10 min)	N.E./↑	
		Male mice – late (45 min)	N.E.	
		Female mice – early (10 min)	↑	
		Female mice – late (45 min)	N.E./↑	
		Splenic lymphocyte mitogen responsiveness:		Lawrence, 1981
		Spontaneous proliferation	↑/N.E./↓	
		To LPS	↑/N.E.	
		To 2-ME	↑	
		To ConA	↓/↑	
		To PHA	N.E./↓	
		Splenic lymphocyte PFC formation	↑/N.E./↓	
		Splenic lymphocyte MLR reactivity:		
		Ni-treated stimulator cells	N.E.	
		Ni-treated responder cells	↑	

Untreated stimulator and responder cells:		
Allogeneic stimulators	↑	Smith and Lawrence, 1988
Syngeneic stimulators	N.E.	
APC-induced DO-11.10 T-lymphocyte hybridoma IL-2 production:		
Pb present in reaction	N.E./↑	
Pb pretreatment of APC	N.E.	
ConA-induced splenocyte IL-2 production	N.E.	
IL-2-induced HT-2 lymphocyte cell line proliferation	N.E.	
DO-11.10 T-lymphocyte hybridoma viability	N.E.	
ConA-induced DO-11.10 T-lymphocyte hybridoma proliferation	N.E.	
Pb-induced DO-11.10 T-lymphocyte hybridoma proliferation	N.E.	
PEM (elicited) phagocytosis of Saccharomyces cerevisiae	N.E.	Tam and Hindsill, 1984
PEM (elicited) intracellular killing of Saccharomyces cerevisiae	N.E.	
Splenic lymphocyte PFC formation (vs SRBC)	↑	Warner and Lawrence, 1986a
Splenic lymphocyte spontaneous proliferation	↑	
Splenic lymphocyte spontaneous proliferation	↑	Warner and Lawrence, 1986b
Numbers of splenic lymphocytes cell cycling	↑	

Table 7 (cont.)

In vitro exposures

Compound	Host cell	Immune parameter analyzed	Effects	Citation
		Numbers of Thy$^+$, L3T4$^+$, and sIgM$^+$ cells	↑	Warner and Lawrence, 1988
		IL-2-induced HT-2 lymphocyte cell line proliferation	N.E./↓	
		ConA-induced splenocyte IL-2 production	N.E.	
		Ni-induced splenocyte IL-2 production	↑	
		Ni-induced splenocyte IL-2 receptor expression	↑	
		Ni-induced splenocyte proliferation:		
		Ni only	↑	
		Ni/anti-IL-2 antibody (vs Ni only)	→	
		Ni/anti-IL-2 receptor antibody (vs Ni only)	→	
		Ni/anti-IFN-γ antibody (vs Ni only)	→	
		Ni/anti-Thy1.2 antibody (vs Ni only)	N.E.	
	Rabbit	Zymosan-induced PAM O$_2^{-}$ production	N.E.	Graham *et al.*, 1975a
		Zymosan-induced PAM O$_2^{-}$ production	N.E.	Labedzka *et al.*, 1989
		Zymosan-induced PAM H$_2$O$_2$ production	N.E.	
		Zymosan-induced PAM chemiluminescence	N.E.	
		Lactate dehydrogenase release	N.E.	
		PAM adherence	N.E.	Lundborg *et al.*, 1987

	Endpoint	Result	Reference
	PAM lysozyme release (cultured post-harvest)	↓	
	BAL fluid lysozyme activity	N.E.	Lundborg and Camner, 1982
	PAM viability	↓	Waters et al., 1975
	PAM lysis	↑	
	PAM acid phosphatase activity	↓	
Rat	PAM viability	↓	Benson et al., 1986a
	Resting PAM O$_2$ consumption	↓	Castranova et al., 1980
	Resting PAM glucose metabolism	↓	
	Phagocytosing PAM O$_2$ consumption	↓	
	Phagocytosing PAM ROI formation/release	↓	
	PAM membrane integrity	N.E.	
	Liver and lung tissue peroxide levels	N.E.	Sunderman et al., 1985
Human	Vascular endothelial cell ICAM-1, VCAM-1, and E-selectin expression	↑	Goebeler et al., 1993
	Vascular endothelial cell NF-κB activation	↑	Goebeler et al., 1995
	Vascular endothelial cell IL-6 production	↑	
	Vascular endothelial cell IL-6 mRNA levels	↑	
	T-lymphoblast (HuT-78) cell line viability	↓	Malinin et al., 1992
	HuT-78 cell cycle progression	N.E.	
	Osmiophilia	↑	
Nickel-copper oxide			
Dog	PAM viability	↓	Benson et al., 1988b
Mouse	PAM viability	↓	Benson et al., 1988b
Rat	PAM viability	↓	Benson et al., 1988b

331

Table 7 *(cont.)*

In vitro exposures

Compound	Host cell	Immune parameter analyzed	Effects	Citation
Nickel disulfide	Rat	PEM viability	N.E.	Kuehn *et al.*, 1982
Nickel ferrosulfide	Rat	PEM viability	↓	Kuehn *et al.*, 1982
Nickel hydroxycarbonate	Guinea pig	Vacuolization in PAM	N.E./↑	Arsalane *et al.*, 1992
		PAM ATP content	↑/↓	
		PAM chemiluminescence:		
		Spontaneous	↑/N.E.	
		PMA-induced	N.E./↓	
		PAM total glutathione (GSH + GSSG) content	↓	
		PAM GSH content	↓	
		PAM GSSG content	N.E.	
		PAM catalase activity	N.E./↓	
		PAM superoxide dismutase activity:		
		Total activity	N.E./↓	
		Cu-Zn-dependent activity	N.E.	
		Mn-dependent activity	N.E.	
Nickel monoarsenide	Rat	PEM viability	N.E.	Kuehn *et al.*, 1982
Nickel monoselenide	Rat	PEM viability	N.E.	Kuehn *et al.*, 1982
Nickel oxide	Dog	PAM viability	↓	Benson *et al.*, 1988b
	Mouse	PAM viability	N.E.	Benson *et al.*, 1988b
	Rabbit	Zymosan-induced PAM $O_2^{.-}$ production	↑/N.E.	Labedzka *et al.*, 1989
		Zymosan-induced PAM H_2O_2 production	N.E./↓	

Substance	Species	Parameter	↓/N.E.	Reference
Nickel oxide (black)	Rat	Zymosan-induced PAM chemiluminescence	N.E.	Benson et al., 1988b
		Lactate dehydrogenase release	N.E.	Benson et al., 1988b
		PAM viability	N.E.	Kuehn et al., 1982
		PEM viability	N.E.	Kuehn et al., 1982
		PAM lactate dehydrogenase release	→	Takahashi et al., 1992
Nickel oxide (green)	Dog	PAM viability	↑	Benson et al., 1986a
	Rat	PAM viability	→	Benson et al., 1986a
		PAM viability	N.E./↓	Takahashi et al., 1992
		PAM lactate dehydrogenase release	N.E./↑	Takahashi et al., 1992
Nickel subarsenide	Rat	PEM viability	→	Kuehn et al., 1982
Nickel subselenide	Rat	PEM viability	N.E.	Kuehn et al., 1982
Nickel subsulfide	Cow	PAM viability	→	Finch et al., 1987
		PAM degranulation	↑	
		PAM membrane tearing	↑	
	Rat	PAM phagocytic activity	→	Fisher et al., 1986
		PAM viability and PAM adherence	N.E.	
	Rat	PEM viability	N.E.	Kuehn et al., 1982
	Human	PBL viability	N.E./↓	Zeromski et al., 1995
		CD4+ and NK cell levels	N.E./↓	
		CD3+, CD8+, CD11a+, and CD20+ cell levels	N.E.	
		PBL NK cell activity (K562 target)	→	
Nickel subsulfide (crystalline)	Dog	PAM viability	→	Benson et al., 1986a
	Rat	PAM viability	→	Benson et al., 1986a

Table 7 (cont.)

In vitro exposures

Compound	Host cell	Immune parameter analyzed	Effects	Citation
Nickel sulfarsenide	Rat	PEM viability	N.E.	Kuehn et al., 1982
Nickel sulfate	Dog	PAM viability	→	Benson et al., 1986a
	Rat	PAM viability	→	Benson et al., 1986a
	Human	Keratinocyte IL-1α and IL-1β production	↑	Gueniche et al., 1994a
		Keratinocyte ICAM-1 expression	↑	
		Keratinocyte TNF-α production	↑	Gueniche et al., 1994b
		Keratinocyte ICAM-1 expression	↑	
		Vascular endothelial cell ICAM-1, VCAM-1, and E-selectin expression	↑	Wildner et al., 1992
		Vascular endothelial cell HLA-DR expression	N.E.	
		PBL viability	N.E./↓	Zeromski et al., 1995
		CD4+ and NK cell levels	N.E./↓	
		CD3+, CD8+, CD11a+, and CD20+ cell levels	N.E.	
		PBL NK cell activity (K562 target)	→	
Nickel sulfide (amorphous)	Mouse	Embryo fibroblast polyI:C-induced IFN-α and IFN-β production: Without GSH present	N.E./↓	Sonnenfeld et al., 1983
		PolyI:C-induced L-929 fibroblast cell line IFN-α and IFN-β production	N.E.	Jaramillo and Sonnenfeld, 1989

Compound	Species	Endpoint	Effect	Reference
	Rat	ConA-induced splenic lymphocyte IL-2 and IFN-γ production	N.E.	Jaramillo and Sonnenfeld, 1989
		PEM viability	N.E.	Kuehn et al., 1982
Nickel sulfide (crystalline)	Mouse	Embryo fibroblast polyI:C-induced IFN-α and IFN-β production: Without GSH present / With GSH present	N.E./↓ / N.E./↓	Sonnenfeld et al., 1983
		PolyI:C-induced L-929 fibroblast cell line IFN-α and IFN-β production	N.E./↓	Jaramillo and Sonnenfeld, 1989
		PolyI:C-induced L-929 fibroblast cell line IFN-α and IFN-β production	N.E./↓	Jaramillo and Sonnenfeld, 1992a
		IFN-α/-β-induced L-929 antiviral activity	N.E.	
	Rat	ConA-induced splenic lymphocyte IL-2 and IFN-γ production	N.E.	Jaramillo and Sonnenfeld, 1989
		NiS-induced spleen cell proliferation	↑	
		ConA-induced splenic lymphoproliferation	↑	Jaramillo and Sonnenfeld, 1992b
		LPS-induced splenic lymphoproliferation	↑	
		NiS-induced splenic CD4+ T-lymphocyte proliferation	↑	
		NiS-induced splenic CD8+ T-lymphocyte proliferation	N.E.	

Table 7 (cont.)

In vitro exposures

Compound	Host cell	Immune parameter analyzed	Effects	Citation
		NiS-induced splenic B-lymphocyte proliferation	N.E.	Jaramillo and Sonnenfeld, 1993
		NiS-induced spleen cell IL-1/IL-2 production	↑	
		ConA-induced splenic lymphocyte IL-2 production	N.E.	
		ConA-induced PEM IL-1 production	N.E.	
		PEM phagocytic activity:		
		IgG-opsonized SRBC/PEM	N.E.	
		Phagocytic index	N.E./↓	
		Binding index	N.E./↑	
		PEM viability	N.E.	Kuehn et al., 1982
Nickel telluride	Rat	PEM viability	→	Kuehn et al., 1982
Nickel titanate	Rat	PEM viability	→	Kuehn et al., 1982
Nickel (unspecified)	Rabbit	PAM rosetting capacity	→	Reichrtova et al., 1986
	Rat	PEM viability	N.E.	Kuehn et al., 1982
		APC MHC–peptide complex transformation	↑	Sinigaglia, 1994

336

Table 8 Immunomodulating effects of platinum compounds

In vivo exposures

Compound	Means of exposure	Species	Immune parameter analyzed	Effects	Citation
Ammonium hexachloroplatinate	IP	Rat	IgE production (2° response to ovalbumin; Pt administered concurrently with antigen during 1° immunization):		Murdoch and Pepys, 1984b
			With *Bordetella pertussis* adjuvant	↑	
			Without *Bordetella pertussis* adjuvant	N.E.	
Ammonium tetrachloroplatinate	ID	Rat	IgE production (1° and 2° responses to ovalbumin–Pt conjugate or Pt compound):		Murdoch and Pepys, 1984a
			With *Bordetella pertussis* adjuvant	N.E.	
			Without *Bordetella pertussis* adjuvant	N.E.	
			With aluminum hydroxide adjuvant	N.E.	
			Without aluminum hydroxide adjuvant	N.E.	
	IM		IgE production (1° and 2° responses to ovalbumin–Pt conjugate or Pt compound):		Murdoch and Pepys, 1984a
			With *Bordetella pertussis* adjuvant	N.E.	
			Without *Bordetella pertussis* adjuvant	N.E.	
			With aluminum hydroxide adjuvant	N.E.	
			Without aluminum hydroxide adjuvant	N.E.	

Table 8 (cont.)

In vivo exposures

Compound	Means of exposure	Species	Immune parameter analyzed	Effects	Citation
	IP		IgE production (1° and 2° responses to ovalbumin–Pt conjugate or Pt compound):		Murdoch and Pepys, 1984a
			With *Bordetella pertussis* adjuvant	N.E.	
			Without *Bordetella pertussis* adjuvant	N.E.	
			With aluminum hydroxide adjuvant	N.E.	
			Without aluminum hydroxide adjuvant	N.E.	
			IgE production (2° response to ovalbumin; Pt administered concurrently with antigen during 1° immunization):		Murdoch and Pepys, 1984b
			With *Bordetella pertussis* adjuvant	↑	
			Without *Bordetella pertussis* adjuvant	N.E.	
			Allergic sensitization to ovalbumin:		Murdoch and Pepys, 1986
			With *Bordetella pertussis* adjuvant	↑	
			Without *Bordetella pertussis* adjuvant	N.E.	
			With aluminum hydroxide adjuvant	N.E.	
			Without aluminum hydroxide adjuvant	N.E.	
	IT		IgE production (1° and 2° responses to ovalbumin–Pt conjugate or Pt compound):		Murdoch and Pepys, 1984a

Compound	Species	Route	Parameter	Result	Reference
		SC	With *Bordetella pertussis* adjuvant	N.E.	Murdoch and Pepys, 1984a
			Without *Bordetella pertussis* adjuvant	N.E.	
			With aluminum hydroxide adjuvant	N.E.	
			Without aluminum hydroxide adjuvant	N.E.	
			IgE production (1° and 2° responses to ovalbumin–Pt conjugate or Pt compound):		
			With *Bordetella pertussis* adjuvant	N.E.	
			Without *Bordetella pertussis* adjuvant	N.E.	
			With aluminum hydroxide adjuvant	N.E.	
			Without aluminum hydroxide adjuvant	N.E.	
Carboplatin	Mouse	IP	Splenic NK cell activity (YAC-1 target)	N.E.	Lichtenstein and Pende, 1986
			Peritoneal NK cell activity (YAC-1 target)	N.E.	
Cesium trichloronitroplatinate	Rat	IP	IgE production (2° response to ovalbumin; Pt administered concurrently with antigen during 1° immunization):		Murdoch and Pepys, 1984b
			With *Bordetella pertussis* adjuvant	↑	
			Without *Bordetella pertussis* adjuvant	N.E.	
Cytoplatam	Mouse	IP	PEM phagocytic activity	↑	Brin and Uteshev, 1990

Table 8 *(cont.)*

In vivo exposures

Compound	Means of exposure	Species	Immune parameter analyzed	Effects	Citation
Diaminodichloroplatinate (CDDP; PDD; cisplatin)	IP	Mouse	Splenic PFC formation	↓	Berenbaum, 1971
			Splenic PFC formation	↓	Khan and Hill, 1971a
			GVH reaction	↓	Khan and Hill, 1972
			NK cell cytolytic activity:		Lichtenstein and Pende, 1986
			Splenic cell (YAC-1 or K562 target)	↑	
			Splenic cell (P815, EL4, or RAJI target)	N.E.	
			Peritoneal cell (YAC-1 or K562 target)	↑	
			Peritoneal cell (P815, EL4, or RAJI target)	N.E.	
			Host resistance to P815 or SL2 tumor cells	↑	Bernsen *et al.*, 1993
			IL-2-enhancement of tumor resistance	↑	
			Using tumor-bearing hosts:		Andrade-Mena *et al.*, 1985
			Splenic PFC formation	↓	
			DTH reaction	↓	
			Splenocyte Winn assay	↑	Bahadur *et al.*, 1984
			PEC Winn assay	↑	
	IP	Rat	Thymic size/cellularity	↓	Thompson and Gale, 1971
			Circulating leukocyte levels	↓	
			IgE production (2° response to ovalbumin; Pt administered	→	Murdoch and Pepys, 1984b

Compound	Route	Species	Parameter	Effect	Reference
			concurrently with antigen during 1° immunization):		
			With *Bordetella pertussis* adjuvant	N.E.	
			Without *Bordetella pertussis* adjuvant	N.E.	
Diaminotetrachloroplatinate	IV	Human	PHA-induced PBL blastogenesis	→	Khan and Hill, 1973
	IP	Mouse	Splenic PFC formation	→	Berenbaum, 1971
Disodium hexachloroplatinate	SC	Mouse	Antinuclear antibody levels	↑/N.E.	Schuppe et al., 1991
			Serum IgE levels	↑	
Ethylenedichloroplatinate	IP	Mouse	Splenic PFC formation	N.E./↓	Berenbaum, 1971
Ethylenetetrachloroplatinate	IP	Mouse	Splenic PFC formation	N.E./↓	Berenbaum, 1971
Iproplatin	IP	Mouse	Splenic NK cell activity (YAC-1 target)	↑	Lichtenstein and Pende, 1986
			Peritoneal NK cell activity (YAC-1 target)	↑	
Oxoplatinum	IP	Mouse	PEM phagocytic activity	↑	Brin and Uteshev, 1990
			Ex vivo T-dependent humoral response	→	Brin et al., 1990
			PHA-induced mitogenic responsiveness	↑	
			T_C-cell activity	↑	
Platinum sulfate	SC	Guinea pig	Allergic reactivity – skin test	N.E.	Rosner and Merget, 1990
	IV	Mouse	Allergic reactivity – footpad test	N.E.	Rosner and Merget, 1990
	IV	Rabbit	Allergic reactivity – skin test	N.E.	Rosner and Merget, 1990

Table 8 *(cont.)*

In vivo exposures

Compound	Means of exposure	Species	Immune parameter analyzed	Effects	Citation
Sodium ammonium hexachloroplatinate	SC	Mouse	Popliteal lymph node weight	↑	Schuppe *et al.*, 1992
			Popliteal lymph node cellularity	↑	
Sodium chloroplatinate	IV	Guinea pig	Serum histamine levels	↑	Saindelle and Ruff, 1969
			Ex vivo basophil histamine release	↑	
	IV	Dog	Serum histamine levels	↑	
			Ex vivo basophil histamine release	↑	
	IV	Rat	Serum histamine levels	↑	
			Ex vivo basophil histamine release	↑	
Tetraaminochloroplatinate	IP	Rat	IgE production (2° response to ovalbumin) (Pt administered concurrently with antigen during primary immunization):		Murdoch and Pepys, 1984b
			With *Bordetella pertussis* adjuvant	N.E.	
			Without *Bordetella pertussis* adjuvant	N.E.	
Transplatin	IP	Mouse	Splenic NK cell activity (YAC-1 target)	↑	Lichtenstein and Pende, 1986
			Peritoneal NK cell activity (YAC-1 target)	↑	

In vitro exposures

Compound	Host cell	Immune parameter analyzed	Effects	Citation
Carboplatin	Mouse	PHA-induced splenocyte blastogenesis	→	Vancurova et al., 1989
		Splenocyte PFC formation	→	
	Human	PHA-induced PBM blastogenesis	→	
		PWM-induced PBM Ig production	→	
Diaminodichloroplatinate (CDDP; PDD; cisplatin)	Mouse	PHA-induced splenocyte blastogenesis	→	Vancurova et al., 1989
		Splenocyte PFC formation	→	
	Human	PHA-induced PBL DNA synthesis	→	Howle et al., 1971
		PHA-induced lymphocyte blastogenesis	→	Khan and Hill, 1971b
		PBM tumor cell killing activity	↑	Kleinermann et al., 1980
		PBL-induced PBM tumoricidal activity	N.E.	
		PHA-induced PBM blastogenesis	→	Vancurova et al., 1989
		PWM-induced PBM Ig production	→	
Oxoplatinum	Mouse	PHA-induced splenocyte blastogenesis	→	Vancurova et al., 1989
		Splenocyte PFC formation	→	
	Human	PHA-induced PBM blastogenesis	→	
		PWM-induced PBM Ig production	→	

Table 9 Immunomodulating effects of vanadium compounds

In vivo exposures

Compound	Means of exposure	Species	Immune parameter analyzed	Effects	Citation
Ammonium metavanadate	IP	Mouse	PEM G6PDH enzyme activity	→	Cohen and Wei, 1988
			PEM GSHPX and GSHRX enzyme activities	→	
			PEM ROI production	→	
			PEM GSSG levels	↑	
			Resistance to *Listeria monocytogenes*	→	Cohen *et al.*, 1986
			Thymic index	→	
			Ex vivo PEM phagocytic activity	→	
			Splenic lymphocyte EAC rosette formation	↑/↓	
			Resistance to *Escherichia coli* endotoxin	↑	
			Splenic megakaryocyte levels	↑	
			Clearance of viable *Listeria* (spleen/liver)	→	Cohen *et al.*, 1989
			PEM uptake and killing of viable *Listeria*	→	
			PEM acid phosphatase activity	→	Vaddi and Wei, 1991a
			PEM lysosomal enzyme activity	N.E.	
			PEM $F_{c\gamma2a}/F_{c\gamma2b}$ expression	→	Vaddi and Wei, 1991b
			PEM $F_{c\gamma2a}$-mediated binding	N.E.	
			PEM $F_{c\gamma2b}$-mediated binding	→	
			LPS-inducible PEM Ca^{2+} influx	N.E.	Vaddi and Wei, 1996
			PEM resting Ca^{2+} levels	↑	

Compound	Route	Species	Parameter	Effect	Reference
	SC		Lymphoid necrosis	↑	Al-Bayati et al., 1992
			Lymphoid necrosis	↑	Wei et al., 1982
			Hepatic and splenic indices	↑	
	Inhalation	Rat	Neutrophil and PAM levels in BAL	↑	Cohen et al., 1996c
			PAM ROI production	↑	
			IFN-γ-primed PAM ROI production	→	
			PAM TNF-α production (inducible)	→	
			IFN-γ-inducible PAM Ia expression	→	
	Per os		Blood neutrophil and lymphocyte levels	↑	Zaporowska and Wasilewski, 1992
			Ex vivo neutrophil phagocytic activity	→	
	SC		Blood monocyte levels	↑	Al-Bayati et al., 1990
			Blood lymphocyte/eosinophil levels	N.E./↓	
			Blood neutrophil levels	N.E./↓	
			Total white blood cell levels	↑/↓/↑	
Sodium orthovanadate	Per os	Mouse	Splenic antibody-producing cell levels	→	Sharma et al., 1981
			DTH reaction (vs SRBC)	N.E.	
			Serum Ig levels	N.E.	
			Splenic lymphocyte mitogen responsiveness	N.E.	
Sodium vanadate	IP	Rat	PAM G6PDH, GSHPX, and GSHRX enzyme activities	N.E./↑	Kacew et al., 1982
Vanadium pentoxide	Inhalation	Monkey	Lung neutrophil recovery in BAL	↑	Knecht et al., 1985, 1992
			Lung PAM/lymphocyte recovery in BAL	N.E.	
	IP	Rat	PAM G6PDH, GSHPX, and GSHRX enzyme activities	N.E./↑	Kacew et al., 1982

Table 9 *(cont.)*

In vitro exposures

Compound	Host cell	Immune parameter analyzed	Effects	Citation
Ammonium metavanadate	Mouse	WEHI-3 macrophage cell line:		Cohen *et al.*, 1993
		TNF-α production	$\rightarrow$	
		IL-1α production	N.E./$\downarrow$	
		Spontaneous PGE$_2$ production	N.E./$\uparrow$	
		WEHI-3 macrophage cell line:		Cohen *et al.*, 1996b
		ROI production	$\leftarrow$	
		Resting Ca^{2+} levels	$\leftarrow$	
		IFN-γ-inducible Ca^{2+} influx	$\rightarrow$	
		Binding of IFN-γ	$\rightarrow$	
		IFN-γ receptor expression	$\rightarrow$	
		IFN-γ-primed ROI production	$\rightarrow$	
		IFN-γ-inducible Ia expression	$\rightarrow$	
		J774 macrophage cell line:		Vaddi and Wei, 1996
		Resting Ca^{2+} levels	$\leftarrow$	
		LPS-induced Ca^{2+} influx	N.E.	
	Rabbit	PAM phagocytic activity	N.E.	Fisher *et al.*, 1978
Sodium orthovanadate	Mouse	Splenic lymphocyte mitogen responsiveness	$\rightarrow$	Sharma *et al.*, 1981
	Rat	Neutrophil morphologic changes	$\leftarrow$	Bennett *et al.*, 1993
		Neutrophil cell spreading	$\rightarrow$	
	Human	PBL (T-lymphocyte) mitogen responsiveness	$\leftarrow$	Evans *et al.*, 1994
		PBL (T-lymphocyte) IFN-γ production	$\leftarrow$	

Compound	Source	Endpoint	Effect	Reference
Sodium vanadate		Jurkat T-lymphocyte cell line IL-2 production	↑	Imbert et al., 1994
		Jurkat T-lymphocyte cell line IL-2 receptor expression	↑	
		Jurkat T-lymphocyte cell line Ca^{2+} mobilization	↑	
	Rabbit	PAM ROI production	↓	Geertz et al., 1994
		Zymosan-induced PAM O_2^{-} production	N.E./↓	Labedzka et al., 1989
		Zymosan-induced PAM H_2O_2-production	N.E./↓	
		Zymosan-induced PAM chemiluminescence	→	
		Lactate dehydrogenase release	N.E./↑	
Vanadium pentoxide	Cow	PAM phagocytic activity	N.E./↓	Fisher et al., 1986
		PAM viability	→	
		PAM adherence	→	
	Mouse	WEHI-3 macrophage cell line: ROI production	↑	Cohen et al., 1996b
		Resting Ca^{2+} levels	↑	
		IFN-γ-inducible Ca^{2+} influx	→	
		IFN-γ-primed ROI production	→	
		PEM ROI production	N.E.	Daum et al., 1993
		PEM plasminogen activator levels	N.E.	
		PAM phagocytic activity	→	Fisher et al., 1978
		PAM adherence	N.E.	
	Rabbit	PAM acid phosphatase activity	→	Waters et al., 1974
		PAM phagocytic activity	→	
		PAM lysosomal enzyme activity	N.E.	

Commonly Used Abbreviations

ADCC	Antibody-dependent cell cytotoxicity
AFC	Antibody-forming cell
APC	Antigen-presenting cell
ATP	Adenosine triphosphate
β-Glu	β-Glucuronidase
β-NAG	β-*N*-Acetyl-β-D-glucosaminidase
BAL	Bronchoalveolar lavage
cAMP	Cyclic adenosine monophosphate
CD3	T-lymphocyte marker
CD4	T-helper lymphocyte surface marker
CD8	T-suppressor lymphocyte surface marker
CD11	Leukocyte function-associated antigens (LFA-1, Mac-1, and p150/95)
CD20	B-lymphocyte maturation marker
CD23	Antibody receptor for IgE
CD25	Interleukin-2 receptor
CD69	Activated leukocyte/lymphocyte surface marker
ConA	Concanavalin A (T-lymphocyte mitogen)
C3	Complement component (3)
DNP	Dinitrophenyl conjugate
DTH	Delayed-type hypersensitivity
E	Erythrocyte
EAC	Erythrocyte antibody complement
F_c	Antibody receptor (for Fc segment)
$F_{c\gamma 2a}R$	$IgG_{2a}F_c$ receptor
$F_{c\gamma 2b}R$	$IgG_{2b}F_c$ receptor
fMLP	Formyl-methionine-leucine-phenylalanine
G6PDH	Glucose-6-phosphate dehydrogenase
GSH	Reduced glutathione
GSHPX	Glutathione peroxidase activity
GSHRX	Glutathione reductase activity
GSHPX/GSHRX	Glutathione peroxidase/glutathione reductase activity ratio
GSSG	Oxidized glutathione
GTP	Guanidine triphosphate
H_2O_2	Hydrogen peroxide
HLA	Human histocompatibility antigen on antigen-presenting cells
HMP	Hexose monophosphate (shunt)
Ia	Mixed histocompatibility antigens on the antigen-presenting cell
ICAM	Intercellular adhesion molecule on antigen-presentng cells

ID	Intradermal injection
IFN	Interferon
Ig	Immunoglobulin (IgG, IgM, IgD, IgA, IgE)
IL	Interleukin
IM	Intramuscular injection
IT	Intratracheal instillation
KLH	Keyhole limpet hemocyanin
LDH	Lactate dehydrogenase
LFA-1	Natural killer cell surface adhesion molecule
LPS	Lipopolysaccharide (B-lymphocyte mitogen)
L3T4	Murine CD4 surface marker
Lyt2	Murine T-suppressor lymphocyte surface antigen
LTB_4	Leukotriene B_4
MAC-1,-2,-3	Macrophage and NK cell surface adhesion molecules (β-integrins)
MAF	Macrophage activating factor
MHC	Major histocompatibility complex
MIF	Macrophage inhibitory factor
MLR	Mixed lymphocyte response
NBT	Nitroblue tetrazolium
NK	Natural killer cell
O_2	Oxygen
$\cdot O_2^-$	Superoxide anion
PAM	Pulmonary alveolar macrophage
PBL	Peripheral blood lymphocyte
PBM	Peripheral blood monocyte
PEC	Peritoneal exudate cell
PEM	Peritoneal exudate macrophage
PFC	Plaque-forming cell
PGE_2	Prostaglandin E_2
PHA	Phytohemagglutinin (T-lymphocyte mitogen)
PMA	Phorbol myristate acetate
PolyI:C	Polyinosinic:polycytidilic acid
PPD	Purified protein derivative
PWM	Pokeweed mitogen (B- and T-lymphocyte mitogen)
RBC	Red blood cells
ROI	Reactive oxygen intermediates
SC	Subcutaneous injection
sIg	Surface immunoglobulin
SRBC	Sheep red blood cells
T_C	Cytotoxic T-lymphocyte
T_H1	T-helper lymphocyte subset
T_H2	T-helper lymphocyte subset
Thy	Surface marker on murine thymocytes/T-lymphocytes
Thy1.2	Allelic form of Thy antigen

TNF-α	Tumor necrosis factor-α
TNP	Trinitrophenyl conjugate
VCAM	Intercellular adhesion molecule on antigen-presenting cells
2-ME	2-Mercaptoethanol
6PGDH	6-Phosphogluconate dehydrogenase

Additional References

References that do not appear in the relevant chapter are cited below.

Arsenic references

BENCKO, V., WAGNER, V., WAGNEROVA, M., and BATORA, J. (1988). Immunological profiles in workers of a power plant burning coal rich in arsenic content. *J. Hyg. Epidemiol. Microbiol. Immunol.* **32**, 137–146.

GONSEBATT, M.E., VEGA, L., HERRERA, L.A., MONTERO, R., ROJAS, E., CEBRIAN, M.E., OSTROSKY-WEGMAN, P. (1992). Inorganic arsenic effects on human lymphocyte stimulation and proliferation. *Mutat. Res.* **283**, 91–95.

MENG, Z. (1993). Effects of arsenic on DNA synthesis in human lymphocytes. *Arch. Environ. Contam. Toxicol.* 25, 525–528.

YOSHIDA, T., SHIMAMURA, T., and SHIGETA, S. (1987). Enhancement of the immune response *in vitro* by arsenic. *Int. J. Immunopharmacol.* **9**, 411–415.

YU, H.S., CHANG, K.L., WANG, C.M., and YU, C.L. (1992). Alterations of mitogenic responses of mononuclear cells by arsenic in arsenical skin cancers. *J. Dermatol.* **19**, 710–714.

ZELIKOFF, J.T., REYNOLDS, C., BOWSER, D., and SNYDER, C.A. (1989). Modulation of tumor surveillance mechanisms by *in utero* exposure to arsenic. *Proceedings of Eastern Regional Symposium on Mechanisms of Immunotoxicity* **4**, 12.

Beryllium references

BENCKO, V., VASILIEVA, E.V., and SYMON, K. (1980). Immunological aspects of exposure to emissions from burning coal of high beryllium content. *Environ. Res.* **22**, 439–449.

CHIAPPINO, G., CIRLA, A., and VIGLIANI, E.C. (1969). Delayed-type hypersensitivity reactions to beryllium compounds. *Arch. Pathol.* **87**, 131–140.

FINCH, G.L., VERBURG, R.J., MEWHINNEY, J.A., EIDSON, A.F., and HOOVER, M.D. (1988). The effect of beryllium compound solubility on *in vitro* canine alveolar macrophage cytotoxicity. *Toxicol. Lett.* **41**, 97–105.

FINCH, G.L., LOWTHER, W.T., HOOVER, M.D., and BROOKS, A.L. (1991). Effects of beryllium metal particles on the viability and function of cultured rat alveolar macrophages. *J. Toxicol. Environ. Health* **34**, 103–114.

HALEY, P.J., FINCH, G.L., HOOVER, M.D., and CUDDIHY, R.G. (1990). The acute toxicity of inhaled beryllium metal in rats. *Fundam. Appl. Toxicol.* **15**, 767–778.

HART, B.A. and PITTMAN, D.G. (1980). The uptake of beryllium by the alveolar macrophage. *J. Reticuloendothel. Soc.* **27**, 49–58.

JONES, J.M. and AMOS, H.E. (1975). Antigen formation in metal contact sensitivity. *Nature* **256**, 499–500.

KANG, K., BICE, D., HOFFMAN, E., D'AMATO, R., and SALVAGGIO, J. (1977). Experimental studies of sensitization to beryllium, zirconium, and aluminum compounds in the rabbit. *J. Allergy Clin. Immunol.* **59**, 425–436.

KANG, K., BICE, D., D'AMATO, R., ZISKIND, M., and SALVAGGIO, J. (1979). Effects of asbestos and beryllium on release of alveolar macrophage enzymes. *Arch. Environ. Health* **34**, 133–140.

KRIVANEK, N. and REEVES, A.L. (1972). The effect of chemical forms of beryllium on the production of the immunologic response. *Am. Ind. Hyg. Assoc. J.* **33**, 45–52.

MARX, J.J. and BURRELL, R. (1973). Delayed hypersensitivity to beryllium compounds. *J. Immunol.* **111**, 590–598.

MOATAMED, F., KARNOVSKY, M.J., and UNANUE, E.R. (1975). Early cellular responses to mitogens and adjuvants in the mouse spleen. *Lab. Invest.* **32**, 303–312.

MORITA, K., INOUE, S., MURAI, Y., WATANABE, K., and SHIMA, S. (1982). Effects of the metals Be, Fe, Cu, and Al on the mobility of immunoglobulin receptors. *Experientia* **38**, 1227–1228.

PFEIFER, S., BARTLETT, R., STRAUSZ, J., HALLER, S., and MULLER-QUERNHEIM, J. (1994). Beryllium-induced disturbances of the murine immune system reflect some phenomena observed in sarcoidosis. *Int. Arch. Allergy Immunol.* **104**, 332–339.

SEKO, Y., KOYAMA, T., ICHIKI, A., SUGAMATA, M., and MIURA, T. (1982). The relative toxicity of metal salts to immune hemolysis in a mixture of antibody-secreting spleen cells, sheep red blood cells, and complement. *Res. Commun. Chem. Pathol. Pharmacol.* **36**, 205–213.

SENDELBACH, L.E., WITSCHI, H.P., and TRYKA, A.F. (1986). Acute pulmonary toxicity of beryllium sulfate inhalation in rats and mice: cell kinetics and histopathology. *Toxicol. Appl. Pharmacol.* **85**, 248–256.

Chromium references

CASTRANOVA, V., BOWMAN, L., MILES, P.R., and REASOR, M.J. (1980). Toxicity of metal ions to alveolar macrophages. *Am. J. Indust. Med..* **1**, 349–357.

CHRISTENSEN, M.M., ENRST, E., and ELLERMANN-ERIKSEN, S. (1992). Cytotoxic effects of hexavalent chromium in cultured murine macrophages. *Arch. Toxicol.* **66**, 347–353.

FIGONI, R.A. and TREAGAN, L. (1975). Inhibitory effect of nickel and chromium upon antibody response of rats to immunization with T-1 phage. *Res. Commun. Chem. Pathol. Pharmacol.* **11**, 335–338.

GALVIN, J.B. and OBERG, S.G. (1984). Toxicity of hexavalent chromium to the alveolar macrophage *in vivo* and *in vitro*. *Environ. Res.* **33**, 7–16.

GRAHAM, J.A., GARDNER, D.E., WATERS, M.D., and COFFIN, D.L. (1975). Effect of trace metals on phagocytosis by alveolar macrophages. *Infect. Immun.* **11**, 1278–1283.

GRAHAM, J.A., MILLER, F.J., DANIELS, M.J., PAYNE, E.A., and GARDNER, D.E. (1978). Influence of cadmium, nickel, and chromium on primary immunity in mice. *Environ. Res.* **16**, 77–87.

HAHON, N. and BOOTH, J.A. (1984). Effect of chromium and manganese particles on the interferon system. *J. Interferon Res.* **4**, 17–27.

HOOFTMAN, R.N., ARKESTEYN, C.W., and ROZA, P. (1988). Cytotoxicity of some types of welding fume particles to bovine alveolar macrophages. *Am. Occup. Hyg.* **32**, 95–102.

JOHANSSON, A., ROBERTSON, B., CURSTEDT, T., and CAMNER, P. (1986b). Rabbit lung after inhalation of hexa- and trivalent chromium. *Environ. Res.* **41**, 110–119.

KINNAERT, P., DESMUL, A., TAGNON, A., and TOUSSAINT, C. (1970). Effect of ^{32}P chromium phosphate on immune reactions and phagocytic activity in rats. *Transplantation* **9**, 457–462.

KUCHARZ, E.J. and SIERAKOWSKI, S.J. (1987). Immunotoxicity of chromium compounds: effect of sodium dichromate on the T-cell activation *in vitro. Arh. Hig. Rada. Toksiokol.* **38**, 239–243.

LAZZARINI, A., LUCIANI, S., BELTRAME, M., and ARSLAN, P. (1985). Effects of chromium(VI) and chromium(III) on energy charge and oxygen consumption in rat thymocytes. *Chem. Biol. Interact.* **53**, 273–281.

PASANEN, J.T., GUSTAFSSON, T.E., KALLIOMAKI, P.L., TOSAVAINEN, A., and JARVISALO, J.O. (1986). Cytotoxic effects of four types of welding fumes on macrophages *in vitro*: a comparative study. *J. Toxicol. Environ. Health* **18**, 143–152.

PRIBYL, D. and TREAGAN, L. (1977). A comparison of the effect of metal carcinogens chromium, cadmium, and nickel on the interferon system. *Acta Virol.* **21**, 507.

VAN HOOGSTRATEN, I.M., BODEN, D., VON BLOMBERG, B.M., KRAAL, G., and SCHEPER, R.J. (1992). Persistent immune tolerance to nickel and chromium by oral administration prior to cutaneous sensitization. *J. Invest. Dermatol.* **99**, 608–616.

Lead references

BINCOLETTO, C. and QUEIROZ, M.L. (1996). The effect of lead on the bone marrow stem cells of mice infected with *Listeria monocytogenes. Vet. Human Toxicol.* **38**, 186–190.

BINGHAM, E., BARKLEY, W., ZERWAS, M., STEMMER, K., and TAYLOR, P. (1972). Responses of alveolar macrophages to metals. *Arch. Environ. Health* **25**, 406–414.

BUCHMULLER-ROUILLER, Y., RANSIJN, A., and MAUEL, J. (1989). Lead inhibits oxidative metabolism of macrophages exposed to macrophage-activating factor. *Biochem. J.* **260**, 325–332.

DENTENER, M.A., GREVE, J.W., MAWSSEN, J.G., and BUURMAN, W.A. (1989). Role of tumor necrosis factor in the enhanced sensitivity of mice to endotoxin after exposure to lead. *Immunopharmacol. Immunotoxicol.* **11**, 321–334.

DEVRIES, C.R., INGRAM, P., WALKER, S.R., LINTON, R.W., GUTKNECHT, W.F., and SHELBURNE, J.D. (1983). Acute toxicity of lead particulates on pulmonary alveolar macrophages. Ultrastructural and microanalytical studies. *Lab. Invest.* **48**, 35–44.

GOVERNA, M., VALENTINO, M., VISONA, I., and SCIELSO, R. (1988). Impairment of chemotaxis of polymorphonuclear leukocytes from lead acid battery workers. *Sci. Total Environ.* **71**, 543–546.

GUILLARD, O. and LAUWERYS, R. (1989). *In vitro* and *in vivo* effect of mercury, lead, and cadmium on the generation of chemiluminescence by human whole blood. *Biochem. Pharmacol.* **38**, 2819–2823.

JAREMIN, B. (1983). Blast lymphocyte transformation (LTT), rosette (E-RFC), and leukocyte migration inhibition (MIF) tests in persons exposed to action of lead during work. *Bull. Inst. Mar. Trop. Med. Gdansk* **34**, 189–197.

JIAN, Z., YING-HAN, X., and HONG-FU, C. (1985). The effects of lead ion on immune function of rabbit alveolar macrophages: quantitation of immune phagocytosis and rosette formation by ^{51}Cr *in vitro. Toxicol. Appl. Pharmacol.* **78**, 484–487.

KAMINSKI, E.J., FISCHER, C.A., KENNEDY, G.L., and CALANDRA, J.C. (1977). Response of pulmonary macrophages to lead. *Br. J. Exp. Path.* **58**, 9–12.

KIMBER, I., STONARD, M.D., GIDLOW, D.A., and NIEWOLA, Z. (1986). Influence of chronic low-level exposure to lead on plasma immunoglobulin concentration and cellular immune function in man. *Int. Arch. Occup. Environ. Health* **57**, 117–125.

LABEDZKA, M., GULYAS, H., SCHMIDT, N., and GERCKEN, G. (1989). Toxicity of metallic ions and oxides to rabbit alveolar macrophages. *Environ. Res.* **48**, 255–274.

LUSTER, M.I., FAITH, R.E., and KIMMEL, C.A. (1978). Depression of humoral immunity in rats following chronic developmental lead exposure. *J. Environ. Pathol. Toxicol.* **1**, 397–402.

MUDZINSKI, S.P., RUDOFSKY, U.H., MITCHELL, D.G., and LAWRENCE, D.A. (1986). Analysis of lead effects on *in vivo* antibody-mediated immunity in several mouse strains. *Toxicol. Appl. Pharmacol.* **83**, 321–330.

NEILAN, B.A., O'NEILL, K., and HANDWERGER, B.S. (1983). Effect of low-level exposure on antibody-dependent and natural killer cell-mediated cytotoxicity. *Toxicol. Appl. Pharmacol.* **69**, 272–275.

QUEIROZ, M.L., PERLINGEIRO, R.C., BINCOLETTO, C., ALMEIDA, M., CARDOSO, M.P., and DANTAS, D.C. (1994b). Immunoglobulin levels and cellular immune function in lead-exposed workers. *Immunopharmacol. Immunotoxicol.* **16**, 115–128.

REIGART, J.R. and GRABER, C.D. (1976). Evaluation of the humoral immune response of children with low level lead exposure. *Bull. Environ. Contam. Toxicol.* **16**, 112–117.

STOURNARAS, C., SPANAKIS, E., PERRAKI, M., ATHANASIOU, M., THANOS, D., and GEORGOULIAS, V. (1990). Triethyl-lead-induced inhibition of proliferation of normal human lymphocytes through decreased expression of the Tac chain of interleukin-2 receptor. *Int. J. Immunopharmacol.* **12**, 349–358.

TALCOTT, P.A. and KOLLER, L.D. (1983). The effect of inorganic lead and/or a polychlorinated biphenyl on the developing immune system of mice. *J. Toxicol. Environ. Health* **12**, 337–352.

Mercury references

BRIDGER, M.A. and THAXTON, J.P. (1983). Humoral immunity in the chicken as affected by mercury. *Arch. Environ. Contam. Toxicol.* **12**, 45–49.

BROWN, D.L., REUHL, K.R., BORMANN, S., and LITTLE, J.E. (1988). Effects of methyl mercury on the microtubule system of mouse lymphocytes. *Toxicol. Appl. Pharmacol.* **94**, 66–75.

BRUNET, S., GUERTIN, F., FLIPO, D., FOURNIER, M., and KRZYSTYNIAK, K. (1993). Cytometric profiles of bone marrow and spleen lymphoid cells after mercury exposure in mice. *Int. J. Immunopharmacol.* **15**, 811–819.

CASTRANOVA, V., BOWMAN, L., REASOR, M.J., and MILES, P.R. (1980). Effects of heavy metal ions on selected oxidative metabolic processes in rat alveolar macrophages. *Toxicol. Appl. Pharmacol.* **53**, 14–23.

CHRISTENSEN, M., MOGENSEN, S.C., and RUNGBY, J. (1988). Toxicity and ultrastructural localization of mercuric chloride in cultured murine macrophages. *Arch. Toxicol.* **62**, 440–446.

CHRISTENSEN, M.M., ELLERMANN-ERIKSEN, S., RUNGBY, J., and MOGENSEN, S.C. (1993a). Comparison of the interaction of methyl mercury and mercuric chloride with murine macrophages. *Arch. Toxicol.* **67**, 205–211.

CHRISTENSEN, M.M., ELLERMANN-ERIKSEN, S., RUNGBY, J., MOGENSEN, S.C., and DANSCHER, G. (1993b). Histochemical and functional evaluation of mercuric chloride toxicity in cultured macrophages. *Prog. Histochem. Cytochem.* **23**, 306–315.

DIETER, M.P., LUSTER, M.I., BOORMAN, G.A., JAMESON, C.W., DEAN, J.H., and COX, J.W. (1983). Immunological and biochemical responses in mice treated with mercuric chloride. *Toxicol. Appl. Pharmacol.* **68**, 218–228.

GEERTZ, R., GULYAS, H., and GERCKEN, G. (1994). Cytotoxicity of dust constituents towards alveolar macrophages: interactions of heavy metal compounds. *Toxicology* **86**, 13–17.

HIROKAWA, K. and HAYASHI, Y. (1980). Acute methyl mercury intoxication in mice. Effect on the immune system. *Acta Pathol. Japan* **30**, 23–32.

HULTMAN, P., BELL, L.J., ENESTROM, S., and POLLARD, K.M. (1993). Murine susceptibility to mercury. II. Autoantibody profiles and renal immune deposits in hybrid, backcross, and H-2d congenic mice. *Clin. Immunol. Immunopathol.* **68**, 9–20.

ILBACK, N.G. (1991). Effects of methyl mercury exposure on spleen and blood natural killer (NK) cell activity in the mouse. *Toxicology* **67**, 117–124.

JANSSON, G. and HARMS-RINGDAHL, M. (1993). Stimulating effects of mercuric- and silver ions on the superoxide anion production in human polymorphonuclear leukocytes. *Free Rad. Res. Commun.* **18**, 87–98.

KANEKO, H., KAKKAR, V.V., and SCULLY, M.F. (1994). Mercury compounds induce a rapid increase in procoagulant activity of monocyte-like U937 cells. *Br. J. Haematol.* **87**, 87–93.

KIMATA, H., SHINOMIYA, K., and MIKAWA, H. (1983). Selective enhancement of human IgE production *in vitro* by synergy of pokeweed mitogen and mercuric chloride. *Clin. Exp. Immunol.* **53**, 183–191.

KOLLER, L.D., EXON, J.H., and ARBOGAST, B. (1997a). Methylmercury: effect on serum enzymes and humoral antibody. *J. Toxicol. Environ. Health* **2**, 1115–1123.

KOLLER, L.D., EXON, J.H., and BRAUNER, J.A. (1977b). Methylmercury: decreased antibody formation in mice. *Proc. Soc. Exp. Biol. Med.* **155**, 602–604.

KOLLER, L.D., ISAACSON-KERKVLIET, N., EXON, J.H., BRAUNER, J.A., and PATTON, N.M. (1979). Synergism of methylmercury and selenium producing enhanced antibody formation in mice. *Arch. Environ. Health* **34**, 248–252.

KUDO, N. and WAKU, K. (1994). Mercuric chloride induces the production of leukotriene B_4 by rabbit alveolar macrophages. *Arch. Toxicol.* **68**, 179–186.

LABEDZKA, M., GULYAS, H., SCHMIDT, N., and GERCKEN, G. (1989). Toxicity of metallic ions and oxides to rabbit alveolar macrophages. *Environ. Res.* **48**, 255–274.

LISON, D., DUBOIS, P., and LAUWERYS, R. (1988). *In vitro* effect of mercury and vanadium on superoxide anion production and plasminogen activator activity of mouse peritoneal macrophages. *Toxicol. Lett.* **40**, 29–36.

NAKATSURU, S., OOHASHI, J., NOZAKI, H., NAKADA, S., and IMURA, N. (1985). Effect of mercurials on lymphocyte functions *in vitro*. *Toxicology* **36**, 297–305.

355

NORDLIND, K. (1982). Effect of metal allergens on the DNA synthesis of unsensitized guinea pig lymphoid cells cultured *in vitro*. *Int. Arch. Allergy Appl. Immunol.* **69**, 12–17.

NORDLIND, K. (1983). Stimulating effect of mercuric chloride and nickel sulfate on DNA synthesis of thymocytes and peripheral lymphoid cells from newborn guinea pigs. *Int. Arch. Allergy Appl. Immunol.* **72**, 177–179.

PATIERNO, S.R., COSTA, M., LEWIS, V.M., and PEAVY, D.L. (1983). Inhibition of LPS toxicity for macrophages by metallothionein-inducing agents. *J. Immunol.* **130**, 1924–1929.

ROBINSON, C.J., BALAZS, T., and EGOROV, I.K. (1986). Mercuric chloride-, gold sodium thiomalate-, and D-penicillamine-induced antinuclear antibodies in mice. *Toxicol. Appl. Pharmacol.* **86**, 159–169.

ROMAN-FRANCO, A.A., TURIELLO, M., ALBINI, B., OSSI, E., MILGROM, F., and ANDRES, G.A. (1978). Anti-basement membrane antibodies and antigen–antibody complexes in rabbits injected with mercuric chloride. *Clin. Immunol. Immunopathol.* **9**, 464–481.

SAEGUSA, J., KUBOTA, H., and KIUCHI, Y. (1991). Antinucleolar autoantibody induced in mice by mercuric chloride. A genetic study. *Indust. Health* **29**, 167–170.

SHENKER, B.J., BERTHOLD, P., DECKER, S., MAYRO, J., ROONEY, C., VITALE, L., and SHAPIRO, I.M. (1992a). Immunotoxic effects of mercuric compounds on human lymphocytes and monocytes. II. Alterations in cell viability. *Immunopharmacol. Immunotoxicol.* **15**, 555–577.

SHENKER, B.J., MAYRO, J.S., ROONEY, C., VITALE, L., and SHAPIRO, I.M. (1993b). Immunotoxic effects of mercuric compounds on human lymphocytes and monocytes. IV. Alterations in cellular glutathione content. *Immunopharmacol. Immunotoxicol.* **15**, 273–290.

TAM, P.E. and HINDSILL, R.D. (1984). Evaluation of immunomodulatory chemicals: alteration of macrophage function *in vitro*. *Toxicol Appl. Pharmacol.* **76**, 183–194.

Nickel references

ARSALANE, K., AERTS, C., WALLAERT, B., VOISIN, C., and HILDEBRAND, H.F. (1992). Effects of nickel hydroxycarbonate on alveolar macrophage functions. *J. Appl. Toxicol.* **12**, 285–290.

BENCKO, V., WAGNER, V., WAGNEROVA, M., and REICHRTOVA, E. (1983). Immunobiochemical findings in groups of individuals occupationally and non-occupationally exposed to emissions containing nickel and cobalt. *J. Hyg. Epidemiol. Microbiol. Immunol.* **27**, 387–394.

BENSON, J.M., HENDERSON, R.F., and McCLELLAN, R.O. (1986). Comparative cytotoxicity of four nickel compounds to canine and rodent alveolar macrophages *in vitro*. *J. Toxicol. Environ. Health* **19**, 105–110.

BENSON, J.M., HENDERSON, R.F., and PICKRELL, J.A. (1988b). Comparative *in vitro* cytotoxicity of nickel oxides and nickel-copper oxides to rat, mouse, and dog pulmonary alveolar macrophages. *J. Toxicol. Environ. Health* **24**, 373–383.

BENSON, J.M., BURT, D.G. CHANG, Y.S., HAHN, F.F., HALEY, P.J., HENDERSON, R.F., HOBBS, C.H., PICKRELL, J.A., and DUNNICK, J.K. (1989). Biochemical responses of rat and mouse lung to inhaled nickel compounds. *Toxicology* **57**, 255–266.

BINGHAM, E., BARKLEY, W., ZERWAS, M., STEMMER, K., and TAYLOR, P. (1972). Responses of alveolar macrophages to metals. *Arch. Environ. Health* **25**, 406–414.

BRAATHEN, L.R., HAAVERLSRUD, O., and THORSBY, E. (1983). HLA antigens in patients with allergic contact sensitivity to nickel. *Arch. Dermatol. Res.* **275**, 355–356.

CHORVATOVICOVA, D. and KOVACIKOVA, Z. (1992). Inhalation exposure of rats to metal aerosol. II. Study of mutagenic effect on alveolar macrophages. *J. Appl. Toxicol.* **12**, 67–68.

DUMONT-FRUYTIER, M., VAN NESTE, D., DE BRUYERE, M., TENNSTEDT, D., and LACHAPELLE, J.M. (1980). Nickel contact sensitivity in women and HLA antigens. *Arch. Dermatol. Res.* **269**, 205–208.

EVERNESS, K.M., GAWKRODGER, D.J., BOTHAM, P.A., and HUNTER, J.A. (1990). The discrimination between nickel-sensitive and non-nickel-sensitive subjects by an *in vitro* lymphocyte transformation test. *Br. J. Dermatol.* **122**, 293–298.

FIGONI, R.A. and TREGAN, L. (1975). Inhibitory effect of nickel and chromium upon antibody response of rats to immunization with T-1 phage. *Res. Commun. Chem. Pathol. Pharmacol.* **11**, 335–338.

FINCH, G.L., MCNEILL, K.L., HAYES, T.L., and FISHER, G.L. (1987). *In vitro* interactions between pulmonary macrophages and respirable particles. *Environ. Res.* **44**, 241–253.

FISHER, G.L., MCNEILL, K.L., and DEMOCKO, C.J. (1986). Trace element interactions affecting pulmonary macrophage cytotoxicity. *Environ. Res.* **39**, 164–171.

HERNANDEZ, M., MACIA, M., CONDE, J.L., and DE LA FUENTE, M. (1991). Cadmium and nickel modulation of adherence capacity of murine peritoneal macrophages and lymphocytes. Intersexual comparisons. *Int. J. Biochem.* **23**, 541–544.

JARAMILLO, A. and SONNENFELD, G. (1992a). Characteristics of the inhibition of interferon-α/β production after crystalline nickel sulfide exposure. *Environ. Res.* **57**, 88–95.

JARAMILLO, A. and SONNENFELD, G. (1992b). Potentiation of lymphocyte proliferative responses by nickel sulfide. *Oncology* **49**, 396–406.

JOHANSSON, A., LUNDBORG, M., SKOG, S., JARSTRAND, C., and CAMNER, P. (1987). Lysozyme activity in ultrastructurally-defined fractions of alveolar macrophages after inhalation exposure to nickel. *Br. J. Indust. Med.* **44**, 47–52.

KARVONEN, J., SILVENNOINEN-KASSINEN, S., ILONEN, J., JAKKULA, H., and TIILIKAINEN, A. (1984). HLA antigens in nickel allergy. *Ann. Clin. Res.* **16**, 211–212.

KASPRZAK, K.S., WARD, J.M., POIRIER, L.A., REICHARDT, D.A., DENN, A.C., and REYNOLDS, C.W. (1987). Nickel–magnesium interactions in carcinogenesis: dose effects and involvement of natural killer cells. *Carcinogenesis* **8**, 1005–1011.

KUEHN, K., FRASER, C.B., and SUNDERMAN, F.W., JR (1982). Phagocytosis of particulate nickel compounds by rat peritoneal macrophages *in vitro*. *Carcinogenesis* **3**, 321–326.

KUSAKA, Y., NAKANO, Y., SHIRAKAWA, T., FUJIMURA, N., KATO, M., and HEKI, S. (1991). Lymphocyte transformation test with nickel in hard metal asthma: another sensitizing component of hard metal. *Ind. Health* **29**, 153–160.

LABEDZKA, M., GULYAS, H., SCHMIDT, N., and GERCKEN, G. (1989). Toxicity of metallic ions and oxides to rabbit alveolar macrophages. *Environ. Res.* **48**, 255–274.

357

LAWRENCE, D.A. (1981). Heavy metal modulation of lymphocyte activities. I. *In vitro* effects of heavy metals on primary humoral immune responses. *Toxicol. Appl. Pharmacol.* **57**, 439–451.

LUNDBORG, M. and CAMNER, P. (1982). Decreased level of lysozyme in rabbit lung lavage fluid after inhalation of low nickel concentrations. *Toxicology* **22**, 353–358.

MALININ, G.I., HORNICEK, F.J., LO, H.K., and MALININ, T.I. (1992). Cytometric and electron microscopic studies of the direct interaction of divalent nickel with intact and chemically-modified HuT-78 lymphoblasts. *Cell Biol. Toxicol.* **8**, 27–41.

MIGALLY, N., MURTHY, R.C., DOYE, A., and ZAMBERNARD, J. (1982). Changes in pulmonary alveolar macrophages in rats exposed to oxides of zinc and nickel. *J. Submicrosc. Cytol.* **14**, 621–626.

MILICEVIC, N.M. and MILICEVIC, Z. (1989). Histochemistry of the acutely-involved thymus in nickel chloride-treated rats. *J. Comp. Path.* **101**, 143–150.

MURTHY, R.C. and NIKLOWITZ, W.J. (1983). Ultrastructural changes in alveolar macrophages of rats exposed to nickel oxide by inhalation. *J. Submicrosc. Cytol.* **15**, 655–660.

REICHRTOVA, E., TAKAC, L., and KOVACIKOVA, Z. (1986). The effect of metal particles from a nickel refinery dump on alveolar macrophages. Part 2. Environmental exposure of rabbits. *Environ. Pollution (Series A)* **40**, 101–107.

REICHRTOVA, E., ORAVEC, C., PALENIKOVA, O., HORVATHOVA, J., TAKAC, L., and BENCKO, V. (1989). Bioaccumulation of metals from a nickel smelter waste in P and F_1 generations of exposed animals. II. Modulation of immune processes. *J. Hyg. Epidemiol. Microbiol. Immunol.* **33**, 245–251.

SMITH, K.L. and LAWRENCE, D.A. (1988). Immunomodulation of *in vitro* antigen presentation by cations. *Toxicol. Appl. Pharmacol.* **96**, 476–484.

SUNDERMAN, F.W., JR, MARZOUK, A., HOPFER, S., ZAHARIA, O., and REID, M.C. (1985). Increased lipid peroxidation in tissues of nickel chloride-treated rats. *Ann. Clin. Lab. Sci.* **15**, 229–236.

TAKAHASHI, S., YAMADA, M., KONDO, T., SATO, H., FURUYA, K., and TANAKA, I. (1992). Cytotoxicity of nickel oxide particles in rat alveolar macrophages cultured *in vitro*. *J. Toxicol. Sci.* **17**, 243–251.

TAM, P.E. and HINDSILL, R.D. (1984). Evaluation of immunomodulatory chemicals: alteration of macrophage function *in vitro*. *Toxicol. Appl. Pharmacol.* **76**, 183–194.

WAKSVIK, H., BOYSEN, M., BROGGER, A., and KLEPP, O. (1981). Chromosome aberrations and sister chromatid exchanges in persons occupationally-exposed to mutagens/carcinogens, in Seeberg, E. and Kleppe, K. (Eds) *Chromosome Damage and Repair*, pp. 563–566, New York: Plenum Press.

WALTON, S., KECZKES, K., LEAROYD, P.A., and RAJAH, S.M. (1986). HLA-A, -B, and -DR antigens in nickel-sensitive females. *Clin. Exp. Dermatol.* **11**, 636–640.

WARNER, G.L. and LAWRENCE, D.A. (1986a). Stimulation of murine lymphocyte responses by cations. *Cell. Immunol.* **101**, 425–439.

WARNER, G.L. and LAWRENCE, D.A. (1986b). Cell surface and cell cycle analysis of metal-induced murine T cell proliferation. *Eur. J. Immunol.* **16**, 1337–1347.

WARNER, G.L. and LAWRENCE, D.A. (1988). The effects of metals on IL-2-related lymphocyte proliferation. *Int. J. Immunopharmacol.* **10**, 629–637.

WASEEM, M., BAJPAI, R., and KAW, J.L. (1993). Reaction of pulmonary macrophages exposed to nickel and cadmium *in vitro*. *J. Environ. Pathol. Toxicol. Oncol.* **12**, 47–54.

Platinum references

HOWLE, J.A., THOMPSON, H.S. STONE, A.E., and GALE, G.R. (1971). *cis*-Dichlorodiammineplatinum(II): inhibition of nucleic acid synthesis in lymphocytes stimulated with phytohemagglutinin. *Proc. Soc. Exp. Biol. Med.* **137**, 820–825.

LICHTENSTEIN, A.K. and PENDE, D. (1986). Enhancement of natural killer cytotoxicity by *cis*-diamminedichloroplatinum(II) *in vivo* and *in vitro*. *Cancer Res.* **46**, 639–644.

MURDOCH, R.D. and PEPYS, J. (1984b). Immunological responses to complex salts of platinum. II. Enhanced IgE antibody responses to ovalbumin with concurrent administration of platinum salts in the rat *Clin. Exp. Immunol.* **58**, 478–485.

MURDOCH, R.D. and PEPYS, J. (1986). Enhancement of antibody production by mercury and platinum group metal halide salts. Kinetics of total and ovalbumin-specific IgE synthesis. *Int. Arch. Allergy Appl. Immunol.* **80**, 405–411.

VANCUROVA, M., JILEK, P., PROCHAZKOVA, J., and CHYLKOVA, V. (1989). Use of *in vitro* models in the immunotoxicologic evaluation of drugs. *Cesko. Farmacie.* **38**, 269–271.

Vanadium references

DAUM, J.R., SHEPHERD, D.M., and NOELLE, R.J. (1993). Immunotoxicology of cadmium and mercury on B-lymphocytes. I. Effects on lymphocyte function. *Int. J. Immunopharmacol.* **15**, 383–394.

EVANS, G.A., GARCIA, G.G., ERWIN, R., HOWARD, O.M., and FARRAR, W.L. (1994). Pervanadate simulates the effects of interleukin-2 (IL-2) in human T-cells and provides evidence for the activation of two distinct tyrosine kinase pathways by IL-2. *J. Biol. Chem.* **269**, 23407–23412.

FISHER, G.L., MCNEILL, K.L., and DEMOCKO, C.J. (1986). Trace element interactions affecting pulmonary macrophage cytotoxicity. *Environ. Res.* **39**, 164–171.

GEERTZ, R., GULYAS, H., and GERCKEN, G. (1994). Cytotoxicity of dust constituents towards alveolar macrophages: interactions of heavy metal compounds. *Toxicology* **86**, 13–27.

Index